T0328273

# Advanced Driver Intention Inference

Theory and Design

# Advanced Driver Intention Inference

## Theory and Design

YANG XING, PHD
Research Fellow
School of Mechanical and Aerospace Engineering
Nanyang Technological University
Singapore

CHEN LV
Assistant Professor
School of Mechanical and Aerospace Engineering
Nanyang Technological University
Singapore

DONGPU CAO
Canada Research Chair in Driver Cognition and Automated Driving
Department of Mechanical and Mechatronics Engineering
University of Waterloo
Canada

ELSEVIER

Elsevier
Radarweg 29, PO Box 211, 1000 AE Amsterdam, Netherlands
The Boulevard, Langford Lane, Kidlington, Oxford OX5 1GB, United Kingdom
50 Hampshire Street, 5th Floor, Cambridge, MA 02139, United States

**Notices**
Knowledge and best practice in this field are constantly changing. As new research and experience broaden our
understanding, changes in research methods, professional practices, or medical treatment may become
necessary.

Practitioners and researchers must always rely on their own experience and knowledge in evaluating and using
any information, methods, compounds, or experiments described herein. In using such information or
methods they should be mindful of their own safety and the safety of others, including parties for whom they
have a professional responsibility.

To the fullest extent of the law, neither the Publisher nor the authors, contributors, or editors, assume any
liability for any injury and/or damage to persons or property as a matter of products liability, negligence
or otherwise, or from any use or operation of any methods, products, instructions, or ideas contained in the
material herein.

**Library of Congress Cataloging-in-Publication Data**
A catalog record for this book is available from the Library of Congress

**British Library Cataloguing-in-Publication Data**
A catalogue record for this book is available from the British Library

ISBN: 978-0-12-819113-2

For information on all Elsevier publications visit our website at
https://www.elsevier.com/books-and-journals

Publisher: Matthew Deans
Acquisitions Editor: Carrie Bolger
Editorial Project Manager: Gabriela D. Capille
Production Project Manager: Sreejith Viswanathan
Cover designer: Alan Studholme

Typeset by TNQ Technologies

Working together
to grow libraries in
developing countries

www.elsevier.com • www.bookaid.org

# List of Abbreviations

| | |
|---|---|
| ABS | Antilock Braking System |
| ACC | Adaptive Cruise Control |
| ACP | Artificial Society, Computational Experiments, and Parallel Execution |
| ACT-R | Adaptive Control of Thought-Rational |
| ADAS | Advanced Driver Assistance System |
| A/D | Analog/Digital |
| AIOHMM | Autoregressive Input-Output HMM |
| ANN | Artificial Neural Network |
| AVM | Around View Monitoring |
| BCI | Brain-Computer Interface |
| BF | Bayesian Filter |
| BN | Bayesian Network |
| CAN | Controller Area Network |
| CBN | Causal Bayesian Network |
| CDI | Comprehensive Decision Index |
| CHMM | Continuous Hidden Markov Model |
| CLNF | Conditional Local Neural Fields |
| CLR | Constant Learning Rate |
| CPS | Cyber-Physical Space |
| CPSS | Cyber-Physical-Social Space |
| CNN | Convolutional Neural Network |
| CSI | Channel State Information |
| DBSCAN | Density-based Spatial Clustering of Application with Noise |
| DBN | Dynamic Bayesian Network |
| DII | Driver Intention Inference |
| DWT | Discrete Wavelet Transform |
| ED | Edge Distribution |
| EEG | Electroencephalograph |
| ECG | Electrocardiogram |
| EMG | Electromyography |
| EOG | Electrooculography |
| ERRC | Error Reduction Ratio-Causality |
| EV | Electric Vehicle |
| FFNN | Feedforward Neural Network |
| FIR | Field Impedance Equalizer |
| FPR | False-Positive Rate |
| GA | Genetic Algorithm |
| GAN | Generative Adversarial Networks |
| GMM | Gaussian Mixture Model |
| GNSS | Global Navigation Satellite System |
| GOLD | Generic Obstacle and Lane Detection |
| GPS | Global Positioning System |
| GP | Gaussian Process |
| HMM | Hidden Markov Model |
| HHMM | Hierarchical Hidden Markov Model |
| HOG | Histogram of Oriented Gradients |
| HRI | Human-Robot Interaction |
| HT | Hough Transform |
| IDDM | Intention-driven Dynamic Model |
| IMM | Interactive Multiple Model |
| IMU | Inertial Measurement Unit |
| IOHMM | Input-Output Hidden Markov Model |
| IPM | Inverse Perspective Mapping |
| JTSM | Joint Time-Series Modelling |
| KRR | Kernel Ridge Regression |
| LADAR | Laser Detection and Ranging |
| LANA | Lane Finding in Another Domain |
| LCII | Lane Change Intention Inference |
| LDA | Lane Departure Avoidance |
| LDW | Lane Departure Warning |
| LIDAR | Light Detection and Ranging |
| LKA | Lane Keeping Assistance |
| LOO | Leave-One-Out |
| LSV | Lane Sampling and Voting |
| LSTM | Long Short-Term Memory |
| MIC | Maximal Information Coefficient |
| MHMM | Modified Hidden Markov Model |
| MLR | Multisteps Learning Rate |
| NGSIM | Next-Generation Simulation |
| NFIR | Nonlinear Finite Impulse Response |
| NN | Neural Networks |
| OLS | Orthogonal Least Squares |
| OOB | Out-of-Bag |
| PCA | Principal Component Analysis |
| PRM | Revolutions Per Minute |
| RANSAC | Random Sample Consensus |
| RF | Random Forest |
| RGB | Red-Green-Blue |
| RGB-D | Red-Green-Blue Depth |
| RMSE | Root Mean Square Error |
| RNN | Recurrent Neural Network |
| ROC | Receiver Operating Characteristic |
| ROI | Region of Interest |
| RVM | Relevance Vector Machine |
| SA | Situation Awareness |
| SAE | Society of Automobile Engineers |
| SBL | Sparse Bayesian Learning |

| | | | |
|---|---|---|---|
| SF | Steerable Filter | TS | Time Sliced |
| SCR | Skin Conductance Response | TPR | True-Positive Rate |
| SWA | Side Warning Assistance | TTI | Time to Intersection |
| SVM | Support Vector Machine | TTC | Time to Collision |
| SVR | Support Vector Regression | TTCCP | Time to Critical Probability |
| TDV | Traffic-Driver-Vehicle | V2V | Vehicle-to-Vehicle |
| THW | Time Headway | WHO | World Health Organization |

# Abstract

Longitudinal and lateral control of the vehicle on the highway are highly interactive tasks for human drivers. The intelligent vehicles and the advanced driver-assistance systems (ADAS) need to have proper awareness of the traffic context as well as the driver to make an appropriate assistant to the driver. The ADAS also need to understand the potential driver intent correctly since it shares the control authority with the human driver. This book provides research on the driver intention inference, particular focus on the two typical vehicle control maneuvers, namely, lane change maneuver and braking maneuver on highway scenarios. A primary motivation of this book is to propose algorithms that can accurately model the driver intention inference process. Driver's intention will be recognized based on the machine learning methods due to its good reasoning and generalizing characteristics. Sensors in the vehicle are used to capture context traffic information, vehicle dynamics, and driver behavior information.

This book is organized in sections and chapters, where each chapter is corresponding to a specific topic. Chapter 1 introduces the motivation, human intention background, and general methodological framework used in this book. Chapter 2 includes the literature survey and the state-of-the-art analysis of driver intention inference. Chapters 3 and 4 contain the techniques for traffic context perception that focus on sensor integration, sensor fusion, and road perception. A review of lane detection techniques and its integration with a parallel driving framework is proposed. Next, a novel integrated lane detection system is designed. Chapters 5 and 6 discuss the driver behavior issues, which provide the driver behavior monitoring system for normal driving and secondary tasks detection. The first part is based on the conventional feature selection method, while the second part introduces an end-to-end deep learning framework. Understanding the driver status and behaviors is the preliminary task for driver intention inference. The design and analysis of driver braking and lane change intention inference systems are proposed in Chapters 7 and 8. Machine-learning models and time-series deep-learning models are used to estimate the longitudinal and lateral driver intention. Finally, discussions and conclusions are made in Chapter 9.

## KEYWORDS

ADAS, Computer vision, Driver behaviors, Driver intention inference, Intelligent vehicles, Automated driving, Machine learning.

# Contents

# CHAPTER 1

# Introduction

Worldwide traffic departments have reported that more than 1.2 million traffic-related injuries happen each year. Among these traffic accidents, more than 80% were caused by human errors [1]. The World Health Organization (WHO) reported that traffic accidents each year cost around €518 billion worldwide and on average, 1%—2% of the world gross domestic product [2,3]. In the past, in-vehicle passive safety systems such as airbags and seat belts have played a significant role in the protection of drivers and passengers. These technologies have saved millions of lives. However, they are not designed to prevent accidents from happening but just try to minimize the injuries after the accidents happen [4]. Therefore recent efforts have been devoted to the development of safer and intelligent systems toward the prevention of accidents. These systems are known as the Advanced Driver Assistance Systems (ADAS).

ADAS is a series of fast-developing techniques that are designed for improving driver safety and increasing the driving experience [5]. ADAS relies on a multimodal sensor fusion technique to integrate multiple sensors such as light detection and ranging (lidar), radar, camera, and GPS into a holistic system. The sensors' working ranges are shown in Fig. 1.1. Most of the current ADAS techniques such as lane departure avoidance, lane keeping assistance, and side warning assistance can help the driver to make the right decision and reduce the workload.

It is predicted that the shipment of ADAS in the future has great potential and can generate a huge amount of commercial benefit based on many automotive market analyzers. One example is shown in Fig. 1.2 according to the prediction of Grand View Research, Inc. ADAS products will show a significant increase in the next 5 years. Therefore the utilization of ADAS products will become more accessible to the public, although this can bring a series of problems. For example, the financial cost will increase. Also, as most of the automotive companies are developing their ADAS products, safety insurance and product quality can be different from each other. Once the drivers are getting familiar with these products, they will heavily rely on these systems. A very famous example is the Tesla car crashes that are caused by their autopilot ADAS products, as shown in Fig. 1.3. The autopilot products of Tesla are one of the most successful commercial driver assistance and semi-automated driving assistance system in the world. The product is set of intelligent computing, perception, and control units, which can significantly increase

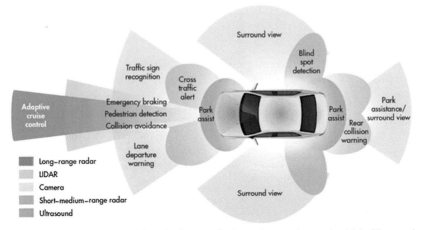

FIG. 1.1 Distribution of Advanced Driver Assistance Systems in an advanced vehicle (deepscale.ai/adas. html). *Lidar*, light detection and ranging.

Advanced Driver Intention Inference. https://doi.org/10.1016/B978-0-12-819113-2.00001-4

driving safety issues. However, even such a smart system can be reported for car crashes worldwide. One of the most common reasons for a crash is the driver overtrusting the autopilot when the system is activated, which is a problem in the future.

The reasons why ADAS cannot be 100% trusted are multifold. Currently, most of the reasons are due to immature techniques; however, a deeper reason is that the driver and the automation lack mutual understanding. The inputs of current ADAS are mainly based only on the vehicle dynamic states and traffic context information. Most of the systems ignore the most critical

factor, the driver. Vehicles are working in a three-dimensional environment with continuous driver-vehicle-road interactions. Drivers are the most essential part of this system, who control the vehicle based on the surrounding traffic context perception. Therefore allowing ADAS to understand driver behaviors and follow driver's intention is of importance to driver safety, vehicle drivability, and traffic efficiency.

Human driver intention inference is an ideal way to allow ADAS to obtain the ability of reasoning. The reasons for developing driver intention inference technique are multifold: first of all, the most important

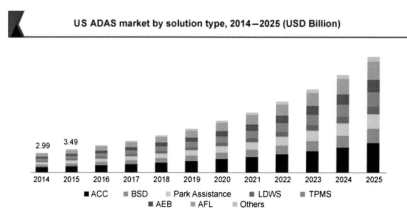

FIGURE 1.2 Advanced Driver Assistance Systems market prediction (Grand View Research, Inc. https://www.grandviewresearch.com). *ACC*, adaptive cruise control; *AEB*, automatic emergency braking; *AFL*, adaptive front light; *BSD*, blind spot detection; *LDWS*, lane departure warning systems; *TPMS*, tire pressure monitoring system.

FIGURE 1.3 A Tesla car has crashed into a parked police car in California, USA. (https://www.bbc.com/news/technology-44300952).

and significant motivation is to improve driver and vehicle safety. Accordingly, two different driving scenarios require inferring the driver's intention. The first one is to better assess the risk in the future based on the driver's interesting region. The second one is to avoid making decisions that are opposite to the driver's intent. For the first case, there is evidence that a large number of accidents are caused by human error or misbehavior, including cognitive (47%), judgment (40%), and operational errors (13%) [6]. Therefore monitoring and correcting driver intention and behavior seem to be crucial in the effectiveness of a future ADAS. Meanwhile, the increasing use of in-vehicle devices and information systems tend to distract the driver from the driving task. For the design of future ADAS, it is therefore beneficial to integrate intended driver behaviors from the early design stages [7,8].

ADAS usually automatically intervene in the vehicle dynamics and share the control authority. To ensure cooperation, it is crucial that ADAS is aware of driver intention and does not operate against the driver's willingness. For example, in complex traffic conditions such as intersection and roundabout, it is crucial not to interrupt the driver making decisions, especially not to suspend the driver with misleading instructions. This makes it reasonable or even necessary for ADAS to have the ability to accurately understand the driver's driving intention. On the other hand, intention information enables for more accurate prediction of future trajectories, which would be beneficial to risk assessment systems [9,10]. Driver intention inference will benefit the construction of the driver model, which

can act as the guidance to design an automatic decision-making system.

Moreover, in terms of the level 3 automated vehicle (according to the SAE International standard on the classification of automated vehicles), accurate driver intention prediction enables a smoother and safer transition between the driver and the autonomous controller [11,12]. When the level 3 automated vehicles are operating in an autonomous condition, all the driving maneuvers are handled by the vehicle. However, once the vehicle cannot deal with an emergent situation, it has to give the driving authority to the driver. This process is known as disengagement [13]. In such a case, the vehicle can assess the takeover ability of the driver according to the continuously detected intention. If the driver is focusing on the driving task at that moment and has an explicit intention, the vehicle can warn the driver and pass the driving authority to the driver. This will make sure the transition between driver and controller is as smooth as possible. However, if the driver is believed to be unable to handle the driving task, the autonomous driving unit should help the driver gain situation awareness as soon as possible and take emergency action immediately.

Another essential reason to develop the driver intention inference system is it will contribute to the development of automated vehicles. As shown in Fig. 1.4, each level of understanding about the driver can be mapped into the corresponding intelligent level of an autonomous vehicle. Comprehensive research on each layer will contribute to the development of the relative layer in the autonomous vehicle. Driver intention

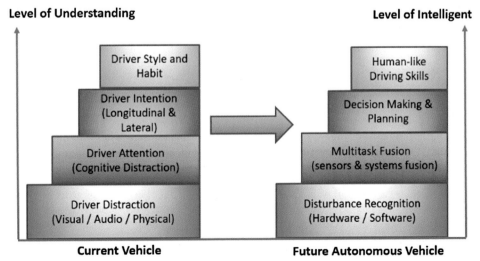

FIGURE 1.4 The evolution from the current vehicle to future autonomous vehicle.

recognition is a relative higher-level understanding of human drivers and related to the decision-making layer of autonomous vehicles. Modeling driver intention mechanisms is critical to the construction of automated decision-making algorithms. Human drivers are the teacher of automated drivers. The automated drivers can learn when and how to make the decision based on the driver's intention knowledge. Once the human driver becomes the passenger in the automated driving vehicles, it will be easier for the passengers to accept that the automated driving systems is such systems that can remember the driving habit of the passenger. Therefore a good study about when and how drivers generate their intentions will benefit the design of the decision-making module for intelligent vehicles. Based on such a design method, the vehicles will be more similar to human drivers, which will make it much easier for humans to accept these highly intelligent vehicles.

As discussed earlier, teaching ADAS to understand driver intention is essential as well as challenging to enhance the safety of the driver-vehicle-traffic close-loop system. To focus more, this book will target two of the most popular driving scenarios in both longitudinal and lateral directions, namely, the braking and lane change maneuvers. For example, during a standard lane change maneuver, the driver is expected to perform a series of behaviors (e.g., mirror checking and turn the steering wheel). The driver's lane change intention can be inferred in an early stage by recognizing driver behaviors and traffic context situations. A driver lane change intention system facing next-generation ADAS is developed in this study. Based on this, four main objectives are determined:

1. Driver intention process analysis: To predict driver lane change intention, it is vital to understand the human intention mechanism, such as how the intention is generated and what is the stimuli of the intention. The nature behind driver intention is the first question that needs to be answered.

2. Traffic context perception: The driver is in the middle of the traffic-driver-vehicle loop. Traffic context is the input to the driver perception system, which makes it act as the stimuli of the driver's intention. Therefore understanding the current traffic situation will benefit the intention inference system.

3. Understanding driver behaviors: Driver behaviors, such as mirror checking, are the most important clues before the driver makes a lane change. The driver has to perform a series of checking action to have a full understanding of the surrounding context before he/she decides to change the lane. Therefore

driver behavior analysis is of importance to infer driver intention.

4. Driver lane change intention inference algorithms: Based on the specific traffic context and driver behaviors, the next task is to infer driver intention properly. The algorithms for intention inference should have the ability to capture the long-term dependency between the temporal sequences. Moreover, the intention inference algorithms should predict the intention as early as possible.

The driver lane change intention platform requires the integration of software and hardware systems. Driver intention inference has to take the traffic context, driver behaviors, or dynamic vehicle information into consideration, which will fuse multimodal signals and mining the long-term dependency between different signals based on machine learning methods. In terms of the hardware system, the sensors, included in this book, contain RGB and RGB-D cameras and vehicle navigation system. Besides, all the sensors are tested and mounted on a real vehicle in this case to collect naturalistic data. Specifically, the traffic context such as lane positions and front vehicle position will be processed with image-processing methods. One web camera is mounted inside the cabinet. The driver behavior dynamics will be evaluated within a steady and dynamic vehicle. The RGB-D camera (Microsoft Kinect V2.0) will be used for the steady vehicle, while another web camera will be used to record the driver behavior during the highway driving task. These signals are recorded with one laptop for further processing and analyzing. The algorithms used in this project are mainly focused on machine learning methods, which include supervised learning, unsupervised learning, and deep learning models. All the algorithms are written in MATLAB and C++.

The driver's intention inference task described in this book relies on machine learning algorithms to work in real time. The reasons for using machine learning can be multifold. First, the real-time traffic context and driver behavior data can be high dimensional and of large volume, and very few mathematic models can deal with such data. However, machine learning algorithms are useful for high-dimensional multimodal data processing. Second, the utilization of a machine learning algorithm enables learning the long-term dependency between driver behaviors and traffic context, which significantly increases the inference accuracy for the lane change intention. Finally, it is hard to find the intention generation and inference pattern based on observation and modeling. The machine learning algorithms provide an efficient way to learn knowledge from naturalistic data. With some advanced deep learning techniques, it is even possible to achieve an end-to-end learning process. Although machine learning algorithms are very powerful in dealing with the tasks described in this book, they do have limitations. The major limitation of using machine learning algorithms is data collection. Data is the heart of the machine learning algorithms. To obtain an accurate intention inference results, several experiments need to be designed and data need to be collected. Insufficient data volume will lead to overfitting and bad inference results. Besides, most of the data used in the book need manual labeling, which is time-consuming. Finally, the training and testing of machine learning algorithms give rise to a higher computational burden both for the financial and temporal costs.

## WHAT IS HUMAN INTENTION?

This section describes the human intention mechanism based on existing studies. Driver intention is a subset of human intention that particularly occurs during driving. The human intention has been theoretically discussed and studied by several studies in the past three decades. From the cognitive psychology perspective, intention refers to the thoughts that one has before producing an action [14]. In the theory of reasoned action given by Fishbein, the intention is in the center of the theory, which is to perform a given behavior. Three aspects determine intentional behavior: the attitude toward the behavior, subjective norm, and perceived behavior control [15] (Fig. 1.5).

Human behavior is directly influenced by intention. Intention can be determined by the three aspects mentioned earlier [15]. Specifically, attitude toward the behavior describes how much willingness does one have to take the behavior; a strong level of attitude can give a strong intention of taking actions in a certain task. Human beliefs determine the attitude toward a behavior about how much outcome it brings after the behavior is taken. Second, the subjective norm reflects the pressure from the surrounding social life of a human. It evaluates how much the family, friends, and the society expect from a person to make certain behavior. Finally, perceived behavior control is developed from the self-efficacy theory given in [16]. It describes the confidence of an individual to perform the behavior. For example, if there are two subjects with the same intention, the one who is more confident in the task can perform better behavior toward finishing a certain task. Based on [16], the planned behavior, perceived behavior control, and the intention can be used directly to predict the behavior performance.

Bratman also pointed out that intention is the main attitude that directly influences plans [17]. Also, Heinze [18] described a triple-level description of intentional behavior, namely, intentional level, activity level, and state level. According to Tahboub, in the human-machine interface scope, intention recognition is generally defined as understanding the intention of another agent. More technically, it is the process of inferring an agent's intention based on its actions [19]. Elisheva proposed a cognitive model with two core components: intention detection and intention prediction. Intention

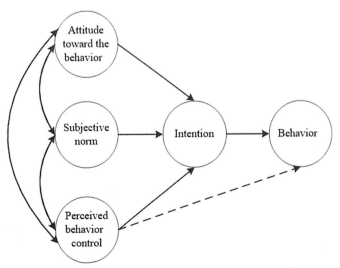

FIGURE 1.5 Architectural diagram of human intention [15].

detection refers to detect whether or not a sequence of actions has any underlying intention at all. Intention prediction, on the other hand, refers to predict and extend the intentional goal by a set of incomplete sequence of actions [20]. In terms of driving intention inference, it equals to the part of intention prediction that is mentioned earlier because we assume drivers always have an intention during a short period (e.g., lane keeping and following can be viewed as driver intention).

The inference and reasoning process make people clever and easier to take part in the social community. A human can recognize other's intentions based on the observation and the knowledge stored in the brain. However, it is a difficult task to make the intelligent machine to infer human intention easily and accurately. To some extent, only when a robot detects human intention based on human observation can it be viewed as an intelligent agent. Based on the study by Meltzoff and Brooks [21], self-experience plays an important role in making an inference of the intended state of other agents. In terms of robots and intelligent vehicles, self-experiences were obtained from learning a large amount of relevant events data.

As mentioned earlier, human intention inference has been widely studied in the past decades. One of the most significant applications of human intention inference is human-robot interface design. Thousands of service robots were designed to assist humans in completing their work in both daily life and a dangerous situation. The traditional robots were designed from a robot's point of view rather than from a human point of view, which reduces the interaction level between humans and robots. A robot should have the ability to learn and infer human intentions and obtain basic reasoning intelligence in order to improve the efficiency of human-robot interaction (HRI) as well as its intelligence. A widely accepted method of classification of human intentions in HRI scope is to classify the human intentions into explicit and implicit intentions. The explicit intention is much clearer than the implicit intention and hence easier to recognize. Explicit intention means humans directly transmit their intention to the robot by language or directly command through the computer interface. On the contrary, implicit intention reflects those human mental states that cannot be communicated to the robot. The robots have to observe and understand human behaviors first, and estimate the human intention based on the gained knowledge and the on-going human actions. Implicit intentions usually can be further separated into informational and navigational intentions. Human implicit intention researches have been done in various areas.

For example, Jang et al. [22] used the eyeball movement pattern as the input to recognize human's implicit intention. The intention recognition task was viewed as a classification problem. They divided implicit human intention into informational and navigational groups and nearest neighbor algorithms, as well as support vector machines (SVMs) to train the classifier. It is also confirmed that the fixation length, fixation count, and pupil size variation were the main factors to discriminate human intention. Kang et al. [23] proposed a method of human implicit intention recognition based on electroencephalographic (EEG) signals. This algorithm focused on service web queries. Three kinds of classification methods were adopted, which were SVM, Gaussian mixture model (GMM), and naïve Bayes. The implicit human intention was classified into two types called navigational and informational intentions. An eye-tracking system was used to help track the subjects' eye movement in the experiment. Results showed that SVM gave the best classification result with 77.4% accuracy. Wang et al. [24] determined the user's intention based on eye-tracking systems. The fuzzy inference was used to infer the user's intention, with eye gazing frequency and gazing time as the input. The fuzzy logic controller outputs the probability of the user's intention on one particular region of the computer screen.

Generally speaking, the human intention inference problem contains a large amount of uncertainty and noise exists in the measurement device. Therefore probability-based machine learning methods are a powerful tool in solving this kind of problem, and it has been successfully applied in many cases. In terms of human intention inference task, which is a work to infer human mental hidden states, the hidden Markov model (HMM) and the dynamic Bayesian theory are two very popular ways of inferring the human mental state. In Ref. [25], a living assistance system for elder and disabled people was developed. The HMM was proposed based on the hand gesture information. Five basic hand gestures were defined to represent human intention, which was come, go fetching, go away, sit down, and stand up. The features of hand movement data were extracted and converted to an observable symbol for HMM. The results of the experiment showed the effects of the intention inference system. In Ref. [26], an intention recognition method for an autonomous mobile robot was developed. The HMMs were trained for five kinds of human actions, such as following, meeting, passing by, picking an object up, and dropping an object off. After this stage, different models were constructed to represent different behaviors, then the robot moved autonomously and can be regarded as an observer to recognize the human intention of the five types of maneuvers and rejustify its model structure. The recognition accuracy can reach 90%−100%. In Ref. [27], a symbiotic human-robot cooperation scheme was proposed. A robot was controlled wirelessly by a human to reflect a human's idea, and other robots were used to infer this robot's intention and help it work. A vague timescale method was used to help the robot infer the target robot's intention with historical data rather than instantaneous information only. Given a simple task, the human behavior model can be constructed with fuzzy automata, and the transition of the human intention is determined by the fuzzy rules using the qualitative expression.

Rani et al. [28] aimed to study and recognize human status from their physiologic cues by using machine learning methods. Four kinds of machine learning methods were adopted, which are SVM, K-nearest neighbor, Bayesian dynamic model, and regression tree. The models were trained to classify five human mental states (anxiety, engagement, boredom, frustration, and anger) based on human physiologic cues. Then a systematic comparison of the performance of machine learning was evaluated. The results showed that SVM gives the best classification result with 85.81% accuracy. In Ref. [29], a human-robot system was designed to reflect a human intention on the path of the robot arm

and assist his/her power. A human-will model was used to explain human intention when cooperating with the robot. A modified hidden Markov Model (HMM) was proposed to infer human path intention on the robot arm, and a filed impedance equalizer (FIE) was used to assist humans in merging their arm force to the robot system. The experimental result showed that the MHMM enables the intended path recognition in an early stage of human motion, and through the FIE, the desired impedance pattern is merged through the proposed assistance system.

Kulić and Croft [30] developed a human intention inference unit aiming to assist a controller of the HRI system. The intention signal was used in a planning and control strategy to enhance the safety and intuitiveness of the interaction. Different physiologic data, which were blood volume pressure, skin conductance, chest cavity expansion/contraction, and corrugator muscle activity, were collected and featured extracted before being fed into the intention recognition unit. Then a fuzzy inference mechanism was used to infer human emotional states such as its valence and arousal level. The final estimation of arousal level achieved 94% accuracy within four subjects and the performance of valence estimation is 80%. The intention of an HRI monitoring system was separated into two components: attention and approval (Fig. 1.6). Human attention was determined by both physical and cognitive processes focused on the robot. Human approval to robot's

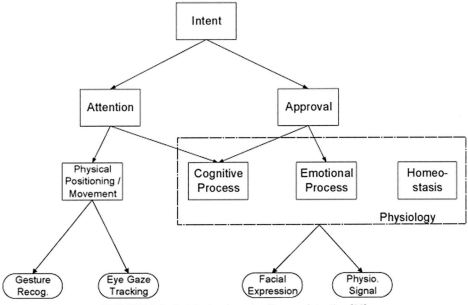

FIGURE 1.6 Architectural graph of human intention [30].

work should not only be determined based on physiologic signals but also be validated physically based on user attention information.

In Ref. [31], the authors proposed an HRI method for a LEGO assembling task. The human upper body type robot was designed to assist the human finish the assembling job by estimated human intention. Eye gaze information was collected and used for intention inference. Three kinds of potential cooperation actions for the robot were defined, which were taking over, settlement of hesitation, and simultaneous execution. The humanoid robot recognizes and classifies the human state into one of the three states and takes appropriate actions to help humans finish the assembling task. The authors introduced a human intention recognition method among human-robot collaboration tasks [32]. The intention was recognized by using a probabilistic state machine to classify explicit and implicit intentions. A human interacted with a robot arm in the experiment that was executed in an interactive workshop. Five explicit intentions were introduced: picking and placing the intention of an object, passing an object to the robot, placing the object, picking and holding an object, and giving the pointed object to the human. Two implicit intentions were piling up the objects and unpiling the objects. The proposed probabilistic state machine for the robot is working efficiently on both explicit and implicit human intention recognition.

Bien et al. [33] pointed out that from the point of intention reading, a human-friendly interface can be classified into three classes according to the autonomy level, namely, a fully manual interface, semiautonomous interface, and fully autonomous. Given the intelligence level, a human-friendly interface can be classified into two classes. In the first class, computational intelligence was used and in the second class, the higher level, the machine can predict human intention on the job, decide whether it can be done or not, and interact with a human. The author also proposed two different kinds of systems designed for elderly and disabled people and that have the ability to read both intentions. Pereira [34] proposed a novel intention recognition method based on a causal Bayesian network and plan generator. Logic programming technique was used to compute the probability of intentions, and those with lower values were filtered out. This made the recognizing agent focus on the most important events, especially when making a quick decision. Then the plan generator generates conceivable plans that can achieve the most likely intentions given by the first step. The plan generator guides the recognizing process concerning hidden actions and unobservable effects. In Ref. [35], an intention recognition method based on the observation of human behaviors, relevant goals, and current context was proposed. The author introduced an extra level between actions and intention, which was called the goal level. The definitions of intention and goal were given as goal was something that humans want to achieve, while intention was a mental state of what a human was going to do (see Fig. 1.7).

To achieve the intention recognition, an intention graph method was used. There were five elements in one intention graph, namely, state, action, goal, intention, and edges. In the goal recognition step, a graph was constructed, and the relevant goal was determined based on the actions. Then in the intention recognition part, the determined goal and user profile information from the current context were used to infer the real intention. Therefore the intention inference can be viewed as the backpropagation of human behavior execution. In Ref. [36], the authors proposed a human intention recognition method in a smart assisted living system based on a hierarchic hidden Markov model (HHMM). They used an inertial sensor mounted on one finger to collect finger gestures. Five finger gestures were defined, which are come, go fetching, go away, sit down, and stand up. The final result showed that by using HHMM, the model could achieve 0.9802, 0.8507, 0.9444, 0.9184, and 1.000 classification accuracy for the five kinds of hand gestures, respectively. Wang

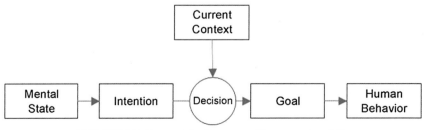

FIGURE 1.7 Human intention recognition procedure [35].

et al. [37] introduced a human intention inference system named the intention-driven dynamic model (IDDM), which is based on the extension of the Gaussian process (GP) model and Bayes theorem. The author extended an intention node in IDDM based on GP and introduced the online algorithm to infer human intention based on movement. The authors made a sufficient comparison with other algorithms such as SVM and GP classification. The performance of the novel method overweighs the traditional ways showing that IDDM is efficient in dealing with human intention problems.

In Ref. [38], a tractable probabilistic model based on the Bayesian network was introduced. The experiment was designed in a virtual kitchen simulation environment. The virtual human intentions are to load the dishwasher, wash dishes, cook and lay the table, get a drink from the robot, and get an object from the robot. An expert provided the mapping of the intention of actions. The user's head and hand motion were tracked, and these observable data were fed into a hidden dynamic Bayesian network (DBN). One of the key benefits of the proposed method is that it can derive the model directly based on expert knowledge. Bascetta et al. [39] developed a human intention estimation method focusing on human intended motion trajectory and body pose prediction for humans interacting with the industrial robot. The proposed method enabled safe cooperation between humans and robots in the industrial area even without protective fences. The robot detected the human through a low-cost camera. Based on the human tracking algorithm, the robot can predict the human's intended working area (four areas were defined, namely, human working area, robot area, inspection area, and cooperation area) before the human comes into a certain area. The intention recognition algorithm was performed by an HMM and results showed that the intention estimation algorithm could successfully predict the interaction area; 92% human intention was correctly recognized. In Ref. [40], a generic human intention method was introduced. The author analyzed the property of human intention and its relationship with actions. Then a DBN method was used because of its flexible characteristics to deal with arbitrary systems or domains. To cover both continuous and discrete intentions, a hybrid DBN method was discussed. To implement the model to the robot, the possible user intention and actions should be first determined; then the model parameters have to be learned from either expert knowledge or data-driven method.

Finally, the measurement nodes have to be modeled based on the given sensor system.

One interesting point should be pointed out that there are some differences between the driver's intention and driver's behavior. Although there are no clear definition and boundary between driver behavior and intention in prior research, the differences between these two concepts should be aware of. Driver intention reflects a driver's mental state, whereas driver behaviors are actions that drivers take. Driver behavior has a much wider scope than driver intention, which is focused not only on driving behaviors but also on some other behaviors including distractions (such as answering phone [41]) and mental and physical fatigue (such as yawning and sleepiness [42]).

In conclusion, driver intention is a subclass of human intention and can be roughly defined as a kind of driver's thoughts during the goal-oriented driving task. The driver's intention will influence the driver's behaviors and a series set of actions will then be executed by the driver based on the strength of driver intention.

## DRIVER INTENTION CLASSIFICATION
Driver intention can be classified into different categories based on a different perspective. It can be classified according to awareness, motivation, timescale, and the direction of driving. Among these, the two most used classifications are based on the timescale of the intention and on the route. In terms of timescope classification, Michon [43] pointed out that the cognitive structure of human behavior in traffic environment is a four-level hierarchic structure, which contains road users, transportation consumers, social agents, and psychobiological organisms. Among these four levels, the road user level is directly connected with drivers and can be further divided into three levels: strategy, tactical, and operational levels (also known as control level)(Fig. 1.8). These three cognitive levels can be viewed as three driver intention levels based on the timescale. The strategy level defines the general plane of a trip that determines the trip route, destination, risk assessment, etc. The time constant will be at least at the minute level or even longer. In this level the driver plans the transport mobility and comfort issues, which formed a long timescale problem. At the tactical level, the driver can make a series of short-term decisions and plans to control the vehicles. The time constant is much shorter than that in the strategical level, which usually lasts for a few seconds. The hierarchic level of intention is shown in Fig. 1.8.

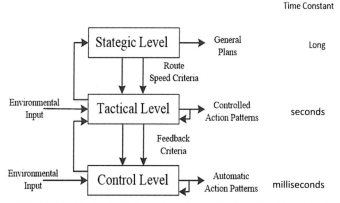

FIGURE 1.8 Three-level architecture of road users to describe driver intention [43].

Tactically planned intentional maneuvers include a sequence of operational maneuvers to fulfill a short-term goal such as turning, lane changing, and braking etc. [44]. All the control commands must meet the criteria from the general goal set at the strategy level. Lastly, the operational level intention is the shortest among the three levels and stands for the willing to remain safe and comfortable in the traffic situation and the driver directly gives control signals to the vehicle. The time constant can be a microsecond level. Most of the existing studies have dedicated to operational level maneuver study and the strategy route plan and prediction based on GPS techniques. However, the main task in this book will focus on tactical level study, particularly lane change maneuvers; as the time constant of tactical maneuvers is a few seconds, it is important to correctly identify and recognize the intention to enhance the functionality of intelligent driver assistance system.

There are also some other ways of classification for the cognition of driver intention. Salvucci [45] developed a driver model in Adaptive Control of Thought-Rational cognitive architecture. Similar to the three-level architecture of the road user model given by Michon, Salvucci developed the integrated driver model into three main components, which were control, monitoring, and decision-making modules. Particularly, the control component was similar to the operational level given by Michon, which gave the charge of perception of the external world and translate any perceptual signals directly to the vehicle. The monitoring component keeps aware of the surrounding situation and the environment by periodically perceiving data and inferring. The decision component, which has the same function as part of Michon's tactical level, makes tactical decisions for individual maneuvers

according to the awareness of the current situation and information that is gathered from the control and monitoring modules. One significant advantage of the cognitive driver model is the incorporation of built-in features that mimic human abilities.

Among cognition scope, Rasmussen [46] developed another three-level architecture model to describe human performance behavior. Three typical levels, namely, skill-based, rule-based, and knowledge-based levels, were defined according to human performance. The three-level model described human performance from a familiar situation to unfamiliar situations. Skill-based behavior represents human's behavior in a familiar situation. All the activities at this level go fast, smooth, and automated, which means a human is very confident to take any maneuver at this level and is very familiar with the task. A more detailed example and how this three-level architecture can be related to the model given by Michon is showed in Table 1.1 [47]. The next level is rule-based behavior. In this level, humans become less familiar and confident with the situation as in the first level. All the behaviors in this relatively familiar situation are based on the stored rule-base in the brain. During unfamiliar situations, it comes to the knowledge base level. In this level, the control of the action moves to a higher conceptual level. The behavior here is goal-oriented and clearly formulated. Humans need to analyze the environment and make an explicit plan to complete the task based on the knowledge they have in their brain. The plan is usually selected from different plans that have been made, and the one that gives the least error should be chosen. This level is similar to the strategy level mentioned earlier. When drivers are dealing with an unfamiliar destination, they have to search the optimal route and transport methods to finally complete this task.

**TABLE 1.1**
**Examples and Relationships Between Models Given by Michon and Rasmussen [47].**

|  | Strategic | Tactical/Maneuver | Operational/Control |
|---|---|---|---|
| Knowledge | Navigation in unfamiliar area | Controlling skid | Novice on first lesson |
| Role | Choice between familiar routes | Passing other vehicles | Driving unfamiliar vehicles |
| Skill | Route used for daily commute | Negotiating familiar intersection | Vehicle handling on curves |

The classification of driver intention toward directional classification, on the other hand, is quite straightforward. As we are considering the underground vehicles, they can only have two directions: longitudinal and lateral. The driver's longitudinal behavior contains braking, acceleration, deceleration, starting, and lane keeping. Lateral behavior usually contains turning, over-taking, and lane changing. In previous research, most researchers pay attention to lateral intention prediction because they are more complicated and more important than longitudinal behaviors owing to the interaction with other vehicles.

In terms of longitudinal intention prediction, most of the previous research focuses on the study on braking intention recognition. Kim et al. proposed a braking intention prediction method by using EEG and electromyographic (EMG) signals and tried to study the neuro-driving simulation framework [48–50]. Three different braking maneuvers were defined in this paper: sharp braking, soft braking, and no braking condition. They applied three algorithms: kernel ridge regression, linear regression and combined linear regression, and classification approach to make a prediction and classification of emergency braking, voluntary braking, and mixed braking. EEG signal and vehicle technical signals (i.e., vehicle brake input) were used as the input of the inference mechanism. The final result showed that the predicted proposed algorithm output can bring the minimum error with the measured value, and linear regression gave the second good result. Moreover, the result indicated that within these three methods, the prediction performance of normal braking is always better than that of emergency braking for the reason that these features are more suitable to be used for normal driving classification. It is considered that the combined EEG features will contribute to the decrease of prediction errors.

Haufe and Kim aimed to prove the ability of correlated neural electrophysiology to improve the prediction of emergency braking situations in the real-world driving environment [51]. Vehicle states as well as EEG and EMG signals were used to train the classification system. The regularized linear discriminant classifier was trained to identify emergency braking intention. Their conclusion suggested that the electrophysiologic method can be used in the braking assistant system. The reasons were multifold. First, the identifying time by using electrophysiology in the real world was even earlier than that in the simulator (can detect braking intention 237 ms before the braking action occurs), and second, the performance of intention identification method was robust even when drivers took a secondary task. Kim et al. also proposed a driver braking intention method by using EEG signals [52]. Particularly, normal braking and emergency braking were defined and classified by the kernel ridge regression model. By analyzing EEG signals, three features were used: event-related potentials, readiness potentials, and event-related desynchronization. The author concluded that, in general, the prediction performance of normal braking could be much better than that of emergency braking. The reasons were the features used in the study were more suitable for normal braking intention analysis and the overfitting problem of the regression model given by the training data. Haufe used EEG and EMG as an input signal to predict the driver's emergency braking intent [53]. The signals after feature extraction were fed into a regularized linear discriminant analysis classifier. The simulated systems with EEG showed a 130 ms earlier detection than those that rely only on brake pedal signals. Khaliliardali proposed a driver intention prediction model to determine whether the driver wants to go ahead or stop [54]. The method was to classify driver go and stop intention based on brain-machine interface. Particularly, EEG, electrooculographic (EOG), and EMG signals from six subjects in a simulation environment were collected, and two classification methods (linear and quadratic discriminant analyses) were used separately to evaluate the classification performance. The final result showed that both the algorithms could achieve above 70% accuracy for driver intention and above 80% for brake intention.

The research mentioned earlier uses the EEG signal as one direct measurement for a human mental state. But other research tends to use surrounding context information, vehicle sensors, and driver information from cameras and the Controller Area Network (CAN) bus. For example, McCall and Trivedi [55] integrated driver intention into an intelligent braking assistance system. A sparse Bayesian learning algorithm was used to infer the driver's intention of braking. Three kinds of data sources were used in this work, which were vehicle data, surrounding traffic information, and driver state. Head and foot monitoring cameras were used to capture the driver's head and foot motion. The system collected data from cameras, GPS, lidar, and radar and CAN bus systems. To capture information about the driver's actions, a color camera was installed to observe the driver's head and a near-infrared camera was used to observe the driver's feet. The system recognized driver actions once the driver removed his/her foot. The results showed that the inference system could detect the driver braking intention 1−3.3 s before the braking action occurs, with a relatively low false alarm rate. The author concluded that by adding foot cameras to the vehicle and from the surrounding data, the inference performance could get a significant increase of 1 s before the braking happens, while the additional head camera showed few influences on the performance of the system. Another interesting research by Tran, Doshi, and Trivedi [56] predicted the driver's braking intention by directly monitoring the foot gesture through cameras. They showed that driver foot gesture is an important factor in vehicle control, and therefore the usage of vision-based foot behavior has its advantage, which is direct and accurate. Also, by monitoring the foot behavior directly, it can help predict a pedal press action before it happens. A new vision-based framework for driver foot behavior analysis was developed using optical flow-based foot tracking and HMM to characterize the instant foot behavior.

Mabuchi and Yamada [57,58] estimated the driver's stop and go intention at the intersection when facing the yellow traffic lights. Two algorithms were used to predict driver intention, which were an SVM and the logistic regression model. Also, two methods were used to model vehicle behavior at the intersection depending on whether or not the driver realizes the yellow light. The experiment was designed in a real-world situation, and 22 subjects participated in the experiment. The final results showed that with a low false-positive rate (5%), the true positive rate of both the classifiers could go

beyond 90%. Takahashi et al. predicted the driver deceleration intent in a downhill road [59]. A gear shift operation will be executed automatically once the driver's mental model outputs a deceleration signal. The results indicated that an automatic gear shift performed by the controller should be finished before the driver forms an intention to shift gears. To introduce the driver's driving experiences and personality traits, the author used Interpretive Structural Modeling (ISM) method to describe these qualitative and quantitative factors. The inference model consists of two types of submodels, which can be regarded as qualitative and quantitative. The deceleration intent can be deduced based on the output of these two submodels.

Kumagai et al. proposed a method to predict the driver's braking intention at intersections (particularly right turns) by using a DBN [60]. The data from the real vehicle, which were vehicle speed, acceleration pedal, and brake pedal, are collected. Two different Bayesian networks, namely, HMM and switching linear dynamic model, were constructed. The results showed that the inference system could predict the future braking intention several seconds prior based on the current observed data. The author also pointed out that the easiest way to make stop prediction is to construct a static table that maps the observed data into the frequency of future stops. However, this method may need more training data to get a convincing result.

In terms of lateral direction intent, most of the driver intention research in a lateral direction focused on lane changing and turning. As lane change analysis is the primary object in the following chapters of this book, this chapter will only focus on turning intention and a few types of research based on steering intention recognition. Cheng and Trivedi [61] developed a vision-based body-pose recovery and behavior recognition algorithms for driver assistance systems. The main task was to focus on a slow turn scenario (when the driver meets an intersection, the driver first stops the car and then starts to turn). It was shown that the classifier can achieve 82% and 96% prediction accuracy for the left and right turns when the prediction was made 1.5 s prior to the intersection. The false-alarm rate is around 20%. When the decision was made 0 s before the vehicles enter the intersections, the classifier performance decreased to 82% and 75% for the left and right turns, respectively. Meanwhile, the author examined the influence of steering data and its combination with pose data; it was found that the classifier trained with both pose and steering data produced the best true-positive rates between −1- and 1.5-s decision times at better than 85%. This means the classifier trained by head and hand pose, along with steering data, gives better performance than head pose only, hand pose, and steering only.

Windridge et al. [62] aimed to find a model that can identify intentional task hierarchies. Martin proposed an architecture that can deal with arbitrary combinations of subsequent maneuvers, and a varying set of available features was also investigated [63]. Right turn maneuver inference was studied as an example of the inference system. In the study, the authors concluded that discriminative approaches were usually used to predict a single type of maneuver, while generative models were more often applied to infer multiple maneuvers. Ohashi et al. designed a system that recognized the turning left, right, and going straight intentions by collecting data of human motions, the velocity of the vehicle, and the distance of host vehicle to the intersection [64]. The intention recognition model was proposed using case-based learning based on the fuzzy associative memory system. The final result showed that the ratio of correctly detecting left turning is 88%, going straight is 95%, and turning right is 86%. The model was proved to be efficient and was able to identify the intention at an early stage when detecting nonlearned data.

Hülnhagen predicted the turn maneuver at the intersection by using probabilistic finite-state machines and fuzzy logic [65]. First, each driving maneuver was modeled by a set of basic elements. Each element was specified by a set of fuzzy rules. Then a Bayes approach was introduced to recognize a driving maneuver. A training method based on the optimization method was used to train the fuzzy rule based on recorded data. The final result showed that the system could achieve a 73% correct rate in the 23 total events. Sato and Akamatsu [66] evaluated the remaining distance of a vehicle to the center of the targeted intersection when the driver was monitored to cover the brake pedal and perform a right turn signal switch maneuver. It was different from most of the existing research, which focused on the time and moment profile. The authors focused on vehicle location to predict driver intention. More specifically, vehicle velocity, relative distance to the leading and following vehicle, the onset locations of covering brake pedal, and turn signals were used to construct a driver structural equation model as the intent inference mechanism.

Liebner et al. [67] aimed to infer driver intention based on an explicit longitudinal driver behavior model and simple Bayesian network. To evaluate the dynamic

vehicle performance at intersections, four different kinds of driver intentions were defined, which were going straight, stop at a red light, turn right, and turn right but stop at a pedestrian crossing. Two particular maneuvers (going straight and turning right only) were evaluated. Based on the different velocity profiles of different driver intentions, different probabilistic distributions for the maximum acceleration parameter were determined. Lastly, a simple Bayesian net was used as the inference mechanism. Measurement data were collected from four different drivers at an inner-city intersection.

Driver tactical maneuvers consist of a series of operational maneuvers. Some of the existing research focus on the analysis of multiple tactics rather than a single tactical task. By using machine learning theory, a single intention inference task can be completed with a discriminative and generative model. However, multiple task inference prefers to use generative models such as Bayesian networks and relevant inference algorithms [44]. Mitrović [68] proposed a driving event recognition scheme based on HMMs. Particularly, seven driving events (right curve, left curve, right on the roundabout, left on the roundabout, right turn, left turn, and straight on the roundabout) were studied and classified by using real vehicle driving data in a normal driving environment. Multiple models were evaluated based on the observation sequences from the training set, and the highest probability decides what kind of events this sequence represents. The left-to-right HMM was selected because of its ability to better characterize the data than general HMM. The driving event with the highest probability was chosen, and only 4 of 238 test events were wrongly classified. Although the system cannot always clarify between left turn and left turn on roundabout, the system can correctly recognize 234 of 238 driving events with about 98.3% accuracy.

In Refs. [69 and 70], the driver assistance system based on the human-machine interface was studied. Oliver and Pentland proposed a driver behavior recognition and prediction method based on dynamic graphic models. Particularly, a dynamic graphic model, HMM, and its extension model, coupled HMM, were used as the prediction controller. The primary task of these studies was to evaluate the influence of contextual information on driver behavior recognition. Thus environmental data were collected and used in the paper, which was surrounding traffic and lane information, vehicle data, driver head position, and viewpoint. All the experiments were designed in a real traffic situation. Seven driver maneuvers were studied, including passing, changing right and left, turning right and left, starting, and stopping. Final results indicated that the predictor could recognize the driver's maneuver 1 s before a significant change in the car and contextual signals take place. Different kinds of situations were studied to test the HMM performance, such as car signal only, car and lane position, car and driver gaze, and car, lane, and driver information. Final results showed that by using a combined data source, the prediction performance of passing and stopping could have 100% accuracy, while the performances of changing lane right and left are much lower than the other maneuvers (6.3% and 23.5%, respectively). It can be concluded that the performance of only use car information will have a plateau. Some driving events, such as passing and lane changing, cannot rely only on car information. The context information is important to identify turning and lane-changing maneuvers. The driver's eye gaze signal, which can reflect the driver's mental state, can be very useful in the prediction of lane changing, passing and turning events, etc.

Liebner inferred the driver's intent based on an explicit model for the vehicle's velocity, and an intelligent driver model was used to represent car-following and turning behavior [71]. The model input data were extracted from real-world data to account for different driving styles. Four kinds of driver intent were defined and recognized, which were going straight, stop at the stop line, turn right, and turn right but stop at a pedestrian crossing. To construct the simple, intelligent driver model, the velocity profile and its derivate were used. The author believed driving maneuvers consist of a linear sequence of actions that were much suitable for the application of HMMs. He and Zong used a double-layer HMM structure to predict driver intention and driving behavior [72]. Specifically, a lower layer Gaussian HMM was designed to infer the short-term driving behavior based on the three main data sources: brake and acceleration pedal, steering wheel angle, and speed. Then, the upper discrete HMM can estimate one of the tactical driver intentions (emergence braking, hill starting, braking in a turn, and obstacle avoidance) based on the recognized driving behaviors. Also, driving behaviors shortly were predicted using the likelihood maximum method. The driving intention was recognized online using Lab-VIEW (Laboratory Virtual Instrument Engineering Workbench). The system using the proposed method can achieve smooth transition of control modes between automated and manual driving.

Jingjing and Yihu [73] proposed a two-dimensional HMM using the Gaussian mixture probability density as driver intention identification. The driver's intention was viewed as a combination of longitudinal and lateral direction intention, and six driving intentions are studied. With a 3-s time interval of input data and eight Gaussian mixture numbers, the identification accuracy can achieve 98.84%. The author believed that speed change and lane change should be considered at the same time rather than separately, which was not enough to reflect the actual situations. Imamura developed a driver intention identification and intention label definition method based on the assumption of compliance with traffic rules [74]. Four subjects participated in the experiment, and the driving behavior data were collected by using a driving simulator with the control of gas, brake, and steering wheel. The GMM was used as the intention identification algorithm, and the expectation-maximization algorithm was used to train GMM parameters. After training the GMM offline, the author constructs an online estimation system. Also, four intention labels, which were normal driving, deceleration, acceleration, and turn left/right are defined. The final result showed an average estimation accuracy of 83.3%.

Liu and Pentland [75] aimed to analyze the sequence of steps within a driving action in their study and to use the first steps to identify the action. The main approach was to model human behavior in a Markov process. The result showed that by using the method developed, the driver's intention can be observed before the maneuver is performed. The experiment was designed in a simulation environment with eight adult male subjects who drove through a citylike virtual environment. The driving command included stop, turn left, turn right, change lanes, and passing a car. A time window method was used to examine its effect on driver intention prediction. In terms of a left turn prediction, the accuracy is between 50% and 60%. For passing, the recognition accuracy is 60%−70% 1 s after the command, and the performance of the right turn can be 60% 2 s after the command. The stopping and following can only achieve 40%−50% 2.5 s after. As can be seen, using multiple tasks for driving intention recognition was much complex than a signal task. It has to define and clear more driving maneuvers and take many more experiments. Moreover, it should pay more attention to the generative inference algorithm, which was supposed to recognize more driving maneuvers correctly and accurately.

## STUDIES RELATED TO DRIVER INTENTION INFERENCE

There are some other research that have not directly studied host driver's tactical intention behavior but have focused on the other aspects of driver mental state and intention relevant areas, for example, the operational driver modeling, strategy level programming, driver type classification, driver's attention and distraction analysis, and driver intention recognition of surrounding vehicles. Among these, the operational and strategic level intention research also have a close relationship to the tactical driver's intention, as they are from the three-level driver mental architecture and can interact with each other. Similarly, studies on driver's attention and distraction can be analyzed with the same methods as it was used in the driver intention study because all these research were preformed to detect and recognize the inner mental state of the driver by observing outer real-world variables. Driving intention, attention, and distraction research all belong to human cognition study, which forms a complete mental research system. Another interesting research area is to infer the surrounding driver's intention. This is also a challenging task and is important to traffic security. Unlike host driver intention inference, surrounding intention inference can only be analyzed based on the measurable parameters of target vehicle, such as its velocity and acceleration estimation and the surrounding environment. As the current technology has not solved the driver's intention inference and vehicle-to-vehicle (V2V) data transition completely, the study on surrounding driver's intention will still focus on the outer vehicle characteristics and analysis of the surrounding environment.

Imamura and Takahashi et al. proposed a driver's intention and driver style estimation method based on the HMM, considering the driver's personality and characteristics [76]. In this paper, the authors defined four kinds of driving state, which are stopping, acceleration, cruise, and deceleration. Besides, the authors evaluated the different driver characteristics of driving in a hurry and driving normally by using a driver's intention ratio index. The driver's intention ratios of hurry driving were always larger than that of normal driving except in the group of cruise state. In the research, the experiment was designed in driving simulator environment and gas, and brake pedal signals, as well as vehicle acceleration, were collected and packed as the input signal. Then the HMM was trained to be the driving state classifier. The final results showed that by considering the input data, the classifier was efficient in recognizing the different driving states and can accurately classify the drivers' intention into a hurry and normal driving group.

In terms of strategic intention study, most of the existing research focus on driving route prediction based on GPS and digital maps to estimate the driver's strategy intention. Rammelt [77] built two types of tactical driver models, which partly connect to the strategy level. There were two different driver models: reactive model and planning model. Both the models were constructed based on DBNs. In a reactive model, the output probability of the intended maneuver was only related to the current sensors' input, and the planning model aims to find a solution to get into a zone of interest. Five index values were introduced to evaluate the performance of the plan, which were a prior, belief in success, benefit, risk, and effort. The planning model, which combines a plan module and probabilistic module, makes a natural formulation of the driving task.

Nakajima proposed a route recommendation method for a car navigation system based on the driver's intention estimation [78]. The method focused on the driver's strategic route plan. A difference-amplification method was used to estimate the driver's intention on route based on the comparison of the newly selected route and the original one given by the Dijkstra algorithm. The difference amplification method updated the driver's current route intention by recalculating the distance cost when the real route was different from the original recommended route. Simmons predicted driver's every day strategic route and destination intention by using the HMM [79]. Drivers' daily driving data were collected from the GPS and map database, and thus their natural driving habit was recorded. The method can predict the driver's destination and route accurately online through the observation of GPS position. The final result showed that approximately 98% accuracy could be achieved. The novel method was designed with the assumption that the driver's driving habits and goals were always similar during weekdays. Data bins were used to characterize the data to construct an extended HMM.

Driver behavior at the intersection also attracted many researchers' efforts for the reason that intersection is one of the most important and challenging parts in real-world driving. The traffic at the intersection can be very complex, which makes it one of the major accidents happening areas. Therefore the development of ADAS, which can predict driver intention and behaviors at the intersection, will largely enhance driving safety and efficiency. Two novel classification methods were introduced by Aoude to identify the driver types at the intersection [80,81]. Specifically, a discriminative model based on the SVM and Bayesian filter and a generative model based on the HMM and expectation-maximization methods were evaluated separately. These classifiers aim to determine whether the oncoming driver was a violator (cannot stop before the stop line of the intersection) or compliant. By observing 10,000 examples in a certain intersection, the classifiers could correctly recognize the driver's stop intention. Under 1 s of time-to-intersection conditions, the SVM with a Bayesian filter classifier achieves 85.4% accuracy and the HMM obtains 80.0% accuracy, both can get a high-level true-positive rate with a relatively low (5%) negative-positive rate. Lefèvre and Javier et al. proposed a Bayesian inference system to recognize the driver behaviors at intersections [82,83]. They introduced contextual information through a digital map to provide informative cues for analyzing driver behaviors. Four kinds of data were used to describe the driver behavior at the intersection, which were entrance road, entrance lane, turn signal, and path in intersections. Finally, a probability density of the intended maneuver was inferred from these data.

Another challenge in research is driver attentive level and distraction monitoring, and their correlation with driver intention inference. The driver's attention reflects a driver's mental state and concentration level while driving, which makes the analysis very similar to driver intention inference. Both can be viewed as a mental inference process based on the observation of human outer physical behavior. McCall and Trivedi [84] described a driver attention monitoring system design on intelligent human-centered vehicle test bed. To monitor and recognize driver attention status, a variety of data cues were captured and synchronized. For instance, facial expression analysis was obtained through the thin-plate splines algorithm. The lane position analysis was proposed by applying one-lane tracking system. The data for the steering wheel and pedal movement analyses were obtained from the vehicle CAN bus; time headway was calculated with data collected via the radar system. All the cues were proved to be effective in the determination of driver attention level. Similarly, Tawari et al.introduced an urban intelligent assistance system to deal with the complex and unique environment in an urban area [85]. Two specific subsystems were developed, which were driver attention guard and merge and lane change assist system. The intelligent driver assistance system was fully mounted on a real-world vehicle to collect natural data while driving. The data sources covered the real-time situation of driver, vehicle, and surrounding environment based on the internal vision system; vision system;

radar, lidar, and GPS; and vehicle dynamic sensors. The merge and lane change assist system can recommend the driver when and how to make the lane change based on a dynamic probabilistic drivability map. Then the real on-road test, which is open to the public and press, was performed to prove the efficiency of this system. Lastly, Doshi and Trivedi evaluated the head and eye gaze dynamics under various conditions leading to the driver's distraction [86]. The experiment was designed in a simulation environment to make sure the safety of the subject. The authors finally concluded that the eye gaze and head pose information were essential features for the recognition of driver distraction and attention states.

## CONCLUSION

Driver intention inference is a very complex task that relies heavily on knowledge from a variety of research area such as psychology, cognition, human-machine interface, and automotive engineering. Driver intention study is a subtask for human intention cognition. As mentioned earlier, human intention can be described as a hierarchic mental architecture and will influence the actions that are going to occur. Therefore by observing human actions, it is possible to infer human intention with this given observation. Similar to human intention, driver intention reflects the driver's thoughts related to the driving task and will influence the driving action and maneuvers based on the strength of the intention. It should also be pointed out that there are some differences between driver maneuver and driver intention. The driver's intention here is closer to the driver's mental research and inference than driver maneuver.

Driver intention can be classified based on different criteria. For example, it can be classified according to the timescale as well as the direction of the action that is going to take place. In terms of intention timescale, there are three typical levels, which are strategy, tactical, and operational levels. Strategy intention determines the driving planning level and route selection task, which is the longest task among the three. Operational level intention focuses on the instance and directly controls the vehicle, such as steering and pedal control. These maneuvers can be finished in a very short period, usually at the millisecond level. The tactical intention, which determines the action of the next few seconds, is more important in the real-world task. Most of the important decision during driving can be classified into the tactical maneuver. Therefore the study of driver's intention at this level is of importance to the improvement of driving security. The driver will get a clearer idea about when and how to optimally execute their thoughts based on the information given by the ADAS.

It is also reasonable to classify driver intention from the perspective of direction. Many previous types of research focus on longitudinal and lateral vehicle dynamics. Lateral dynamics are usually more complex than longitudinal dynamics. Thus driver intention in the lateral direction has been the research hot spot for a few years. A lot of studies have focused on lane changing and lane keeping, and some other lateral intention focus on analyzing driver behavior at intersections, such as turning, stopping, or passing. On the other hand, most of the longitudinal driver intention studies focus on intelligent braking systems development. By predicting and recognizing the braking intention, the assistance system gives better guidance to the driver and improves cooperation efficiency.

Apart from the direct tactical driver intention study mentioned earlier, many other types of research have a close relationship to driver intention. Some research focus on driver behavior at intersections by inferring the other vehicle's driving intention based on sensor fusion and vehicle communication. Moreover, to recognize the driver intention online at a strategic level is another interesting aspect. Most of the research determine the driver strategic intention based on GPS, digital map information, and weekly driving task recordings. From a cognitive point of view, driver attentive level and distraction recognition are very similar to the driver intention study, which also can be viewed as a cognition process by observing the driver's outer observable behaviors.

In summary, driver intention is a mental process during driving and can be identified by the following driving actions. There are many ways to classify driver intention based on different criteria, and the tactical level intention is the most important aspect that we are concerned about.

## CHAPTER OUTLINES
The main content of this book is divided into chapters according to different functional modules within the driver intention system. The book starts with an overview of the current issues related to the driver intention inference, intelligence, and automated vehicles. This chapter serves as a motivation to the following research and states the application area of the proposed technology. Chapter 2 provides the analysis of both human intention and driver intention. Literature surveys about

driver lane change focus on the state of the art of the intention inference and the challenges. Chapters 3 and 4 cover a literature review of lane detection techniques and the integration methodologies for the lane detection system.

Meanwhile, a parallel driving framework for lane detection, namely, the parallel lane detection system is introduced. Chapters 5 and 6 illustrate the driver behavior recognition studies concerning the lane change maneuver prediction based on conventional machine learning and deep learning methods, respectively. Chapter 7 introduces braking intention recognition and braking intensity prediction. Chapter 8 covers the lane change intention inference framework and the prediction results using naturalistic data. Finally, Chapter 9 concludes the whole book and discuss the challenges as well as future works.

The main content of each chapter is summarized as follows.

Chapter 1: Introduction and motivation for the research detailed throughout this book are proposed.

Chapter 2: This chapter provides studies about human intention as well as driver intention. Driver intention is classified into different categories according to different criteria. The lane change intention is also surveyed from the sensory level, algorithm level, and the evaluation level.

Chapter 3: As traffic context is the major reason for lane change maneuvers, one of the most important aspects, lane detection, is studied in this chapter. This chapter covers a sufficient survey of the lane detection techniques and the integration with other onboard systems.

Chapter 4: A novel algorithm-level integrated lane detection framework is introduced. The integrated lane detection and evaluation system are designed to improve the robustness of the lane detection system for further driver intention inference system construction.

Chapter 5: Driver behaviors are the major clues for driver intention inference because driver behaviors carry valuable information that can reflect the mental state of the driver. This chapter provides a driver behavior and secondary tasks recognition algorithm based on the driver's head and body feature selection. The feature engineering and conventional machine learning methods for driver behavior recognition are introduced.

Chapter 6: This chapter moves a step further in driver behavior monitoring, which use an end-to-end deep learning approach to avoid complex feature extraction and selection procedure.

Chapter 7: This chapter introduces an unsupervised machine learning method based on braking intention recognition system. The braking intention is identified and compared to two different unsupervised learning methods. Then the feedforward neural network is used to estimate the braking intensity based on different braking styles. Lastly, the feedforward neural network is used to construct an integrated braking intention learning and braking intensity estimate system.

Chapter 8: Driver intention inference algorithm should be able to learn the dependency roles among the temporal sequence data. In this chapter, a driver lane change intention inference algorithm based on the deep learning method is proposed. Multimodal naturalist data that are collected on highways are fed into the driver intention inference system. The algorithm can predict driver lane change intention a few seconds before the lane change maneuver initiates. Then regarding the surrounding driver intention inference tasks, a time-series trajectory prediction model is proposed to predict the track of the leading vehicle in real time.

Chapter 9: This chapter provides the final remarks, discussions, and the future work required for driver lane change intention inference.

## REFERENCES

[1] L.S. Angell, J. Auflick, P.A. Austria, D.S. Kochhar, L. Tijerina, W. Biever, T. Diptiman, J. Hogsett, S. Kiger, Driver Workload Metrics Task 2 Final Report, 2006. No. HS-810 635.

[2] M. Peden, World Report on Road Traffic Injury Prevention, World Health Organization, Geneva, 2004.

[3] Fundación Instituto Tecnológico para la Seguridad Del Automóvil (FITSA), El valor de la seguridad vial. Conocer los costes de los accidentes de tráfico para invertir más en su prevención (in Spanish), Report funded by the Spanish General Directorate of Traffic, Universidad Politécnica de Madrid, Madrid, 2008.

[4] T. Kowsari, Vehicular Instrumentation and Data Processing for the Study of Driver Intent, Dissertation, The University of Western Ontario, 2013.

[5] D. Geronimo, et al., Survey of pedestrian detection for advanced driver assistance systems, IEEE Transactions on Pattern Analysis and Machine Intelligence 32 (7) (2010) 1239−1258.

[6] M.G. Ortiz, Prediction of Driver Behaviour, Ph.D. dissertation, Universitätsbibliothek Bielefeld, 2013.

[7] J.C. McCall, Human Attention and Intent Analysis Using Robust Visual Cues in a Bayesian Framework, Dissertation, The University of California, San Diego, 2006.

[8] J.C. McCall, M.M. Trivedi, Human behavior based predictive brake assistance, in: 2006 IEEE Intell. Veh. Symp., 2006, pp. 8−12, 200.

[9] H. Berndt, K. Dietmayer, Driver intention inference with vehicle onboard sensors, in: 2009 IEEE Int. Conf. Veh. Electron. Saf., 2009, pp. 102−107.

[10] H. Berndt, S. Wender, K. Dietmayer, Driver braking behavior during intersection approaches and implications for warning strategies for driver assistant systems, in: 2007 IEEE Intell. Veh. Symp., 2007, pp. 245−251.

[11] A. Eriksson, N.A. Stanton, Takeover time in highly automated vehicles: noncritical transitions to and from manual control, Human Factors 59 (4) (2017) 689−705.

[12] J. Nilsson, P. Falcone, J. Vinter, Safe transitions from automated to manual driving using driver controllability estimation, IEEE Transactions on Intelligent Transportation Systems 16 (4) (2015) 1806−1816.

[13] C. Lv, et al., Analysis of autopilot disengagements occurring during autonomous vehicle testing, IEEE/CAA Journal of Automatica Sinica 5 (1) (2018) 58−68.

[14] P. Carruthers, The illusion of conscious will, Synthese 159 (2) (2007) 197−213.

[15] R. Netemeyer, M. Van Ryn, I. Ajzen, The theory of planned behavior, Orgnizational Behaviour Human Decision Processes 50 (1991) 179−211.

[16] Bandura, Toward a unifying theory of behavioral change, Psychological Review 84 (2) (1977) 191−215.

[17] A.M. Treisman, A. Treisman, A feature-integration theory of attention, Cognitive Psychology 12 (1980) 97−136.

[18] C. Heinze, Modelling Intention Recognition for Intelligent Agent Systems, No. DSTO-RR-0286, Defence Science and Technology Organisation Salisbury (Australia) Systems Sciences Lab, 2004.

[19] K. a Tahboub, Intelligent human-machine interaction based on dynamic Bayesian networks probabilistic intention recognition, Journal of Intelligent and Robotic Systems 45 (2005) (2006) 31−52.

[20] E. Bonchek-Dokow, Cognitive Modeling of Human Intention Recognition, Dissertation, Bar Ilan University, Gonda Multidisciplinary Brain Research Center., 2011.

[21] A.N. Meltzoff, R. Brooks, Training study in social cognition, Developmental Psychology 44 (5) (2008) 1257−1265.

[22] Y.M. Jang, R. Mallipeddi, S. Lee, H.W. Kwak, M. Lee, Human intention recognition based on eyeball movement pattern and pupil size variation, Neurocomputing 128 (2014) 421−432.

[23] J.-S. Kang, U. Park, V. Gonuguntla, K.C. Veluvolu, M. Lee, Human implicit intent recognition based on the phase synchrony of EEG signals, Pattern Recognition Letters 66 (2015) 144−152.

[24] M. Wang, Y. Maeda, Y. Takahashi, Human intention recognition via eye tracking based on fuzzy inference, in: Soft Computing and Intelligent Systems (SCIS) and 13th International Symposium on Advanced Intelligent Systems (ISIS), 2012 Joint 6th International Conference on, IEEE, 2012.

[25] C. Zhu, W. Sun, W. Sheng, Wearable sensors based human intention recognition in smart assisted living systems, Informatica y Automatica 2008 (2008) 954−959.

[26] R. Kelley, A. Tavakkoli, C. King, M. Nicolescu, G. Bebis, Understanding human intentions via hidden Markov models in autonomous mobile robots, in: Proc. 3rd Int. Conf. Hum. Robot Interact., 2008, pp. 367−374.

[27] Y. Inagaki, et al., A study of a method for intention inference from human's behavior, in: Robot and Human Communication, 1993. Proceedings., 2nd IEEE International Workshop on, IEEE, 1993.

[28] P. Rani, C. Liu, N. Sarkar, E. Vanman, An empirical study of machine learning techniques for affect recognition in human-robot interaction, Pattern Analysis and Applications 9 (1) (2006) 58−69.

[29] Y. Yamada, Y. Umetani, H. Daitoh, T. Sakai, Construction of a human/robot coexistence system based on a model of human will-intention and desire, International Conference on Robotics and Automation 4 (May) (1999) 2861−2867.

[30] D. Kulić, E.A. Croft, Estimating intent for human robot interaction, in: IEEE Int. Conf. Adv. Robot., 2003, pp. 810−815.

[31] K. Sakita, Flexible Cooperation between Human and Robot by Interpreting Human Intention from Gaze Information, 2004.

[32] M. Awais, D. Henrich, Human-robot collaboration by intention recognition using probabilistic state machines, in: 19th Int. Work. Robot. Alpe-Adria-Danube Reg. RAAD 2010 − Proc., 2010, pp. 75−80.

[33] Z.Z. Bien, K.H. Park, J.W. Jung, J.H. Do, Intention reading is essential in human-friendly interfaces for the elderly and the handicapped, IEEE Transactions on Industrial Electronics 52 (6) (2005) 1500−1505.

[34] L.M. Pereira, H.T. Anh, Intention recognition via causal bayes networks plus plan generation, in: Lect. Notes Comput. Sci. (Including Subser. Lect. Notes Artif. Intell. Lect. Notes Bioinformatics), vol. 5816, LNAI, 2009, pp. 138−149.

[35] S. Youn, K. Oh, Intention recognition using a graph representation, Imaging 1 (1) (2007) 2−7.

[36] C. Zhu, Q. Cheng, W. Sheng, Human intention recognition in smart assisted living systems using a hierarchical hidden Markov model, Automation Science and Engineering (2008) 253−258.

[37] Z. Wang, K. Mülling, M. Deisenroth, H. Ben Amor, D. Vogt, B. Schölkopf, J. Peters, Probabilistic movement modeling for intention inference in human-robot interaction, The International Journal of Robotics Research 32 (7) (2013) 841−858.

[38] O.C. Schrempf, D. Albrecht, U.D. Hanebeck, Tractable probabilistic models for intention recognition based on expert knowledge, in: IEEE/RSJ Int. Conf. Intell. Robot. Syst., 2007, pp. 1429−1434.

[39] L. Bascetta, G. Ferretti, P. Rocco, H. Ardo, H. Bruyninckx, E. Demeester, E. Di Lello, Towards safe human-robot interaction in robotic cells: an approach based on visual tracking and intention estimation, in: 2011 IEEE/RSJ Int. Conf. Intell. Robot. Syst., 2011, pp. 2971−2978.

[40] O.C. Schrempf, U.D. Hanebeck, A generic model for estimating user intentions in human-robot cooperation, in: Proc. 2nd Int. Conf. Informatics Control. Autom. Robot. (ICINCO 2005), vol. 3, 2005, pp. 251−256.

[41] H. Alm, L. Nilsson, The effects of a mobile telephone task on driver behaviour in a car following situation, Accident Analysis and Prevention 27 (5) (1995) 707−715.

[42] Q. Ji, Z. Zhu, P. Lan, Real-time nonintrusive monitoring and prediction of driver fatigue, IEEE Transactions on Vehicular Technology 53 (4) (2004) 1052−1068.

[43] J.A. Michon, A critical view of driver behavior models: what do we know, what should we do?, in: Human Behavior and Traffic Safety Springer, US, 1985, pp. 485−524.

[44] A. Doshi, M.M. Trivedi, Tactical driver behavior prediction and intent inference: a review, in: IEEE Conf. Intell. Transp. Syst. Proceedings, ITSC, 2011, pp. 1892−1897.

[45] D.D. Salvucci, Modeling driver behavior in a cognitive architecture, Human Factors 48 (2) (2006) 362−380.

[46] J. Rasmussen, Skills rules and knowledge, other distinctions in human performance models, IEEE Transactions on Systems, Man, and Cybernetics 13 (3) (1983) 257−266.

[47] A.R. Hale, J. Stoop, J. Hommels, Human error models as predictors of accident scenarios for designers in road transport systems, Ergonomics 33 (10−11) (1990) 1377−1387.

[48] I.-H. Kim, et al., Detection of multi-class emergency situations during simulated driving from ERP, in: Brain-Computer Interface (BCI), 2013 International Winter Workshop on, IEEE, 2013.

[49] I.-H. Kim, J.-W. Kim, S. Haufe, S.-W. Lee, Detection of braking intention in diverse situations during simulated driving based on EEG feature combination, Journal of Neural Engineering 12 (1) (2014) 016001.

[50] J.-W. Kim, et al., Brain-computer interface for smart vehicle: detection of braking intention during simulated driving, in: Brain-Computer Interface (BCI), 2014 International Winter Workshop on, IEEE, 2014.

[51] S. Haufe, J.-W. Kim, I.-H. Kim, A. Sonnleitner, M. Schrauf, G. Curio, B. Blankertz, Electrophysiology-based detection of emergency braking intention in real-world driving, Journal of Neural Engineering 11 (5) (2014) 056011.

[52] J. Kim, I. Kim, S. Lee, Decision of Braking Intensity during Simulated Driving Based on Analysis of Neural Correlates, 2014, pp. 4129−4132.

[53] S. Haufe, M.S. Treder, M.F. Gugler, M. Sagebaum, G. Curio, B. Blankertz, EEG potentials predict upcoming emergency brakings during simulated driving, Journal of Neural Engineering 8 (5) (2011) 056001.

[54] Z. Khaliliardali, R. Chavarriaga, L. Andrei Gheorghe, J.D.R. Millan, Detection of anticipatory brain potentials during car driving, in: Proc. Annu. Int. Conf. IEEE Eng. Med. Biol. Soc. EMBS, 2012, pp. 3829−3832.

[55] J.C. McCall, M.M. Trivedi, Driver behavior and situation aware brake assistance for intelligent vehicles, Proceedings of the IEEE 95 (2) (2007) 374−387.

[56] C. Tran, A. Doshi, M.M. Trivedi, Modeling and prediction of driver behavior by foot gesture analysis, Computer Vision and Image Understanding 116 (3) (2012) 435–445.

[57] R. Mabuchi, K. Yamada, "Estimation of driver's intention to stop or pass through at yellow traffic signal, Electronics and Communications in Japan 98 (4) (2015) 35–43.

[58] R. Mabuchi, K. Yamada, Study on driver-intent estimation at yellow traffic signal by using driving simulator, IEEE Intelligent Vehicles Symposium, Proceedings (IV) (2011) 95–100.

[59] H. Takahashi, K. Kuroda, A study on mental model for inferring driver's intention, Decision Control. Proceedings of 35th IEEE Conference 2 (December) (1996) 1789–1794.

[60] T. Kumagai, Y. Sakaguchi, M. Okuwa, M. Akamatsu, Prediction of driving behavior through probabilistic inference, in: Eng. Appl. Neural Networks, Eighth Int. Conf., no. September, 2003, pp. 8–10.

[61] S.Y. Cheng, M.M. Trivedi, Turn-intent analysis using body pose for intelligent driver assistance, Pervasive Computing IEEE 5 (4) (2006) 28–37.

[62] D. Windridge, A. Shaukat, E. Hollnagel, Characterizing driver intention via hierarchical perception-action modeling, IEEE Transactions on Human-Machine Systems 43 (1) (2013) 17–31.

[63] M. Liebner, C. Ruhhammer, F. Klanner, C. Stiller, Generic Driver Intent Inference Based on Parametric Models, in: ITSC, 2013, pp. 268–275.

[64] I. Tamai, T. Yamaguchi, O. Kunihiro, Humane automotive system using driver intention recognition, in: SICE, 2004, pp. 4–7.

[65] T. Hülnhagen, I. Dengler, A. Tamke, T. Dang, G. Breuel, Maneuver recognition using probabilistic finite-state machines and fuzzy logic, in: IEEE Intell. Veh. Symp. Proc., 2010, pp. 65–70.

[66] T. Sato, M. Akamatsu, Influence of traffic conditions on driver behavior before making a right turn at an intersection: analysis of driver behavior based on measured data on an actual road, Transportation Research Part F: Traffic Psychology and Behaviour 10 (2007) 397–413.

[67] M. Liebner, F. Klanner, M. Baumann, C. Ruhhammer, C. Stiller, Velocity-based driver intent inference at urban intersections in the presence of preceding vehicles, IEEE Intelligent Transportation Systems Magazine 5 (2) (2013) 10–21.

[68] D. Mitrović, Reliable method for driving events recognition, IEEE Transactions on Intelligent Transportation Systems 6 (2) (2005) 198–205.

[69] N. Oliver, A.P. Pentland, Graphical models for driver behavior recognition in a SmartCar, in: Proc. IEEE Intell. Veh. Symp. 2000 (Cat. No.00TH8511), no. Mi, 2000, pp. 7–12.

[70] N. Oliver, A.P. Pentland, Driver behavior recognition and prediction in a SmartCar, in: AeroSense 2000. International Society for Optics and Photonics, 2000.

[71] M. Liebner, M. Baumann, F. Klanner, C. Stiller, Driver intent inference at urban intersections using the intelligent driver model, in: IEEE Intell. Veh. Symp. Proc., 2012, pp. 1162–1167.

[72] L. He, C. Zong, C. Wang, Driving intention recognition and behaviour prediction based on a double-layer hidden Markov model, Journal of Zhejiang University - Science C 13 (3) (2012) 208–217.

[73] M. Jingjing, W. Yihu, Driving intentions identification based on continuous pseudo 2D hidden Markov model, in: Intell. Comput. Technol. Autom. Int. Conf., 2012, pp. 629–632.

[74] S. Diego, Real-time Implementation of Estimation Method for Driver's Intention on a Driving Simulator, 2014, pp. 1904–1909.

[75] A. Liu, A. Pentland, Towards real-time recognition of driver intentions, in: Proc. Conf. Intell. Transp. Syst., 1997, pp. 236–241.

[76] T. Imamura, T. Takahashi, Z. Zhang, T. Miyake, Estimation for driver's intentions in straight road environment using hidden Markov models, in: IEEE Int. Conf. Syst. Man Cybern., no. i, 2010, pp. 2971–2975.

[77] P. Rammelt, Planning in driver models using probabilistic networks, in: Proceedings. 11th IEEE Int. Work. Robot Hum. Interact. Commun., no. 2048, 2002, pp. 87–92.

[78] S. Nakajima, D. Kitayama, Y. Sushita, K. Sumiya, N.P. Chandrasiri, K. Nawa, Route recommendation method for car navigation system based on estimation of driver's intent, in: 2012 IEEE Int. Conf. Veh. Electron. Safety, ICVES 2012, 2012, pp. 318–323.

[79] R. Simmons, B. Browning, Y.Z.Y. Zhang, V. Sadekar, Learning to predict driver route and destination intent, in: 2006 IEEE Intell. Transp. Syst. Conf., 2006, pp. 127–132.

[80] G.S. Aoude, V.R. Desaraju, L.H. Stephens, S. Member, J.P. How, S. Member, Driver behavior classification at intersections and validation on large naturalistic data set, IEEE Transactions on Intelligent Transportation Systems 13 (2) (2012) 724–736.

[81] G.S. Aoude, V.R. Desaraju, L.H. Stephens, J.P. How, Behavior Classification Algorithms at Intersections and Validation Using Naturalistic Data, no. Iv, 2011, pp. 601–606.

[82] S. Lefèvre, J. Ibañez-Guzmán, C. Laugier, Context-based estimation of driver intent at road intersections, in: Computational Intelligence in Vehicles and Transportation Systems (CIVTS), 2011 IEEE Symposium on, IEEE, 2011.

[83] S. Lefèvre, C. Laugier, J. Ibañez-guzmán, Exploiting Map Information for Driver Intention Estimation at Road Intersections, no. Iv, 2011, pp. 583–588.

[84] J.C. McCall, M.M. Trivedi, Visual context capture and analysis for driver attention monitoring, in: IEEE Intell. Transp. Syst. Conf., 2004, pp. 332–337.

[85] A. Tawari, S. Sivaraman, M.M. Trivedi, T. Shannon, M. Tippelhofer, Looking-in and looking-out vision for Urban Intelligent Assistance: estimation of driver attentive state and dynamic surround for safe merging and braking, in: IEEE Intell. Veh. Symp. Proc., no. Iv, 2014, pp. 115–120.

[86] A. Doshi, M.M. Trivedi, Head and eye gaze dynamics during visual attention shifts in complex environments, J. Vis. 12 (2) (2012) 9-9.

# CHAPTER 2

# State of the Art of Driver Lane Change Intention Inference

## DRIVER INTENTION INFERENCE BACKGROUND

Traffic accident statistics have shown that more than 80% of traffic accidents were caused by driver errors, such as misbehavior, distraction, and fatigue [1−3]. Various passive safety systems such as airbags and seat belts have played a significant role in the protection of the driver and passengers when traffic accidents occur. Although these technologies have saved a large number of lives, they are not designed to prevent traffic accidents from happening but only protect the passengers after the accident happens [4,5]. Therefore many efforts have been devoted to the development of safer and more intelligent systems toward the prevention of accidents instead of only minimizing the impact of the accidents. The most successful active safety system is the advanced driver assistance systems (ADAS), which contains a series of active safety functionalities.

Most of the ADAS techniques such as lane departure avoidance, lane keeping assistance (LKA), and side warning assistance (SWA) can assist the drivers by alarming in time, assisting in decision making, and reducing driver workloads [6−8]. However, the inputs of these systems usually rely on the vehicle's dynamic states such as the steering wheel angle, velocity, yaw angles, Global Positioning System (GPS), and other traffic information. However, most of these systems ignore the most important factor on the road—the human driver. Vehicles operate in a basic three-dimensional environment with continuous driver-vehicle interactions. The three-dimensional road environment consists of road context, human driver, and vehicles. The structured or unstructured road plays a fundamental function for drivers and other road entities. The host driver interacts with other drivers by controlling the vehicle based on the road context. Among the three aspects, drivers are the major component in this system, who control the vehicle to maintain a safe and efficient driving task. Therefore allowing ADAS to understand drivers' intentions and behaviors is important to driver safety, vehicle drivability, and traffic efficiency.

Driver intention inference (DII) is an ideal way of allowing the ADAS to understand the driver. The reasons for recognizing driver intention are multifold. First, DII improves driving safety. Specifically, there are two different driving scenarios that require inferring the driver's intention: to better assess the risk in the future and to avoid making decisions that are opposite to the driver's intent [9]. For the first case, there is a shred of evidence that a large number of accidents are caused by human error, misbehavior, cognitive overload, misjudgment, and operational errors [2]. Monitoring and correcting driver intention in time are crucial to the effectiveness of ADAS. The intelligent vehicles can carry more high-resolution road perception sensors than a human driver, based on the sensor fusion technique. If the vehicle can identify the driver's intention in advance, the intelligent vehicle will pay more attention to the related road and analyze the possibility of such an intention based on the comprehensive assessment of the traffic context. Hence, estimating driver intention prior to the maneuver can largely increase driving safety. Meanwhile, the increasing usage of in-vehicle devices and other entertainment devices can distract the driver from normal driving. The driver may not sufficiently check the surrounding context before executing the control maneuver. Hence, for the design of intelligent ADAS, it is beneficial to understand the driver intentions and execute correct assistance actions [10,11].

In terms of making the right decisions, ADAS intervenes with the vehicle dynamics and shares the control authority with the driver. The current ADAS functions can be designed in a serial and parallel manner. For example, the LKA system (LKAS) can generate feedback control to the vehicle to slightly minimize the error between the vehicle central line and the lane middle position. If the system does not understand the driver lane change intent, conflicts between the driver and vehicle automation will occur, which can significantly decrease the driver's trust to the LKAS. Another example can be estimated in the complex traffic conditions such as at intersections and roundabouts, it is crucial not to interrupt

Advanced Driver Intention Inference. https://doi.org/10.1016/B978-0-12-819113-2.00002-6

the driver, especially not to interrupt the driver with misleading instructions. Moreover, mutual understanding between the driver and vehicle is a basic requirement for driver-vehicle collaboration. To ensure efficient cooperation, it is important for the ADAS to be aware of the driver's intention and not operate against the driver's willingness. This makes it reasonable for the ADAS to accurately understand the driver intention in real time.

Furthermore, the DII system benefits from the development of future automated vehicles. DII can be used to construct the driver model, which can act as the guidance for the design of the automated driver. Analysis of the driver cognitive intention process based on the stimuli classification, control maneuver, and interaction feedback can be used to construct a human-like decision-making model. Moreover, in terms of the level 3 automated vehicles, as determined by the SAE International standard, accurate driver intention prediction enables a smoother and safer transition between the driver and the autonomous vehicle controller [12−14]. When level 3 automated vehicles work at automated conditions, all the driving maneuvers are handled by the vehicle; however, once an emergent situation occurs, the vehicle has to disengage and give the driving authority to the driver. In such a case, the vehicle can determine whether the driver is ready to take over or not by assessing the driver's intention in advance. If the driver has a reasonable driving intention, the vehicle should follow the driver's maneuver and the assistance system should give fewer instructions. The application of DII makes the transition between driver and controller as smooth as possible.

In this section, the literature review of DII will be proposed, and the lane change intention will be used as an exemplary scenario for further explanation. The components of this section can be summarized as follows. First, a state-of-the-art literature review about driver lane change intention is proposed. Then based on the lane change intention inference (LCII) scenarios, the system architecture is categorized from different aspects. Next, the critical time flow of DII will be introduced, which leads to a comprehensive understanding of the architecture of the intention inference system. Lastly, future works of DII will be proposed and discussed to benefit the development of future intelligent vehicles.

## LANE CHANGE MANEUVER ANALYSIS—AN EXEMPLARY SCENARIO

Lane change maneuver occurs when the host vehicle driver is not satisfied with the current vehicle status or surrounding traffic context. To pursue a more comfortable driving experience with comfort zone, the driver may try to overtake the leading vehicle that has a slower speed to maintain their current speed or try to maintain a safe distance with the car behind the vehicle. Sometimes lane change also happens when the driver tries to leave a certain traffic segment. When this situation occurs, the driver will choose to drive thevehicle into the adjacent lane and finish the lane changing process.

When drivers initiate a lane change maneuver, they are responsible for their own safety. A lane change maneuver requires more control actions and environment checking behaviors than normal driving. The driver has to pay attention to the front as well as the rear situation of the target lane. Besides, the driver should still pay attention to the current context and surrounding vehicles around the blind spot area. The driver has to decide the best moment to make a lane change maneuver based on current traffic information. Lee et al. supposed that the driver is more likely to start a lane change when there is a gap of at least 12 m from the front and rear vehicles or the relative velocity with the other vehicle (leading/following) is less than 22 km/h. However, the relative velocity and the distance from other vehicles are not a good measurement for lane change maneuver. The time to collision (TTC) can be a better risk measurement index [120,121]. Lee determines that drivers will choose to change lane if the minimum TTC with the front or rear vehicle is between 4 and 6 s. Wakasugi [122] suggested that the driver usually starts a lane change when the TTC is larger than 6 s and more likely to abort the lane change maneuver if TTC is less than 10 s.

Another important result given from the US traffic statistics is that less than half of the drivers signal when they change the lane during their normal driving. Besides, drivers are more likely to use a turn signal when moving into the adjacent lane rather than when coming back to the original lane when finishing an overtaking [120]. Salvucci and Liu [32] revealed that 50% of the turn signal is activated at the very beginning of a lane change and after 1.5−2 s this proportion increases to 90%. This suggests two different driving patterns: using a turn signal to show the lane changing intent before performing the maneuver and using a turn signal to indicate that the driver is performing the lane change maneuver. They also studied the driver eye movement during the lane change process. The driver will spend more time on the current lane before changing the lane. As the maneuver starts, the driver moves his/her eyes to the mirror. Then during the maneuver, the driver directs his/her gaze to the target lane.

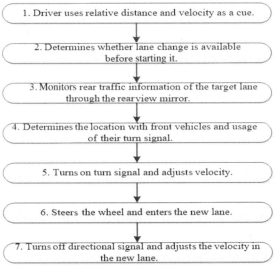

FIG. 2.1 Lane change procedure [123].

According to the author in Ref. [123], lane change in highway can be viewed as a seven-step procedure, which can be summarized as the driver using traffic information as a cue, determining the possibility of making a lane change, monitoring the surrounding traffic, rechecking the current state before lane changing, turning on signals, controlling and steering the vehicle, and finishing the lane change maneuver; the process is shown in Fig. 2.1.

In conclusion, lane change maneuver is a complex task during the driving process. Additional requirements are needed by the drivers, such as mirror checking and risk assessment. They have to be clear about their surrounding traffic information, especially the blind spot area. Sometimes driver's distraction can also have an adverse effect on the lane changing maneuver. The driver has to focus on the traffic and pay attention to the lane change maneuver. Hence, most of the time, when the driver is going to make a lane change, the driver's attention will be highly focused on the driving task and the distraction level is low.

## LANE CHANGE ASSISTANCE SYSTEMS

Currently, some lane change assistance (LCA) systems have been developed to assist the driver in making lane change maneuvers and avoiding the collision. This section first analyzes some recent ADAS products on the market such as lane departure warning, LKA, and LCA system. Then their standard technique, advantages, disadvantages, and the necessity of introducing

lane change intention will be described. According to these limitations, the literature review of LCII will be given in the next section.

### Lane Departure Warning

Lane departure refers to the driver crossing the lane unaware of the surrounding context, and sometimes by distraction, and it can be dangerous. Lane departure warning system (LDWS) is an in-vehicle electronic system that can alert the driver of the unintended lane departure [124]. LDWS has no direct action on and control over the vehicle chassis and steering units, which means LDWS is just a warning system rather than a driving assistance system like LKAS. LDWS usually uses vision sensor systems such as monocular or stereo cameras to detect the lane boundary and vehicle location, and then it can alert the driver if the vehicle is coming into the dangerous situation (crossing lane) (as shown in Fig. 2.2). LDWS is usually operated at speeds of 56–72 km/h for passenger vehicles and 64–80 km/h for trucks. The warning zones represent that the vehicle is going to depart and no corrective steering force is applied to the vehicle. The threshold of the width of the warning zone controls the sensitivity of the LDWS.

Two documents guide the design of LDWS. The first one is the ISO 17361 Lane Departure Warning Systems (ISO 17361:2007) (Fig. 2.3) and the second one is the Lane Departure Warning Systems (FMCSA-MCRR-05-005). Both these systems define the LDWS in a similar way: LDWS is an in-vehicle system that detects the vehicle lane departure situation through sensor systems and alerts the driver when necessary. The driver maintains the vehicle control responsibility to make sure the vehicle stays in the right place, and the LDWS does not interact with the vehicle control of the driver.

According to Fig. 2.3, the status indication represents the current position and state of the vehicle. There are three main components of the LDWS: lateral position detection and lane detection, lane departure warning and system status monitoring, and status warning system. Besides, a few other modules can help the LDWS in making decisions, which are the suppression request, vehicle speed reading, driver behavior, preference reasoning, etc. If the system recognizes a proper driver intention, it may not warn the driver based on the intelligent decision-making module.

Normally, two kinds of sensor systems are used in the LDWS: camera-based and infrared systems. The camera-based LDWS relies on the detection of the line painted on the roadway and passes this information as well as the vehicle status (such as vehicle location, velocity, and heading angle rate) to the processer. If the

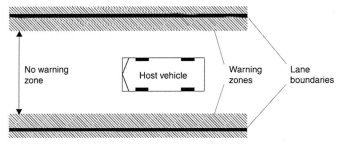

FIG. 2.2 Detection zone definition for lane departure warning system [124].

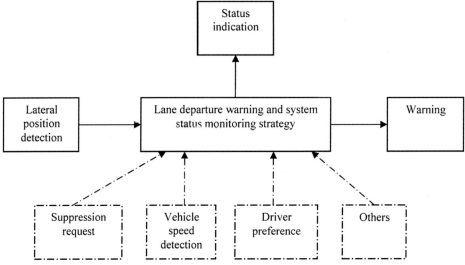

FIG. 2.3 The lane departure warning system architecture defined in ISO 17361 [124].

data suggest the vehicle is leaving its path unintention-
ally, it will alert the driver through the driver-machine
interface. One significant drawback of camera-based
LDWS is it relies heavily on the lane marks. If the lane
mark is not clear and the weather condition is adverse,
such as rainy, foggy, and ice on the carriageway, the system cannot work very well. The infrared system is usually mounted under the front bumper to detect the
lane marks on the road and classifies the different reflec-
tions from the signals. When the vehicle crosses the
lane, the sensor system detects a change and hence alerts
the driver. In terms of the human-machine interface
used in LDWS, there are three typical types, namely,
auditory warning, visual warning, and hepatic feedback.
Usually, a visual warning is suitable when the departure
occurs slowly and an auditory and hepatic warnings are
the preferred choice when an emergency lane departure
happens.

## Lane Keeping Assistance
LKAS can be viewed as an advanced version of LDWS for
the reason that it monitors the lane departure maneuver
and alerts the driver if necessary. Besides, it also assists
the driver to control the vehicle by giving assist actuation signals to the vehicle's chassis and steering control
unit. It supports the driver in keeping to the lane by controlling the vehicle heading and yaw angles. LKAS is an
extension of LDWS, which is designed to assist the
driver in maintaining the lane position [125]. LKAS
uses similar sensor systems as in LDWS, in addition to
the lane keeping controller and the assistance actuator
(Fig. 2.4). The control variable for LKAS is the position
of the vehicle within the lane. LKAS normally works in a
similar operational status as LDWS, i.e., above 65 km/h,
and focuses on the safety issues rather than on comforts.
LKA can also patch with adaptive cruise system to help
the driver steer and maintain the vehicle in the path.

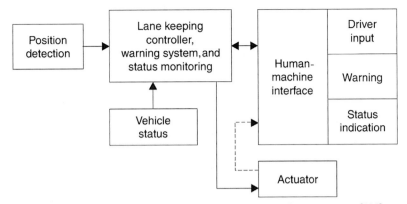

FIG. 2.4 Functional architecture of the lane keeping assistance system [125].

The most significant difference between LKAS and LDWS is whether a controller is used or not. The lane keeping controller calculates an optimal actuator output based on the heading and lateral position of the vehicle within the lane and passes the control signal to the actuator unit. The actuator converts the control signal into an action signal to support the driver. Currently, there are two different assist methods for LKAS: steering input assist and corrective braking assist. The LKAS uses the same lane detection sensor and algorithms as LDWS to identify the lane curves, vehicle position, and relative status. In terms of the steering force assist, when lane departure is detected, the LKAS alerts the driver through the human-machine interface and intervenes by applying a slight countersteering angle. On the other hand, the brake actuators generate the necessary yaw rate of the vehicle by controlling the brake pressure of each wheel separately.

### Lane Change Assistance

Lane change collision usually occurred when the driver takes a lane change maneuver but neglects the adjacent vehicle in the "blind spot" or starts at an inappropriate time. LCA system monitors the adjacent and rear lanes and the traffic to assist the driver when the driver is making a lane change maneuver. If the driver starts to change lane and the LCA detects a vehicle in the adjacent lane, the LCA will warn the driver to stop changing the lane. Currently, this system does not have the function to intervene with the driver, but in the future, LCA is expected to have lane change planning and control functionality. The LCA warns the driver when he/she chooses an inappropriate time to change the lane. Lane change collision usually happens during three situations: the subject vehicle has a side collision with the vehicle in the lane, the subject vehicle experiences a rear

collision with the approaching vehicle, and the vehicle has a collision with the front overtaking car. Usually, lane change collision happens because the driver did not see the vehicle in the blind spot area or carried out a misbehavior during lane changing.

There is one document defined to support the LCA system, ISO 17387 Lane Change Decision Aid System (ISO 17387:2008). The standard defined in ISO 17383 supports the LCA system development for cars, vans, and trucks. In this standard, the system definition is to warn the driver against the collision that may occur when the driver initiates a lane change maneuver [81]. However, the systems just warn the driver through human-machine interface without intervening and controlling the vehicle. The driver should always be responsible for the safe action of the vehicle. The standard requires the LCA to operate following a state diagram (Fig. 2.5) rather than define its functional element, as shown in Fig. 2.5.

There are two key aspects to be considered in LCA, which are coverage zone and the maximum closing speed of the target vehicle. The coverage zone is the entire area that the sensor system monitored and covered outside the vehicle. Two subgroups of the coverage zone are adjacent and behind. This classification method enables the LCA to be separated into three types. Type 1 covers the adjacent region of the subject vehicle and therefore supports blind spot warning function only. Type 2 system, on the other hand, monitors the behind area of the vehicle and only supports closing vehicle warning. Type 3 system has a series of sensors to cover the areas both adjacent to and behind the vehicle and can provide blind spot warning and closing vehicle warning function. The LCA products currently on the market can be separated into the blind spot monitoring system and lane change warning system (monitoring the blind spot and the

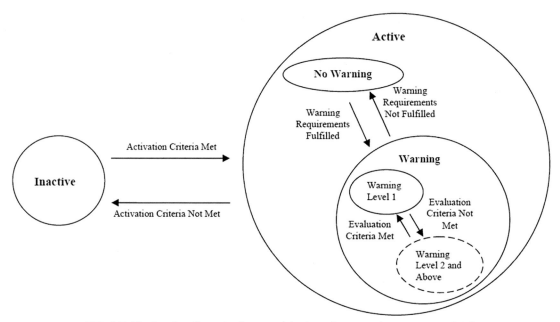

FIG. 2.5 The functionality state diagram of the lane change assistance system [125].

behind area). Currently, four kinds of sensors are used in LCA, which are radar, camera, infrared, and ultrasonic sensors. Among these, radar and the camera can be used in both blind spot and rear closing vehicle monitoring.

Several research have been carried out to study the impact of LDWS and LCA on driver behavior. Portouli [126] pointed out that the usage of LDWS has no effect on the lane change frequency but it increases the use of turn signals, which proved drivers improved their driving to avoid warning. It is also suggested that the driver does not drive faster because the driver feels safer when the LDWS is activated. As LCA does not have that many uses, a few research can clearly point out how exactly it influences the driver's behavior. While Kiefer and Hankey [127] suggest that the driver begins to look at both the side mirrors and the rearview mirror because the LCA increases the driver's awareness of safety. Meanwhile, no adverse effects were found on mirror usage and lane change frequency in the vehicle-mounted LCA.

The drawbacks of the current ADAS mentioned earlier are that most of them rely heavily on sensor systems. If the weather condition is adverse, it is usually difficult to obtain a good lane mark and around vehicle detection, thus it can hardly provide proper assistance to the driver. Moreover, as stated earlier, nearly half of

the drivers do not use turn signals in most conditions. However, some current LCA systems are activated by the turn signals. If the drivers do not use the turn signal, these LCA systems will fail to work. Therefore the DII algorithm is an ideal way to reduce the conflicts between driver intent and ADAS. It does not heavily rely on the outer sensor system but can determine the driver's mental status according to inner vehicle sensors. If a driver lane change intention is estimated, it can communicate with the LCA system in advance to make sure LCA can be executed and can improve the driver safety more significantly. Hence, the next section will give a detailed review of the research on driver lane change intention.

### Limitations and Emerging Requirement

Based on the analysis of the lane change process and the current LCA systems such as LDWS, LKAS, and LCA, it can be found that most of the current systems do not have the capability to fully interact with the human driver, as no driver behavior reasoning is used. These systems are mainly being activated based on the outer traffic and road perception; however, driver behaviors are a critical clue to the lane change maneuver prediction. If driver lane change intention can be detected in advance, the driver assistance system can be activated in advance such that the potential driving interest region and the blind spot region can be well monitored. Hence, it is necessary to design a DII unit to better interact with the human driver. Normally the ADAS products are mainly designed to provide driving assistance in the L1 or L2 automated driving vehicles, in which the driver is responsible for vehicle control. For higher level automated vehicles, in which the driver can share the control authority with the automation, the driver intention reasoning is even more important to the automation to efficiently interact and cooperate with the driver.

## HUMAN INTENTION MECHANISMS

The human intention has been theoretically studied in the past two decades. From a cognitive psychology perspective, intention refers to the thoughts that one has before performing the actions [15]. Similarly, in this study, intention (particular focus on the tactical driving intention) is the attitude toward performing a series of vehicle control maneuvers. The intention is determined by three aspects, namely, the attitude toward the behavior, subjective norm, and the perceived behavior control [16]. In Ref. [16], human behaviors are found to be the response of the intention. The attitude toward the behavior describes how willing is the human and how much effort the human puts to take the behavior; a strong level of attitude can give a strong willingness of taking actions in a certain task. Second, the subjective norm reflects the pressure from the surrounding social life of the human. Finally, the perceived behavior control was developed from the self-efficacy theory. It describes the confidence of an individual to perform the behavior.

Bratman [17] pointed out that intention is the main attitude that directly influences future plans. In addition, Heinze [18] described a triple-level description of the intentional behavior, which contained intentional level, activity level, and state level. In the human-machine interface scope, according to Ref. [19], intention recognition is the process of understanding the intention of another agent. More technically, it is the process of inferring an agent's intention based on its actions. Elisheva [20] proposed a cognitive model with two core components, which were intention detection and intention prediction. Intention detection refers to detect whether a sequence of actions has any underlying intention. Intention prediction, on the other hand, refers to the prediction of the intentional goal based on a set of incomplete sequence of actions. The intention inference and reasoning process makes people clever and enables them to take part in

the social community. Humans can recognize others' intentions based on their observation and social skill knowledge. However, it is difficult to make an intelligent machine, such as a smart vehicle, learn how to infer human intention accurately. To some extent, only when a robot can detect human intention based on its own observation can it be viewed as an intelligent agent.

Human intention inference has been widely studied in the past decades. One of the most significant applications of human intention inference is human-robot interface design [21−24]. Thousands of the service robots were designed to assist humans in completing their works either in daily life or in a dangerous workspace. Traditional robots were designed from a robot's perspective rather than from a human's point of view, which reduces the interaction level between humans and robots. To improve the efficiency of human-robot interaction (HRI) and the intelligence of the robot, the robot should have the ability to learn and infer human's intention and obtain basic reasoning intelligence.

A widely accepted method of classification for the human intention in HRI scope is to classify the human intention into explicit and implicit intentions. Implicit human intention can be further separated into informational and navigational according to Ref. [26]. The explicit intention is clearer than the implicit intention and, hence, is easier to be recognized. Explicit intention means humans can directly transmit their intention to the robot by language or direct command through the computer interface, while implicit intention reflects the human mental state without any communication with the robot [25]. The robot has to observe and understand human behavior first and then makes an estimation of the human intention at the right movement based on the knowledge base. Human intention inference problem contains a large amount of uncertainty, and noise exists in the measurement device. Therefore probability-based machine learning methods are powerful tools in solving this kind of problem and have been successfully applied in many cases. The human intention inference can be mathematically modeled as a process of inferring a series of human mental hidden states. The hidden Markov model (HMM) and the dynamic Bayesian theory are two popular methods for the inference of human mental states [27−29].

## DRIVER INTENTION CLASSIFICATION

Driver intention can be classified into different categories from a different perspective. For example, it can be classified according to the motivation, timescale, and direction of driving. Among these, the two most straightforward ways of classifications are based on the timescales of the intention and the driving direction.

### Timescale-Based Driver Intention Classification

In terms of the timescale-based classification method, Michon pointed out that the cognitive structure of human behavior in the traffic environment is a four-level hierarchic structure, which contains road user, transportation consumer, social agents, and psychobiological organisms [30]. Among these, the road user level is directly connected with the drivers and can be further divided into three sublevels: strategy, tactical, and operational levels (also known as control level), as shown in Fig. 2.6. The three cognitive levels can be viewed as three driver intention levels based on the timescale characteristic. Strategy level defines the general plan of a trip such as the trip route, destination, and risk assessment. The time constant will be at least in minutes or even longer. At this moment, the driver will decide the transport mobility and comfort issues, which is a long timescale problem. In terms of the tactical level, in which the time constants are in seconds, the driver will make a short-term decision and control the vehicle to negotiate the prevailing circumstance. The tactically planned intentional maneuver consists of a sequence of operational maneuvers to fulfill the short term goal, such as turning, lane changing, and braking maneuvers [31].

All the control commands must meet the criteria from the general goal that are set at the strategic level. Lastly, the operational intention is the shortest one among the three levels and stands for the willing of the driver to remain safe and comfortable in the traffic situation. The driver directly gives control signals to the vehicle and the time constant is normally in milliseconds. As mentioned earlier, real-time lane change intent inference plays a critical role in the improvement of driving safety. In addition, continuous LCII is a relatively complex and difficult task than some other driving intentions. Salvucci and Liu [32] concluded that lane change not only was a control procedure but also incorporated a set of critical aspects of driving such as lower-level controls. Normally, lane change maneuvers will contain a series of short-term driving behaviors such as the acceleration and deceleration in the longitudinal direction and the steering wheel control in the lateral direction.

There are also some other classification methods for driver intention. Salvucci developed a driver model, namely, Adaptive Control of Thought-Rational cognitive architecture [33]. Similar to the three-level architecture of road user model given by Michon, Salvucci developed the integrated driver model into three main

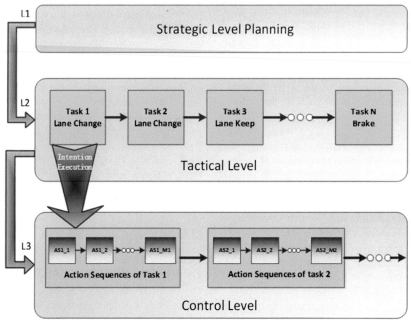

FIG. 2.6 Driver intention classification based on the time constant.

components, which are control, monitoring, and decision-making modules. The control component is like the operational level given by Michon, which is responsible for the perception of the external world and for transferring the perceptual signals directly to the vehicle. The monitoring component keeps aware of the surrounding situation and environment by periodically perceiving data and inferring. The decision component, which has the same function as part of Michon's tactical level, makes tactical decisions for each maneuver according to the awareness of the current situation and the information gathered from the control and monitoring modules. One significant advantage of the cognitive driver model is that incorporation of the built-in features helps mimic human abilities.

## Direction-Based Driver Intention Classification

The direction-based driver intention classification, on the other hand, is quite straightforward. There are two basic directions for the underground vehicle, which are the longitudinal and lateral intentions. The driver's longitudinal behavior includes braking, acceleration, starting, lane keeping, etc. Lateral behaviors usually contain turning, lane changing, and merging. In previous studies, most researchers pay attention to the lateral intention prediction such as the lane change, turning, and overtaking maneuvers. The lateral intentions are more complicated than the longitudinal intention owing to the frequent interaction with surrounding vehicles.

In terms of the longitudinal intention, most of the previous studies focus on braking intention recognition. Haufe et al. proposed a driver braking intention prediction method using electroencephalographic (EEG) and electromyographic (EMG) signals [34,35]. Khaliliardali [36] proposed a driver intention prediction model to determine whether the driver wants to go ahead or stop. The method was to classify the go and stop intention based on the brain-machine interface. The EEG, EMG, and electrooculographic signals from six subjects in the simulation environment were collected and two classification methods (linear and quadratic discriminant analyses) were used separately to evaluate the classification performance. McCall and Trivedi [37] integrated driver intention into an intelligent braking assistance system. A sparse Bayesian learning algorithm was used to infer the driver's intention of braking. Trivedi et al. predicted the driver's braking intention by directly monitoring the foot

gesture through cameras [38,39]. They showed that the driver foot gesture plays an important role in vehicle control. Therefore the usage of vision-based foot tracking is more direct and accurate. Mabuchi and Yamada [40] estimated driver's stop and go intention at intersections when the yellow light occurs. Takahashi and Kuroda [41] predicted the driver deceleration intent in a downhill road. Kumagai et al. [42] proposed a method to predict driver's braking intention at intersections (particularly right turns) by using a dynamic Bayesian network.

As aforementioned, direction-based intention classification is less precise than the timescale-based methods, as the driving maneuvers can be very complex and may contain multiple short-stage actions. For example, the lane change maneuvers can consist of short-period acceleration and turn, and it is less accurate to describe the intention as merely longitudinal or lateral.

## Task-Based Driver Intention Classification

Driver tactical maneuvers consist of a series of operational maneuvers. Some of the existing studies focus on the analysis of multiple tactics rather than a single tactical task. The multitask-based model usually contains both longitudinal and lateral maneuvers compared with the single-task-oriented model. By using machine learning theory, a single intention inference task can be modeled with a discriminative and generative model. However, the multitask inference model prefers to use the generative models such as Bayesian networks and HMM [31]. Oliver and Pentland [43] proposed a driver behavior recognition and prediction method based on the dynamic graphic models. Seven driver maneuvers, namely, passing, changing right and left, turning right and left, starting, and stopping were analyzed. Liebner inferred the driver's intent based on an explicit model for the vehicle's velocity, and an intelligent driver model was used to represent the car-following and turning behaviors [44]. Liu and Pentland [45] aimed to analyze the patterns within a driving action sequence. The primary approach was to model the human behaviors in a Markov process. Imamura developed a driver intention identification and intention labeling method based on the assumption of compliance with traffic rules [46].

As can be seen, multitask driving intention recognition is more complicated than the single-task model. It must define and clarify several driving maneuvers and need more experiments. Moreover, it relies on the design of generative inference algorithms.

## DRIVER INTENTION INFERENCE METHODOLOGIES

### The Architecture of Driver Intention Inference System

DII system requires multiple techniques such as vision-based perception system, data fusion and synchronization, and model training based on machine learning methods. According to the previous literature, LCII system mainly contains the following modules: road and traffic perception module, vehicle dynamic measurement module, driver behavior recognition module, and DII module, as shown in the bounding box in Fig. 2.7.

As shown in Fig. 2.7, the traffic information is first captured by the environment perception block. Similar to the previous literature that uses cameras, light detection and ranging (lidar), radar, and GPS signals to detect the surrounding traffic situation, this block will process the current road and traffic information with machine vision techniques and output the position of ego-vehicle and the velocity. After the traffic context is detected, the relative distance and velocity between the ego-vehicle and the front vehicles can be obtained by collecting the vehicle status information captured through the Controller Area Network (CAN) bus.

The traffic and vehicle data will be fed into the intention inference model, along with the driver behavior signals. The driver behavioral signals include the driver head rotation, eye gaze, body movement, etc. Next, the intention inference model will calculate the probability of a lane change intention based on the fused information. Once the decision is made, the lane change decision module outputs a binary signal to indicate a lane change maneuver. After the lane change decision is activated, the interaction module models the driver's hand and foot dynamics and the interaction with the vehicle control interface. A taxonomy of driver intention modules is depicted in Fig. 2.8.

The relationship between tactical intention and operational intention is illustrated in Fig. 2.9. Fig. 2.9

contains three parts, namely, the traffic context perception unit, level 2 tactical intention unit, and level 3 control units. Specifically, in the third level, three layers are defined. The upper layer is driver dynamics, which represents the checking and monitoring behavior of the driver. The second interface layer will be activated once the lane change decision is made. Finally, the control signals are fed into the lowest vehicle control layer. In the driver dynamics module, the most common dynamics are brain dynamics, which can be measured by EEG, eye gaze behavior, head movement, and body movement (includes hand, body, and foot dynamics).

From Fig. 2.9, we can clearly define the time flow of the driver intention procedure. The driver first captures the traffic context and then generates the corresponding intention according to the traffic. Next, the driver will check the surrounding traffic by performing a series of checking behaviors to make sure safety control. Once the driver is confident with the lane change decision, he/she will control the vehicle through the steering wheel and the pedal. Finally, the vehicle responds to relative control behaviors and vehicle dynamic changes.

### Inputs for Driver Intention Inference System

The driver is in the center of the traffic-driver-vehicle (TDV) loop. The signals from the three parts of the TDV loop can be used to infer human driver intentions. The major signals that can be used as the inputs of LCII system are summarized in Table 2.1. The perception of traffic context enables the ADAS to understand the specific driving environment of the human driver so that a more reasonable intention inference can be detected. The driver behavior information will help determine how long the driver has generated the intention and how the driver checks the surrounding traffic with respect to the corresponding intention. Finally, the vehicle dynamic signals indicate what kind of driving action the driver has taken to realize the intent.

The input data used to infer driver mental intent and predict driver behavior is multimodal signals. It can

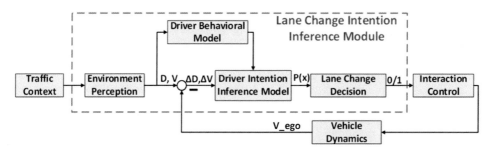

FIG. 2.7 Driver lane change intention inference framework.

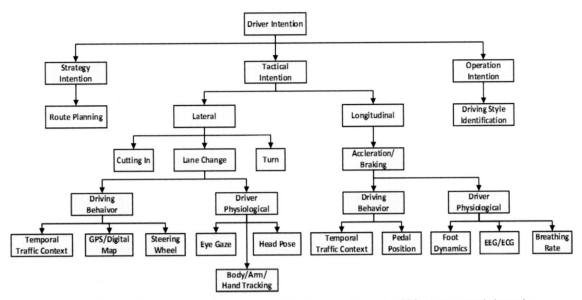

FIG. 2.8 Taxonomy of driver intention systems. *ECG*, electrocardiography; *EEG*, electroencephalography.

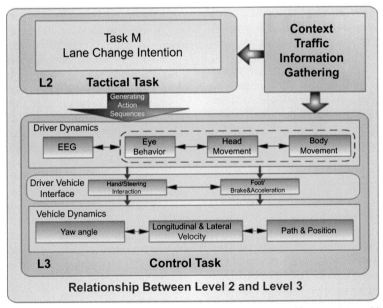

FIG. 2.9 Relationship between tactical intention and operational intention with respect to multimodal inputs. *EEG*, electroencephalography.

contain data from the vehicle CAN bus, radar, and lidar system, as well as the inner driver monitoring system. Basically, there are three main kinds of sources to collect the input data of intention inference system, which are traffic environment, vehicle parameters, and driver behavior information. Besides, to infer different intentions, different input data should be selected. For example, the steering wheel angle can be a significant signal to reflect the vehicle lateral behavior, which is also helpful in the prediction of driver lane change intention. However, in terms of longitudinal intention, a steering wheel angle can give limited information and

**TABLE 2.1**
**Common Input Signals and Sensors Used for Driver Intention Inference.**

| Sensor Sources | Sensor Categories |
|---|---|
| Traffic | Current ego-vehicle position (collected with GPS and digital map), relative distance, velocity, and acceleration with respect to the front and surrounding vehicles (collected with cameras, radar, or lidar) |
| Vehicle | CAN bus signals (including steering wheel angle, steering wheel velocity, brake/gas pedal position, velocity, heading angle, etc.) |
| Driver | Cameras (head rotation, gaze direction, foot dynamics). EEG, EMG, heart rate, etc. |

*CAN*, Controller Area Network; *EEG*, electroencephalography; *EMG*, electromyography; *lidar*, light detection and ranging.

the brake pedal will play an important role at this moment. Selecting the most important and relevant data can increase the accuracy of the prediction rate and decrease the false alarm rate of the system. For the road-vehicle-driver system, the three parts construct a whole driving model. Each part plays its unique role in the design of ADAS and intention inference system. To make the vehicle better understand the driver and to develop an intelligent vehicle toward automated driving, information from each part should be as detailed as possible.

### Traffic context

Traffic context is the major stimuli for driver intention. Traffic environment, also known as the road situation, is the source of stimuli for drivers to take actions during their driving. A better understanding of the surrounding traffic information will improve the performance of the intention inference system. For example, lane changing usually occurs when there is a lower vehicle in front of the host vehicle. To keep comfortable, the driver always chooses to make a lane change. Nowadays, most of the traffic information can be obtained from ADAS.

There are many kinds of sensors that can be used to capture the surrounding traffic context, such as cameras, radar, and lidar systems. The most popular vision-based ADAS are LDW and LKA. Vision-based LDW is able to compute the distance between the host vehicle and the lane boundary, the vehicle lateral velocity and acceleration, yaw angle, and road curvature, etc. [47].

Adaptive cruise control (ACC) is a cruise system that can be divided into laser based and radar based. The relevant distance between the host vehicle and the front vehicles can be detected through the radar of the ACC system [48]. The SWA system uses at least two radars that are mounted under the side mirrors to monitor the rear and side vehicles. The scan area can be up to 50 m behind the vehicle [49,50]. In Ref. [50], the authors evaluated the impact of different sensors on the prediction of driver intention. GPS, a space-based navigation system, gives rough information about vehicle location, road type, and road geometry compared to the abovementioned sensors. Berndt designed an HMM-based intention classifier that extracted the distance to the next turn, the street curvature, and street type from a digital map [51]. Rafael introduced an interactive multiple-model–based approach to predict lane change maneuvers on the highway. The system used the GPS/IMU (inertial measurement unit) sensors to collect the vehicle position data [52]. The advantage of the GPS system is that it is able to give the location and time information in rough weather conditions when the camera and radar system cannot work. The precision of GPS is usually between 3 and 10 m. However, it can give location and time information in all weather conditions even in a dusty situation when the camera and radar system cannot work. Fig. 2.10 illustrates the detection regions given by different context sensors. In Ref. [53], McCall and Trivedi proposed a preliminary work for the study of a driver-centered assistance system, which focuses on the lane change events. A modular scalable architecture was provided to capture the driver behaviors and the surrounding environment. Radar and video devices are used to obtain the forward, rear, and side information. Meanwhile, inner cameras were also used to monitor the driver foot gesture and head movement. The work in Ref. [54] concentrated on the lane change intention prediction according to the sensor data. The data contains the lane information given by a lane tracker, the vehicle velocity, lateral position and its derivation, and the steering wheel angle.

### Vehicle dynamics

Vehicle status information such as steering wheel angle, brake pedal position, and velocity can be viewed as the direct response to the control actions. This information reflects the driver's control actions that are taken on the vehicle. Hence, these signals have been widely used in many works for driver intention identification. Vehicle data can be collected from the vehicle CAN bus, such as velocity, acceleration, and heading angle, which can reflect the driver control to the vehicle and can process a

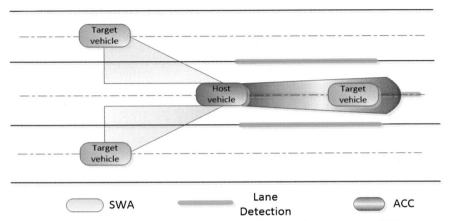

**FIG. 2.10** Traffic information detected through advanced driver assistance systems. *ACC*, adaptive cruise control; *SWA*, side warning assistance.

large amount of data with high transfer speed. In Ref. [42], vehicle speed, acceleration pedal, and brake pedal position are collected to predict the driver braking intention. In terms of the lane change intent, throttle pedal position, brake pressure, cross-acceleration, steering wheel angle, steering wheel angle velocity, yaw rate, and velocity are collected from CAN bus in Refs. [49,51]. These signals are particularly useful to understand the driving intent after the driver has determined to finish the intention. Schmidt and Beggiato proposed a lane change intention recognition method based on the construction of an explicit mathematic model of the steering wheel [55]. In Ref. [56], the authors proposed a driver lane change/keep intention inference method on a driving simulator. The input signals contained the steering wheel angle, acceleration, and the relative speed of the front vehicle. In Ref. [57], the speed, transmission position, steering angle velocity, steering angle, acceleration pedal position, and lateral acceleration data were captured on a simulator for LCII. The authors in Ref. [58] proposed a driver intention recognition method based on artificial neural networks (ANNs). CAN bus data and driver gaze information were collected and fed into the intention model. The experiment was designed in a six-degrees-of-freedom dynamic driving simulator and a total of 284 lane change and lane keeping instances were recorded.

Although vehicle dynamic data reflect the response to the driving actions, it gives delayed information compared to the driver behavior and traffic context information with respect to the intention inference. Vehicle data can reflect the driver's intention only after the maneuver has been initiated. However, the vehicle dynamic status–based mathematic model has limited ability in the prediction of driver intention. Although

the vehicle data gives a limited contribution to the prediction of driver intention, it is still an important data source that can increase the accuracy of the intention identification and can help predict the intent at an early stage after the intended behavior is initiated.

### Driver behaviors

With an increasing demand for ADAS, driver behavior and driver state analysis is getting more and more popular. Unlike the CAN bus data, driver behavioral signals, such as head and eye movement, can give an early clue about the driver's intention. Many studies have evaluated the impact of head/eye movement on intention prediction. In Ref. [59], the authors applied the pupil information as the cognitive signals for the lane change intent prediction. Normally, driver eye movement can be classified as intention guided and nonintention guided. Intention-guided eye movement means the eye fixation or saccades were done on purpose, whereas the nonintention-based movement may be due to distractions. Driver visual fixation will no longer follow the driver's attention when the driver is distracted. At this moment, eye movement cannot reflect the driver's mental purpose and the intention prediction result cannot be trusted [61]. In terms of intention-oriented eye tracking, it can be viewed as a cognitive progress of information gathering, which can reflect the driver's mental state earlier than the vehicle parameters. A CAN bus data–based driver behavior recognition belongs to the cognitive process of action execution. In addition, the driver's intention at an information-gathering step is less likely to change when compared with that in the action execution step [62]. Although head/eye movement can be caused by distraction, most of the time the driver shifts the eye gaze on purpose, which makes eye movement a useful signal for intention decoding and inference [63]. Doshi and Trivedi [64] evaluated the relationship between gaze pattern, surrounding traffic information, and driver intention. They proposed an intention recognition method based on reasoning eye movement first. They tried to find out whether an eye movement should attribute to a specific goal or irrelevant stimuli before using eye movement to infer driver intention by fusing eye gaze with a saliency map. Many prior researchers have focused on applying the eye-tracking technique to predict driver intention [65−67,100,128,129], and it has been proved that eye movement information does improve the intention prediction accuracy and decrease the false alarm rate. Besides, by using eye movement information, normally a driver's intention can be recognized much earlier than by using vehicle parameters only. The eye-tracking system can be classified into intrusive glass type and nonintrusive camera-based system (Fig. 2.11).

Eye movement is detected through cameras mounted on the vehicle dashboard or glass-type driver eye-tracking system. Still many researchers have developed powerful algorithms for eye movement detection [68,130]. Many studies have paid attention to the eye-tracking techniques to predict driver intention [65−67]. It has been proved that the eye movement information does improve the intention prediction accuracy and helps decrease the false alarm rate. A significant challenge to eye movement detection is eye tracking. The eye movement is normally detected with cameras mounted on the dashboard or the wearable eye-tracking system. Owing to the physical characteristics of the eye (small scale and occlusion, etc.), it is not easy to detect the eye and track the pupil robustly.

Moreover, glass, lightness, and even hairs near the eye can influence eye-tracking performance. According to these challenges, some robust algorithms for eye movement detection have been proposed [68−70]. Lethaus et al. [71] evaluated how early the gaze information can reflect the driver intent, and how many gaze features can be used to infer the intent. In their experiment, vehicle data was collected in a driving simulator and the eye data was captured by the SMI eye-tracking system. Five viewing zones are defined, and ANN is used to predict driver intention. They finally concluded that a 10-s window for the eye gaze data is enough for intention prediction, whereas a 5-s window gives better performance because the 10-s window carries more noise. Besides, an efficient distance of 3.5−5 s before the maneuver occurs for classifying lane change left and 2.5−3 s for lane change right detection was observed.

Similar to gaze direction, head motion is another cognitive process for information gathering. Both eye and head movement can be seen as a cognitive process of information gathering. It has been proved in Ref. [94] that head movement is a more important factor than eye movement for driver intention prediction and it has also been widely used in many past DII research [90−92]. The true-positive rate (TRP) of using head movement, along with lane and vehicle information, can achieve 79.5%, whereas with only use lane and vehicle information the rate is only 50%. Another interesting point the authors found out was that head movement information plays a more important role than eye movement information in the prediction of driver intention. Further research is done in Ref. [94], in which the authors claim that both eye and head movement are useful data for the detection of driver distraction, attention, and mental state inference. However, there is a

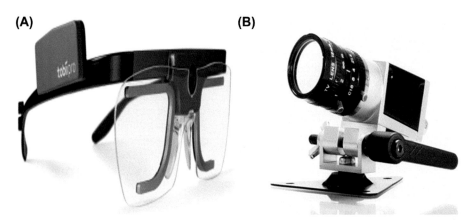

FIG. 2.11 Two different kinds of eye-tracking devices are available on the market. **(A)** An intrusive eye tracking glass by Tobii and **(B)** a nonintrusive camera.

difference between eye and head movement in the classification of driver mental state. Specifically, when the driver is executing a mental goal-oriented task the head moves early, whereas the eye stays or moves later. On the other hand, when outer stimuli occur, the driver's eye will shift first and the head moves later. This is an interesting research because it offers a way to determine whether the ongoing driver behavior is goal oriented or stimulus based.

Murphy and Trivedi [74] concluded that a variety of head pose estimation algorithm can be used to track head movement. They concluded that several head pose estimation algorithms can be used to track head movement. However, the in-vehicle head-tracking system has its own problems. One significant challenge is the online computing ability of the onboard processor. Driver's eye and head movements are usually sampled at 30−60 Hz; for a multiple camera system, a large amount of data will influence the precision of the driver status recognition unit. In Ref. [75], a processing method for in-vehicle driver eye-/head-tracking system was proposed aiming at handling the natural eye-/head-tracking system for driver distraction detection. The data processing method was able to improve the sensitivity and specificity of the eye-tracking system by 10%. Another challenge is the noise issue. Vibration from the road and lightness variation issues exist in the vehicle cabinet, which brings a large amount of noise to the captured images. Sometimes head-tracking data will get lost because of the algorithm problems and head tracking usually needs reinitializing [76]. Besides, according to the current head-tracking algorithms, those developed with a monocular camera show worse performance than the multicameras-based algorithms. However, using multiple cameras in the vehicle will increase the system cost.

In Ref. [60], the authors proposed a driver head-tracking system for driver LCII. The head-tracking system consists of six cameras. Head movement was regarded as a more important factor than eye gaze for driver intention prediction [66]. Head movement was widely adopted for DII [66,72,73]. In Ref. [66], the authors claimed that both eye and head movements are useful data for the detection of driver distraction, attention, and mental state inference. However, there is a difference between the eye and head movement for the classification of driver mental state. Specifically, the head moves earlier than the eye when the driver is executing a mental goal-oriented task. On the other hand, when the outer stimuli occur, the driver's eye will shift first and the head moves later. This is an interesting conclusion because it offers a way to determine

whether the ongoing driver behavior is goal oriented or stimulus based. In addition to camera-based driver eye- and head-tracking systems, some other driver behaviors such as the foot, hand, and body gestures were also recorded through a camera in some research [38,78,131]. A carefully selected driver behavior feature will improve the performance of the driver intent predictor.

### Electroencephalography

EEG is a brain action measuring device that measures the flow of brain electric current with noninvasive electrodes on the scalp. It has been widely used in cognitive neuroscience for the study of brain activities. Meanwhile, EEG is also an important sensor for brain-computer interface (BCI) design. EEG is sensitive to a small change in electric activities, which is suitable to detect a human mental state. Therefore many researchers have studied the impact of introducing EEG on the detection of driver mental activity. A direct application of EEG to on-vehicle BCI design is driver workload detection. EEG has been widely used to monitor the driver workload and other status such as drowsiness, happiness, sadness, mental fatigue, and abnormal conditions [77]. However, a drawback of the EEG signal is it usually contains various noise, hard to acquire, and gets weak when sampled with poor quality [79]. This is because the brain electric current is detected with a noninvasive method, in which the signal has to cross the brain layers, scalp, and skull. Therefore EEG recording systems usually contain electrodes, amplifier units, A/D converter, and noise filter units. EEG signals consist of high-dimensional data with large noise, which leads to the use of a machine learning method as the main solution when dealing with EEG signals. Machine learning algorithms can be used in the feature extraction and classification unit [132]. The general EEG processing procedure is shown in Fig. 2.12.

In addition, EEG has been proved not only suitable for driver attention and status detection but also to be able to predict driver intention. Haufe et al. [35] have used EEG, for the first time, to design a braking assistance system. As EEG measures brain activity, it can be much faster than human muscle reaction. If the braking assistant system can recognize braking intention by EEG rather than by detecting the brake pedal, it will save extra time for the driver and improve the driving safety. It is thus finally proved that, by using EEG, the system can detect a brake intention 130 ms faster than a brake pedal measurement only method [35]. EEG has also been used in driver steering intention prediction [82], with different machine learning-based classification

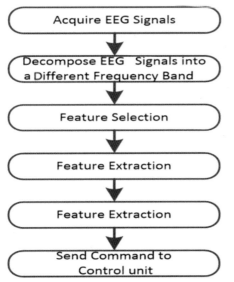

FIG. 2.12 General algorithm for electroencephalographic (EEG) signal processing [79].

methods, and the prediction accuracy of steering intention can achieve 65%−80%. A general process of EEG signal processing is shown in Fig. 2.11. As shown in prior studies, most of the EEG-based BCI is designed to predict driver's instant or operational level intention.

The impact of EEG signals on the intended inference is worthy to be examined.

Table 2.2 illustrates previous research that used EEG to detect driver intentions. Most of the research focus on using EEG to recognize the driver's operational intent such as braking and steering. One concern is that in the real-world driving environment, EEG signals contain lots of noise and can be affected by head movement, which makes it difficult to use. However, EEG signals are still a good measurement of driver statuses such as attention and fatigue.

## ALGORITHMS FOR DRIVER INTENTION INFERENCE

Conventional intention inference algorithms can be roughly divided into the following groups: mathematic model, driver cognitive model, and the widely used machine learning models. Owing to their ability to deal with a high-dimensional feature vector, machine learning methods are widely accepted by the LCII system. As mentioned in Ref. [49], the intelligent vehicle adopted 200 sensor signals. At this moment, machine learning methods are the most suitable tools to fuse the signals and construct the LCII system. The machine learning algorithms can be divided into generative model and discriminative model. The generative model provides a joint probability distribution over the observed and target values, which can generate both

**TABLE 2.2**
**Performance of Previous Driver Intention Research Using EEG Signal.**

| Reference | Target | Inputs | Algorithms | Num[a] | TRP |
|---|---|---|---|---|---|
| [102] | Steering | EEG | Gaussian distribution | 3 | 80 (aver) |
| [133] | Steering | EEG, steer angle, brake and gas pedal | Hierarchic Bayesian and logistic regression | 8 | 70 (aver) |
| [35] | Steering | EEG | Bayesian decision theory | 5 | 65% |
| [134] | Braking | EEG, EMG, gas and brake pedal | Regularized linear discriminant analysis | 20 | Varying with time |
| [81] | Braking | EEG, EMG, brake and gas pedal | Regularized linear discriminant analysis | 15 | 70−90 |
| [36] | Stop and go | EEG, EMG, EOG, steering, brake and gas pedal | Linear and quadratic discriminant analysis | 6 | 0.76 ± 0.12 |
| [135] | Turning | EEG, steer angle, brake and gas pedal | Linear discriminant analysis | 7 | 0.69 ± 0.16 |
| [136] | Alertness, drowsiness | EEG, driving behavior | A wide variety of classification algorithms | 6 | 80%−90% |

EEG, electroencephalography; EMG, electromyography; EOG, electrooculography; TRP, true-positive rate.
[a] Num is short for number, which indicates the number of participants in the experiment.

the inputs and outputs according to some learned hidden states. However, the generative model is less easy to be trained compared with the discriminative model, as it may require training multiple models and providing a model to each class based on the different probability distribution.

On the contrary, the discriminative model only provides the dependence of the target on the observed data. Discriminative models usually can be generated from the specific generative models through the Bayes rule. The most popular generative models include Bayesian networks, HMMs, Gaussian mixture models, etc., and the widely used discriminative models are support vector machine (SVM), neural networks, and linear regression. As mentioned in Ref. [31], discriminative models provide a better result on a single target problem than the generative models, whereas the generative models are more suitable for multitarget problems. Instead of these two typical methods, driver intention can also be modeled based on the cognitive models and the deep learning models. A taxonomy of the algorithms for intention inference is shown in Fig. 2.13, and in Table 2.3 the LCII results based on some articles are illustrated.

### Generative Model

Generative models, such as HMM, are widely used in previous LCII studies [55,56,59,66,83,84]. Pentland and Liu [45] used the HMM to recognize seven kinds of driver intention. Berndt used the HMM to investigate early lane change intention [51]. The aim of the study is

to identify the left and right lane change maneuvers at an early stage after the lane change was started. Final results showed 71% and 74% recognition accuracy for the left and right lane changes, respectively. In Ref. [85], a new feature named comprehensive decision index (CDI) was introduced.

Fuzzy logic was applied to represent the surrounding environment and the lane changing willingness of the driver. The overall performances of the algorithm with different input parameters (a situation, situation and vehicle, and situation, vehicle, and CDI) were analyzed. Li et al. [86] proposed an integrated intention inference algorithm based on HMM and Bayesian filter (BF) technique. A preliminary output from the HMM was further filtered using the BF method to make the final decision. The HMM-BF framework achieved a recognition accuracy of 93.5% and 90.3% for the right and left lane changes, respectively. In Ref. [56], the authors proposed a driver lane change/keep intention inference method based on a dynamic Bayesian network. A four-step framework for the DII was developed and the autoregression was combined with an HMM to take the past driver behaviors into consideration.

In Ref. [57], the authors constructed a lane change intention recognition method based on the continuous hidden Markov model (CHMM). The authors evaluated the CHMM performance with different model structures (3, 6, and 9 hidden states). According to the results, the CHMM with six hidden states and 1.5 s window size (data collected between 0 and 1.5 s prior to the vehicle crossing the lane) gave the best

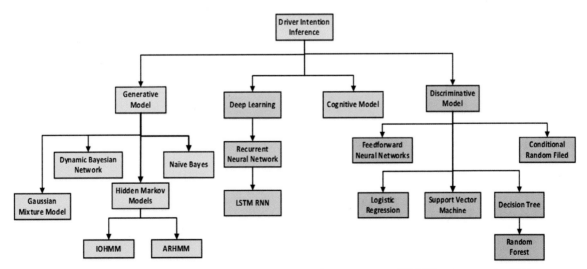

FIG. 2.13 Taxonomy of algorithms for driver intention inference system. *ARHMM*, autoregressive hidden Markov model; *IOHMM*, input-output hidden Markov model; *LSTM RNN*, long short-term memory recurrent neural network.

**TABLE 2.3**
Summary of Various Previous Lane Change Intention Inference Systems.

| Reference | Signals | Algorithm | No. of Subjects | Environment | Performance | Prediction Horizon |
|-----------|---------|-----------|-----------------|-------------|-------------|--------------------|
| [57] | Steering angle, steering force, velocity | CHMM | 10 | Simulator | 100% | 0.5–0.7 s After steering |
| [66] | Lane position, CAN bus, eye and head movement | RVM | 8 | On road | 88.51% | 3 s Prior to the lane change |
| [98] | Lane position, CAN, and head movement | Sparse Bayesian learning | 3 | On road | 90% | 3 s Prior to the lane change |
| [65] | Eye movement | Finish questionnaire | 17 | Simulator | 77% | – |
| [99] | Eye movement (pupil size variation) | SVM | 24 Samples | On road | 73.13% ± 1.25% | – |
| [51] | CAN, digital map | HMM | 50 LCL, 50 LCR | On road | 71% L, 74% R | – |
| [87] | CAN, distance between vehicles | HMM | 20 | On road | 80%–90% | – |
| [50] | CAN, LDW, ACC, head movement, SWA | RVM | 15 | On road | 91% | 1 s Prior to the lane change |
| [43] | CAN, head and eye movement | HMM | 70 | On road | 12.5% LR, 17.6% LL | 1 s Prior the maneuver |
| [40] | CAN, lane style and position, head and eye movement | RVM | 108 Lane changes | On road | 79.20% | – |
| [67] | CAN, eye movement | ANNs, BNs, NBCs | 10 | Simulator | 95.8% (ANN), 93.2% (BNs), 90.6% (NBCs) | – |
| [83] | Steering angle | Queuing network model | 14 | Simulator | LCN 98.61%, LCE 91.67% | – |
| [85] | Steering angle, rate, comprehensive decision index | Fuzzy logic and HMM | 4 (69 samples) | Simulator | 94% | 1.67 s After lane change maneuver starts |
| [100] | CAN, eye movement | State transition diagram | 20 (8576 lane changes) | Simulator | 80% | – |
| [84] | Relative position and velocity, heading, time to collision, time headway | SVM, RF, LR | 5 | Simulator | 89.5% (SVM), 88.9% (RF), 87.2% (LR) | – |

*(continued)*

**TABLE 2.3**
**Summary of Various Previous Lane Change Intention Inference Systems.—cont'd**

| Reference | Signals | Algorithm | No. of Subjects | Environment | Performance | Prediction Horizon |
|---|---|---|---|---|---|---|
| [45] | Steering angle and velocity, vehicle velocity and acceleration | HMM | 8 | Simulator | 88.3% ± 4.4% | – |
| [49] | CAN, ACC, SWA, LDW, head movement | RVM | 15 (500 samples) | On road | 80% | 3 s Prior to the lane change |
| [101] | CAN, GPS, eye movement | Finish questionnaire | 22 | On road | – | – |
| [54] | Steering angle and relative lane position | SVM and Bayesian filter | 2 (139 samples) | On road | 80% | 1.3 s Prior to the lane change |
| [58] | CAN, eye movement | ANN | 10 | Simulator | 95% (L), 85% (R) | – |
| [102] | CAN, lidar, radar, hand, head, and foot dynamics | Latent dynamic conditional random field | 1000 Samples | On road | 90% | 2 s Prior to the lane change |
| [103] | CAN | SVM and BN | 4 | Simulator | 95% (LK), 80% (LC) | – |
| [33] | CAN, eye movement | Computational model based on ACT-R | 11 | Simulator | 90% | 1 s After steering |
| [86] | CAN bus | CHMM and Bayesian filter | 188 LCL, 212 LCR, 242 LK | On road | 93.5% (L), 90.3% (R) | 0.5–0.7 s After steering |
| [94] | GPS, digital map, head movement, CAN bus | LSTM-RNN | 10 Drivers, 1180 miles | On road | 90.5% | 3.5 s Prior to the lane change |

*ACC*, adaptive cruise control; *ACT-R*, Adaptive Control of Thought-Rational; *ANN*, artificial neural network; *BNs*, Bayesian networks; *CAN*, Controller Area Network; *CHMM*, continuous hidden Markov model; *HMM*, hidden Markov model; *LC*, lane changing; *LCL*, lane change left; *LCR*, lane change right; *LDW*, lane departure warning; *LK*, lane keeping; *LR*, logistic regression; *LSTM-RNN*, long short-term memory-recurrent neural network; *NBCs*, Naïve Bayesian classifiers; *RF*, random forest; *RVM*, relevance vector machine; *SVM*, support vector machine; *SWA*, side warning assistance.

classification result (95.48%). The authors also concluded that the most important factors for lane change intention recognition were data representation, size of the sliding window, and the initial sets of model parameters. In Ref. [87], the authors proposed a context-based highway lane change intention system based on HMM. Four different inference systems were defined, which were vehicle state model only, front vehicle distance-based model, rear vehicle distance-based model, and front and rear distance-based model.

From the comparison of the four scenarios, the author pointed out that the classification performance did not show a significant increase with the additional context information. However, they showed that the additional context information leads to a high false-positive result rate and the system performance was worse than the system with vehicle state information only. One possible explanation is that the HMM has limited ability to capture the context information during the lane change process. Therefore a more powerful algorithm such as double-layered HMM and input-output HMM [94] should be used.

Besides the HMM, other generative models such as the Bayesian network and Naive Bayesian classifier are also used for intention detection. For example, in Ref. [67], the authors used three different algorithms, namely, an ANN, Bayesian networks, and Naive Bayesian. With a direct comparison, the result demonstrates that the performance of ANN was the best among the three. In Ref. [49], a driver lane change behavior classifier based on a hybrid model that combines a Bayesian network and SVM was proposed. In Ref. [60], the authors proposed a driver head-tracking system for driver LCII. A Naive Bayesian classifier was trained and used to classify the glance area of the driver. In Ref. [88], the authors proposed a lane change detection method based on the object-oriented Bayesian network. The system was designed according to the modularity and reusability of the Bayesian network, which makes it easy to extend the system according to different requirements. The whole system was constructed by various sub-Bayesian networks with a different function. Seven main driving maneuvers were studied, which were object follow, lane follows, ego-vehicle cut out, object cut out, ego-vehicle cut in, object cut in, and others.

### Discriminative Model

Discriminative models, such as SVM and ANN, are also widely used in LCII because of the rich background theories and the successful application experience. In Refs. [49,50], a Bayesian extension to the SVM algorithm, namely, the relevance vector machine (RVM), was used to classify the driver lane change (right and left) and lane keeping intentions. To decrease the false alarm rate of the system, multiple detection suppression techniques were used, under the assumption that consecutive detection arises from the same intention. The classifier can achieve 80% accuracy with a relatively low false alarm rate. The authors also examined the classification result in a certain time before the maneuver happens using different sources of sensors and testing situations. The conclusion was given that the online classification results were worse in a real-time environment than in the experimental environment.

Campbell identified three kinds of driver intentions: lane keeping, preparing for lane changing, and lane changing [84]. Three classification algorithms were used in the study (SVM, random forest, and logistic regression). The SVM model achieved the best classification performance compared with the other two algorithms. The authors in Ref. [58] proposed a driver intention recognition method based on ANNs. The accuracy for lane change left detection is better than

that for the lane change right. The results showed that head rotation had consistent gains between 1.5 and 2.5 s prior to the lane change maneuver.

The work in Ref. [54] constructed a multiclass classifier by combining the SVM and Bayesian filter (BF). Results showed that the proposed algorithm realized prediction of the intention in an average of 1.3 s in advance and can achieve a maximum prediction horizon of 3.29 s. It was concluded that one of the important tasks in the future is to improve the performance of the lane tracker system by reducing the false alarm rate. The authors in Ref. [89] proposed a driver lane change and lane keeping intention classification method based on SVM. In Ref. [90], the authors introduced and compared three machine learning methods, which were the feedforward neural network, recurrent neural network (RNN), and SVM. To evaluate the classification performance, four evaluation criteria were introduced, which were the mean value of prediction horizon, the number of the correctly recognized lane change, the number of not recognizing the lane change, and the number of false alarms. The results showed that SVM gave the best results followed by RNN. The classifiers were able to predict the lane change 1−1.5 s prior to the vehicle crossing the lane.

### Cognitive Model

Apart from the machine learning algorithms, human cognitive models are also used in some studies. For example, Salvucci introduced a real-time system for detecting driver lane change intention based on a mind-tracking architecture [32,33]. The mind-tracking computational model continually infers the driver's unobserved intention from the observed actions. The computation model was based on the cognitive model implemented in the ACT-R study. The system contains four steps: data collection, model simulation, action tracking, and thought inference. During the simulation, the system ran several models simultaneously. Mind tracking detected the driver's intentions by examining the "thoughts" of the best matching model. The mind-tracking system achieved 85% accuracy with a 4% false alarm rate for lane change intention detection.

In Ref. [83], the authors constructed a queuing network cognitive architecture to model the normal lane change (LCN) and emergency lane change (LCE) maneuvers. The differences between the outputs of the driver model and the measured data are compared. The driver lane change behavior model was built by the queuing network model, which contains three main modules: preview module, prediction module, and control module. The intention was detected based on a threshold of root-mean-square error (RMSE) value.

The intention with the smallest RMSE value is determined as the final output. The proposed method achieved a high accuracy (above 90%) and a low false alarm rate (0.294%). Comparing with those intelligent inference methods based on eye gaze and head movement, this method can be easily extended into real-world applications. However, as the algorithm was based on the steering wheel angle signal only, it cannot infer the driver maneuver at a very early stage or before the maneuver happens.

## Deep Learning Methods

Recently, tremendous achievement has been made in the deep learning area because of the development of deep learning theories, parallel computation hardware, and large-scale annotated dataset. The deep networks have achieved state-of-the-art performance on many computer vision tasks, such as image classification, segmentation, and object detection domains [91,92]. The deep convolutional neural network has been widely used in many intelligent and automated vehicles [93]. Meanwhile, the RNN also achieved significant results on natural language processing and image captioning areas [73,94]. RNN can be used to process the temporal dependence between the dataset, as it allows connection between the previous layers and current layers. To increase the long-term dependency property and overcome the gradient descent, a long short-term memory (LSTM) scheme was proposed [95]. The LSTM largely increases the long-term memory ability of the RNN model and has been successfully applied to many tasks.

As aforementioned, DII usually requires taking the previous driver behaviors and traffic context into consideration. The conventional HMM has a limited ability to capture long-term dependency. However, at this moment, RNN can provide a better prediction of the driver's intention. In Ref. [96], an LSTM-RNN model was designed to infer the driver's intention when the vehicle enters an intersection. The RNN outperforms the quadratic discriminate analysis model. Similarly, a series of studies have been proposed in Ref. [94]. The authors compared the lane-change intention inference performance of the LSTM-based RNN with multiple HMMs. Driver head rotation data, along with the traffic context from the GPS and digital map, were collected. The lane change intent can be detected 3.5 s before the vehicle comes into another lane with a precision and recall of 90.5% and 87.4%, respectively. Based on the review of previous LCII algorithms, the deep learning algorithms show a significant advantage in the prediction accuracy and larger prediction horizon. Although deep learning algorithms lead to a higher computational burden, these algorithms are much powerful than conventional methods. With the development of hardware platforms and software systems, deep learning methods can be more distinct in the future.

Machine learning algorithms are particularly suitable for DII problems because it is difficult or even impossible to build an accurate mathematic model for the human mental state. On the other hand, machine learning methods do not need to build such kind of models manually and most of the algorithms have a rich theory background. Machine learning methods can use observed data to train a model in both supervised and unsupervised ways, which is easy to realize. Machine learning-based models can be roughly classified into discriminative models, such as SVM and RVM, and generalize models, such as dynamic Bayesian network and HMM. The generalized model relies on probability theory and prior knowledge about the work, whereas the discriminative model does not require statistical information. Both these algorithms are used in past research to study human and driver intentions. In Ref. [32], the authors introduce an extended Bayesian network as a tool to infer human intention. In Refs. [60,80,101], the HMM known as a kind of dynamic Bayesian network is accepted to predict driver intention. In Refs. [90−92], SVM and RVM are adopted to classify driver intent. In this work, the two most popular kinds of algorithms (SVM and HMM) will be reviewed and discussed.

## EVALUATION OF DRIVER INTENTION INFERENCE SYSTEM

It is important to evaluate the performance of the driver intention classification system. To have a clear perspective about how the classifier works, the driver intent classification can be evaluated with the TRP, false-positive rate (FRP), and the prediction horizon.

### Detection Accuracy

TRP and FRP are two important factors that describe the performance of the classifier and have been used in a large number of related studies. TRP, also called the hit rate or recall, measures how many times the classifier detects the intent successfully, while the false alarm rate describes how many times the classifier misclassifies the intent into the wrong category. Sometimes the FRP can be more critical than the TRP because the driver normally does not want to be disrupted by the classifier. If a classifier pursues a high TRP at the price of a high FRP, this system will hardly be accepted. However, if a system has a slightly lower TRP and a lower false alarm rate, it is still helpful in some situations. Therefore the main objective of the classifier is to maximize the TRP while minimizing the low FRP. A continent way to visualize the performance of the classifier is by using receiver operating characteristics (ROCs). ROC is a graph technique that is used to visualize, organize, and select classifiers based on their performance [97].

### Prediction Horizon

Besides the TRP and FRP indexes, predicting the horizon of intent is another important factor to evaluate the performance of the classifier. Some of the previous studies report a TRP and FRP without giving a clear prediction horizon, which is unfair. As shown in Fig. 2.14, there are four critical moments for a lane change process. T1 is the moment when the driver generates the lane change intention. T2 is the moment when the driver finishes traffic context checking and begins to make a lane change. T3 represents the moment that the vehicle starts to cross the lane. At T4 the driver finishes the lane change task. Normally, as there is no precise driver intention model that can explain when the driver generates an intention, T1 is very difficult to be determined. Therefore most of the studies use T2 and T3 as the time criteria to evaluate the prediction horizon. The earlier the prediction is made, the more difficult the task will be. After the driver has taken some actions such as steering the wheel and slightly braking based on his/her intent, it will be easier to recognize the driver's intent. However, if the intelligent inference unit tries to recognize the driver's intent before some actions are taken, the task will be much difficult because only limited and uncertain information can be obtained. Moreover, the earlier the prediction is made, the higher the FRP will be. Therefore a trade-off between the FRP and the prediction horizon exists, which needs to be carefully considered. It is mentioned in Ref. [50] that the TRP and FRP can vary with the prediction horizon. These performance indexes will be less powerful in describing the performance of the classifier, without a clear prediction horizon.

Some researchers have paid attention to the prediction horizon issue. Doshi et al. [50] found that the data collected 3 s prior to the maneuver was significant enough to present the lane change intention. Salvucci [33] tested a lane change intent inference system on a driving simulator and detected the lane change at the very start with 65% accuracy. As time goes on, the accuracy increased to 80% after half a second and 90% after 1 s. Berndt identified a lane change maneuver at a very early stage after it was initialized and achieved 71% prediction accuracy for lane change left and 74% for lane change right in a conditional simulation environment [51]. Bi et al. [83] detected a driver's normal and emergency lane change intents within 0.325 and 0.268 s, respectively. Kumar realized an average of 1.3 s prediction of the intent in advance and can achieve a maximum of 3.29 s prediction horizon with a maximum of 82% accuracy [54]. By evaluating the performance of SVM and ANN, Dogan [90] proposed a classifier to predict the lane change 1–1.5 s prior to the vehicle crosses the lane on a driving simulator. The prediction horizon in the simulation environment is always better than that in real-world testing. This is mainly due to the large noise and distraction in the real-world environment. However, the performance of the classifier in the real world gives a real indication and is beneficial to the analysis of driver response in the real world.

## CHALLENGES AND FUTURE WORKS

In this section, challenges and part of the future works are highlighted. Four primary works to enhance the DII system are discussed, which are the design of next-generation ADAS, driver situation awareness (SA) and interaction-aware modeling, autonomous driving, and parallel DII.

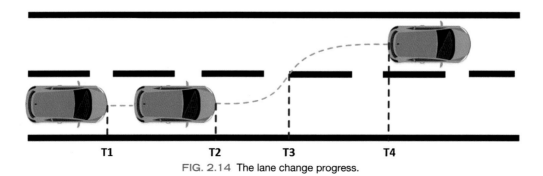

FIG. 2.14 The lane change progress.

## Design Next-Generation Advanced Driver Assistance Systems

Next-generation ADAS require further advances in understanding the driver from outer behaviors, mental status, and sophisticated environment perception. As aforementioned, current ADAS are only equipped with an isolated driver status recognition system, which fails to exploit the relationship between different functions. Meanwhile, a holistic traffic context perception system is required according to the fast development of sensors and an onboard computing device. These concerns give rise to the following discussion.

### Integration of driver monitoring systems

The studies of driver behavior-oriented assistance systems can be partially summarized into the following aspects: driver attention, driver intention, driver workload, driver style, and driver distraction. For each research area, a vast amount of studies have been proposed. However, there are still no explicit connections between these systems. It is believed that driver behavior under the distracted condition and the nondistracted state is different [104,105]. Also, if the driver is overloaded after a long drive, the physical behaviors are also different [106]. In terms of the DII system, how to correctly infer the driver intention with different mental status still needs to be studied. Therefore the construction of a robust DII, which can adapt to different driver status, is expected. Also by considering driver monitoring systems as a whole, the control conflicts between the driver and the vehicle can be reduced.

### The need for a comprehensive environment model

Sensing efficiently and precisely is another emerging requirement for the context perception module. A holistic approach is needed in the future to construct a comprehensive environment model from both the sensor's view and the driver's view. The driver-oriented context perception must process the context data sequence and analyze the potential driving solutions for the human driver. This can be treated as an active guidance that can influence the driver's intention generation process rather than only providing the fused context data to the driver and inferring the intention afterward. Dynamic analysis of the potential driving behaviors concerning the current context will significantly increase the intention prediction horizon and accuracy. However, real-time estimation leads to a more stringent requirement to onboard perception and computing hardware.

### Design cognitive model for driver intention

A more challenging work is to exploit a comprehensive understanding of the intention generation process according to the traffic context and human behaviors. Currently, driver attention and workload can be mathematically modeled, which provides a better explanation for the driver's cognitive attention and workload behaviors [108,109]. However, there are still limited studies on the explicit modeling of driver intention. Describing the intention generation process with more precise cognitive language and a mathematic model would be one of the core studies in the future.

## Situation Awareness and Interaction-Aware Modeling

The prediction of driver maneuver and the vehicle trajectory need to be made according to the driver SA and interaction behaviors. In Ref. [99], three kinds of vehicle motion modeling methods were proposed: the physics-based motion model, maneuver-based motion model, and interaction-aware motion model. The maneuver-based motion model predicts the vehicle trajectory based on the early recognition of the driver's intended maneuvers, which is like the intention inference task described in this study. However, most of the maneuver-based models assume that the surrounding vehicles move independently without interacting with each other, which can be unreasonable in some complicated situations such as in the roundabout or urban area. Therefore the interaction-aware modeling methods with respect to the driver SA should be further studied in the future. This section will discuss this problem from two points, namely, driver SA modeling and interaction-aware modeling.

### Situation awareness modeling

Driver SA can be viewed as the knowledge that was learned and updated from the driving tasks to handle the multifaced situation and guide the driver to make decisions when engaged in real-time multitasking [100]. The perceptual and cognitive process of maintaining the SA can also be divided into three categories, which are automatic (usually unconscious and require no cognitive resources), recognition-primed process (few demands on cognitive resources), and conscious controlled process (requires heavy cognitive resources) [100]. The driver SA model carries the habit, knowledge, and attitude toward the specific driving tasks and is closely related to the DII because the SA knowledge directs how to understand the driver correctly. For example, a driver's intention-oriented SA system at the intersection has been discussed in Ref. [26]. Four significant contributions of the SA system are summarized as avoidance of unnecessary warnings, detection of occluded traffic participants, enhancement of DII, and prediction of future trajectories of other entities.

Regarding the lane change maneuvers, the four factors are also important because the SA model enables the analysis of surrounding traffic flow and provides guidance to the DII system. In Ref. [101], the driver lane change maneuver was classified into five categories based on the different interaction styles with surrounding vehicles. With the analysis of 1000 naturalistic highway lane change data, it was found that 72% of the lane change was self-motivated and had no significant interaction with the surrounding vehicles. However, without the proper SA, drivers may be unable to finish the intended maneuver smoothly when encountering a complex interaction. For example, a low-speed vehicle is in front of the ego-lane and a rear vehicle is fast approaching in the overtaking lane. In this case, the driver may wish to overtake the front vehicle and must control the vehicle according to the SA and the motion prediction of the rear vehicle. If the driver postpones the lane change maneuver and lets the rear vehicle pass first, a conflict will be generated between the desired intention and the actual maneuver.

Most of the previous driver intention studies do not provide enough analysis of this conflict because the driver's intention is unable to be predicted and labeled precisely, especially in such a complex condition. Furthermore, as mentioned in Ref. [101], most of the naturalistic lane change maneuvers have no significant interaction with other vehicles. The complex interaction scenarios are hard to be repeated in the real world so that not enough data can be used to analyze the conflict situations. However, the situation assessment and understanding can be used to predict the dangerous maneuver at intersections so that the conflict between the actual intention and the expected intention can be clarified [102]. Specifically, the intended stop/go maneuver and the expected maneuver of the driver when approaching an intersection were compared to gain a risk assessment of the dangerous maneuver. In Ref. [103], the context information and the corresponding traffic rules were applied with the dynamic Bayesian network so that the expected maneuver of the driver can be estimated. The future motion of the traffic participants is the combination of the tactical intentions and their corresponding risk assessment to perform the maneuver [104]. Therefore it is believed that driver SA and risk assessment can contribute to a better prediction of driver intention.

In summary, traffic SA concerning the assessment of the traffic contexts, traffic rules, road layout, driver behaviors, etc. are critical to the correct prediction of driver intention. A driver may generate a series of checking behaviors and perform the maneuver after the intention. However, the intended maneuver may be postponed or aborted due to the inappropriate situation. Hence, a comprehensive SA model is needed to fully understand the driver's behavior, cognitive process, perception, and interaction habit so that a precise prediction of the driver's intention can be achieved.

### Interaction-aware modeling

The interaction-aware motion prediction assumed traffic entities influence each other and provide a long-term motion prediction of other road users as the mutual dependencies between the drivers' decisions are considered. Regarding the lane change maneuver, a suddenly cut-in maneuver in front of the ego-vehicle can cause a lane change decision to the ego-driver to avoid collision [105]. At this moment, the DII algorithms may become less powerful than the interaction-aware algorithms in the prevention of collision, as DII is mainly designed for the prediction of active intention. Here we roughly define the active intention as a goal-oriented intention and the passive intention is mainly caused by other road entities, and the host driver must finish a specific maneuver in a short period. In Ref. [106], an integrated interaction-aware motion prediction model was proposed based on the combination of model-based intention estimation for surrounding entities and learning-based lateral motion prediction. The proposed method provides a reliable estimation of the future planning of the surrounding vehicles and the average prediction time before the lane change maneuvers can be extended by more than 60%.

In Ref. [107], a unified framework for maneuver classification, trajectory prediction, and interaction-aware motion prediction was proposed. It was shown that the surrounding vehicle motion should be predicted according to the comprehensive analysis of the potential maneuver and the probability of the future trajectory. In Ref. [108], a generic probabilistic interactive situation awareness model is proposed based on a two-layer HMM (TLHMM) framework. The TLHMM was used to model the real-world interaction behaviors in highway entrances, roundabouts, and T-intersections by computing the joint maneuver distribution of the multiple interactive agents. However, the model has a limitation in long-term prediction because the TLHMM cannot precisely remember the long-term dependency and temporal patterns. With the interaction-aware prediction model, the long-term motion and intention of surrounding entities can be estimated and used for host DII. This will lead to a holistic understanding of the current traffic context and enhance the DII system with an even earlier prediction. Moreover, the interaction-aware motion prediction enables the inference of a sudden lane change intention (passive intention) as discussed in the cut-in scenario. However, one of the disadvantages of the interaction-aware model is the computational complexity grows exponentially with the increasing number of vehicles [106]. The interaction-aware method relies on a comprehensive perception of the local traffic context, which increases the overall system cost.

Another interesting point is to predict driver intention and interaction behaviors based on transfer learning. In Ref. [108], the second layer of the TLHMM was trained with virtual data and high-level meta-features instead of traffic context information, which can be quickly applied to the real-world target. The complex interaction behaviors and scenarios are hard to be recorded and duplicated in the real world, although it can be carefully designed and sufficiently tested in the simulation environment. Hence, if the knowledge learned from the simulation can be properly transferred to the real-world scenario, the real-world interaction-aware model can be more precise and robust. This is also a major concern of parallel driving and parallel driver, which will be discussed later.

Despite sensing traffic context with onboard multisensor fusion, the interaction-aware prediction model can also be constructed based on the vehicle-to-vehicle (V2V) techniques [109]. The V2V communication does not rely on high-cost sensors but can provide efficient interactive communication and SA for the local area vehicles. The intention inference for the host driver and surrounding drivers can be detected and shared even earlier with the V2V techniques. The impact of the interaction-aware motion prediction and the V2V technique on the host DII has not been adequately studied in the past. Future works are expected in this area so that a risk-free and highly interactive traffic framework can be built.

### Autonomous Driving

The automated driving technology was divided into different levels based on the SAE standard J3016. With level 3 or higher intelligence, the autonomous vehicle is responsible for the environment perception, decision making, motion planning, and vehicle control. The automotive industry wishes to replace human drivers

with autonomous cars so that human mistakes can be avoided. However, it does not mean that driver modeling is not needed in the future.

DII systems require a comprehensive understanding of the driving environment as well as the driver's behavioral pattern. The process of intention generation and execution reflects the driver SA regarding the traffic context. Current decision-making algorithms for autonomous vehicles are mainly based on optimization, probabilistic models, and reinforcement learning. Neither of the algorithms takes the driver experience into the loop. The autonomous vehicle makes a lane change mainly based on the predefined rule base or a probabilistic model such as the Markov chain. These algorithms usually fail to consider the acceptance of the human passenger. The DII system will provide important guidance to the autonomous vehicle so that the autonomous vehicle can learn how human drivers make a lane change as well as when and where to execute the lane change. Meanwhile, combining DII with driving styles is also considerable [110]. Different drivers have different driving styles, and the intention inference system cannot work uniformly. In some situations, gentle drivers prefer to wait before changing the lane, whereas aggressive drivers like challenging tasks. If the autonomous vehicle takes the different intention patterns from different driver styles, it can minimize the uncomfortable driving experience for passengers.

Another emerging topic for DII toward autonomous decision making and motion planning is to estimate when and where the driver is going to drive [111]. Most of the existing DII algorithms do not pay attention to the intended position. The position should be estimated with the comprehensive environment perception and driver behaviors in the past few seconds. Current intention inference algorithms enable the intention prediction before the maneuver. However, the intended position estimation is still a difficult task. The positioning pattern learned from the driver can be transferred to the autonomous vehicle in a more straightforward manner. The path planning model can take the estimated short-term destination into calculation so that a more reasonable and human-like path can be generated. Therefore transferring the DII knowledge to the autonomous vehicle will bring more naturalistic human-like behaviors in both the decision-making and motion planning stages.

### Parallel Driver Intention Inference System

As aforementioned, the DII system suffers from hardware and algorithm limitations. Also, there is still no explicit model to describe the real mental intention process. One of the emerging challenges is shortness of data for model training and model evaluation. It is hard to collect plenty of real-world data to increase the data diversity because it dramatically increases the temporal and financial cost. Therefore a novel approach is required to sufficiently train and evaluate the intention inference system, and it would be better to have a self-learning ability to exploit the unseen pattern and principles that are behind the driver's intention nature. Fei-Yue Wang first developed the parallel theory in 2004 [112]. The construction of a parallel system requires the ACP approach as the background knowledge, which is a combination of Artificial society, Computational experiments, and Parallel execution [113].

The physical system in the real world can be viewed as a Newton machine, whereas the software defined-artificial world is a Merton machine [114]. In Ref. [115], the parallel system is described in the cyber-physical-social space, which extends the conventional cyber-physical space by integrating an additional dimension of human and social characteristics. Based on the ACP approach, a parallel DII system is proposed in Fig. 2.15.

In the artificial society, a virtual driving environment will be developed based on the modeling of traffic context as well as the driver behaviors such as head, facial, and body features [116]. There is plenty of simulation software that can build the three-dimensional

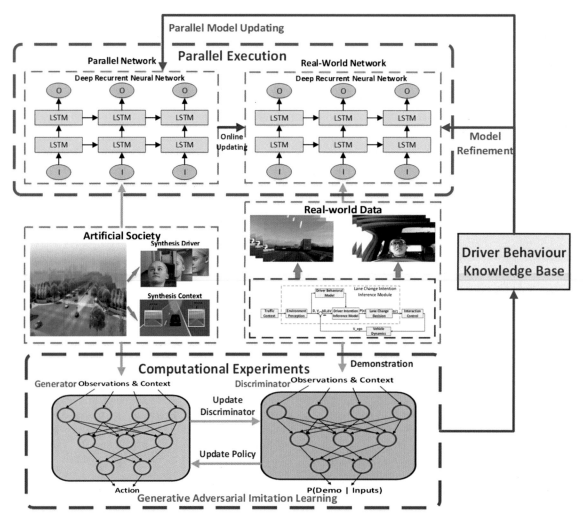

FIG. 2.15 The architecture of the ACP-based parallel driver intention inference system. *LSTM*, long short-term memory.

(3D) driving context, such as *CarSim* or *PanoSim*. The virtual facial images and videos can be generated based on the high-resolution 3D scans as used in Ref. [117]. The driver's facial dynamic model can be trained according to the real driver patterns using deep learning approaches such as generative adversarial networks [118]. Then a generative adversarial imitation learning method can be used to train the virtual driver model [119]. The virtual driver will be sufficiently evaluated with the data from both the artificial world and the real world. Finally, the learned driver behavior knowledge concerning the current traffic context can be used for the training and testing of the DII model. If the virtual model gives better inference accuracy, it will guide the real-world model to deal with challenging tasks and will update the real-world model with online learning methods. With the parallel DII system, the intention inference model can be trained and evaluated with many more scenarios so that a more robust intention inference model is generated.

## CONCLUSIONS

DII is an important function for ADAS and intelligent vehicles. It is an efficient method to avoid conflicts between the human driver and the intelligent units. The understanding of human intention enables a better design of the decision-making algorithms for automated vehicles. Human-like decision making is a major task for current intelligent and automated vehicles. Driver intention can be classified into three levels based on the time constant property. The relationship between level 2 and level 3 is clarified in this study. Based on the framework, the traffic context is viewed as the stimuli for the intention, while the driver's behavioral information and vehicle dynamics are the response to the stimuli. With respect to the LCII, multimodal sensors are used for intention inference. The signal sources can be classified into three parts: the traffic context, driver behavioral dynamics, and vehicle dynamic information. Existing algorithms for intention inference can be summarized into four major groups, which are the generative model, discriminative model, cognitive model, and deep learning models. A comprehensive evaluation method for intention inference should consider both the accuracy and the prediction horizon. A smart intention inference system should try to predict the driver's intention as early as possible. Future works for DII should concentrate on the precise modeling of the intention generation process and its cooperation with the other driver status monitoring system.

## REFERENCES

[1] A. Koesdwiady, et al., Recent trends in driver safety monitoring systems: state of the art and challenges, IEEE Transactions on Vehicular Technology 66 (6) (2017) 4550–4563.

[2] E. Bellis, J. Page, National Motor Vehicle Crash Causation Survey (NMVCCS) SAS Analytical Users Manual, 2008. No. HS-811 053.

[3] C.M. Martinez, et al., Driving style recognition for intelligent vehicle control and advanced driver assistance: a survey, in: IEEE Transactions on Intelligent Transportation Systems, 2017, pp. 1–11. PP.99.

[4] M.M. Michałek, M. Kiełczewski, The concept of passive control assistance for docking manoeuvres with n-trailer vehicles, IEEE/ASME Transactions on Mechatronics 20 (5) (2015) 2075–2084.

[5] F.-Y. Wang, S.-ming Tang, Concepts and frameworks of artificial transportation systems, Complex Systems and Complexity Science 1 (2) (2004) 52–59.

[6] V. Gaikwad, S. Lokhande, Lane departure identification for advanced driver assistance, IEEE Transactions on Intelligent Transportation Systems 16 (2) (2015) 910–918.

[7] Y.S. Son, et al., Robust multirate control scheme with predictive virtual lanes for lane-keeping system of autonomous highway driving, IEEE Transactions on Vehicular Technology 64 (8) (2015) 3378–3391.

[8] J. He, J.S. McCarley, A.F. Kramer, Lane keeping under cognitive load: performance changes and mechanisms, Human Factors 56 (2) (2014) 414–426.

[9] M. Liebner, F. Klanner, Driver intent inference and risk assessment, in: Handbook of Driver Assistance Systems: Basic Information, Components and Systems for Active Safety and Comfort, 2014, pp. 1–20.

[10] F. Qu, et al., A security and privacy review of VANETs, IEEE Transactions on Intelligent Transportation Systems 16 (6) (2015) 2985–2996.

[11] W. Liu, et al., Parking like a human: a direct trajectory planning solution, IEEE Transactions on Intelligent Transportation Systems 18 (12) (2017) 3388–3397.

[12] C. Lv, et al., Characterization of driver neuromuscular dynamics for human-automation collaboration design of automated vehicles, in: IEEE/ASME Transactions on Mechatronics, 2018.

[13] H.J. Kim, J.H. Yang, Takeover requests in simulated partially autonomous vehicles considering human factors, IEEE Transactions on Human-Machine Systems 47 (5) (2017) 735–740.

[14] A. Eriksson, N.A. Stanton, Takeover time in highly automated vehicles: noncritical transitions to and from manual control, Human Factors 59 (4) (2017) 689–705.

[15] P. Carruthers, The illusion of conscious will, Synthese 159 (2) (2007) 197–213.

[16] I. Ajzen, The theory of planned behavior, Organizational Behavior and Human Decision Processes 50 (2) (1991) 179–211.

[17] M. Bratman, Intention, Plans, and Practical Reason, 1987.

[18] C. Heinze, Modelling Intention Recognition for Intelligent Agent Systems, No. DSTO-RR-0286, Defence Science and Technology Organisation Salisbury (Australia) Systems Sciences Lab, 2004.

[19] K.A. Tahboub, Intelligent human-machine interaction based on dynamic Bayesian networks probabilistic intention recognition, Journal of Intelligent and Robotic Systems 45 (1) (2006) 31–52.

[20] E. Bonchek-Dokow, Cognitive Modeling of Human Intention Recognition, Bar Ilan University, Gonda Multidisciplinary Brain Research Center, Diss, 2011.

[21] H. Bösch, F. Steinkamp, E. Boller, Examining psychokinesis: the interaction of human intention with random number generators–A meta-analysis, Psychological Bulletin 132 (4) (2006) 497.

[22] Chadalavada, R. Teja, et al., That's on my mind! Robot to human intention communication through on-board projection on shared floor space, in: Mobile Robots (ECMR), 2015 European Conference on, IEEE, 2015.

[23] J. Huang, et al., Control of upper-limb power-assist exoskeleton using a human-robot interface based on motion intention recognition, IEEE Transactions on Automation Science and Engineering 12 (4) (2015) 1257–1270.

[24] J.-H. Han, S.-J. Lee, J.-H. Kim, Behaviour Hierarchy-based affordance map for recognition of human intention and its application to human–robot interaction, IEEE Transactions on Human-Machine Systems 46 (5) (2016) 708–722.

[25] Y.-M. Jang, et al., Human intention recognition based on eyeball movement pattern and pupil size variation, Neurocomputing 128 (2014) 421–432.

[26] S. Li, X. Zhang, Implicit intention communication in human–robot interaction through visual Behaviour Studies, IEEE Transactions on Human-Machine Systems 47 (4) (2017) 437–448.

[27] T. Takeda, Y. Hirata, K. Kosuge, Dance step estimation method based on HMM for dance partner robot, IEEE Transactions on Industrial Electronics 54 (2) (2007) 699–706.

[28] C. Zhu, C. Qi, W. Sheng, Human intention recognition in smart assisted living systems using a hierarchical hidden Markov model, in: Automation Science and Engineering, 2008. CASE 2008. IEEE International Conference on, IEEE, 2008.

[29] N.M. Oliver, B. Rosario, A.P. Pentland, A Bayesian computer vision system for modeling human interactions, IEEE Transactions on Pattern Analysis and Machine Intelligence 22 (8) (2000) 831–843.

[30] J.A. Michon, A critical view of driver behaviour models: what do we know, what should we do?, in: Human Behaviour and Traffic Safety Springer, Boston, MA, 1985, pp. 485–524.

[31] A. Doshi, M.M. Trivedi, Tactical driver behaviour prediction and intent inference: a review, in: Intelligent Transportation Systems (ITSC), 2011 14th International IEEE Conference on, IEEE, 2011.

[32] D.D. Salvucci, A. Liu, The time course of a lane change: driver control and eye-movement behavior, Transportation Research Part F: Traffic Psychology and Behaviour 5 (2) (2002) 123–132.

[33] D.D. Salvucci, Modeling driver behaviour in a cognitive architecture, Human Factors 48 (2) (2006) 362–380.

[34] S. Haufe, et al., Electrophysiology-based detection of emergency braking intention in real-world driving, Journal of Neural Engineering 11 (5) (2014), 056011.

[35] I.-H. Kim, et al., Detection of braking intention in diverse situations during simulated driving based on EEG feature combination, Journal of Neural Engineering 12 (1) (2014) 016001.

[36] Z. Khaliliardali, et al., Detection of anticipatory brain potentials during car driving, in: Engineering in Medicine and Biology Society (EMBC), 2012 Annual International Conference of the IEEE, IEEE, 2012.

[37] J.C. McCall, M.M. Trivedi, Driver behaviour and situation aware brake assistance for intelligent vehicles, Proceedings of the IEEE 95 (2) (2007) 374–387.

[38] C. Tran, A. Doshi, M.M. Trivedi, Modeling and prediction of driver behaviour by foot gesture analysis, Computer Vision and Image Understanding 116 (3) (2012) 435–445.

[39] E. Ohn-Bar, et al., On surveillance for safety critical events: in-vehicle video networks for predictive driver assistance systems, Computer Vision and Image Understanding 134 (2015) 130–140.

[40] R. Mabuchi, K. Yamada, Study on driver-intent estimation at yellow traffic signal by using driving simulator, in: Intelligent Vehicles Symposium (IV), 2011 IEEE, IEEE, 2011.

[41] H. Takahashi, K. Kuroda, A study on mental model for inferring driver's intention, in: Decision and Control, 1996., Proceedings of the 35th IEEE Conference on, vol. 2, IEEE, 1996.

[42] T. Kumagai, et al., Prediction of driving behaviour through probabilistic inference, in: Proc. 8th Intl. Conf. Engineering Applications of Neural Networks, 2003.

[43] N. Oliver, A.P. Pentland, Graphical models for driver behaviour recognition in a smart car, in: Intelligent Vehicles Symposium, 2000. IV 2000. Proceedings of the IEEE, IEEE, 2000.

[44] M. Liebner, et al., Driver intent inference at urban intersections using the intelligent driver model, in: Intelligent Vehicles Symposium (IV), 2012 IEEE, IEEE, 2012.

[45] A. Liu, A. Pentland, Towards real-time recognition of driver intentions, in: Intelligent Transportation System, 1997. ITSC'97., IEEE Conference on, IEEE, 1997.

[46] T. Imamura, et al., Estimation for driver's intentions in straight road environment using hidden Markov models, in: Systems Man and Cybernetics (SMC), 2010 IEEE International Conference on, IEEE, 2010.

[47] S.S. Beauchemin, et al., Portable and scalable vision-based vehicular instrumentation for the analysis of

driver intentionality, IEEE Transactions on Instrumentation and Measurement 61 (2) (2012) 391−401.

[48] A. Vahidi, A. Eskandarian, Research advances in intelligent collision avoidance and adaptive cruise control, IEEE Transactions on Intelligent Transportation Systems 4 (3) (2003) 143−153.

[49] B. Morris, A. Doshi, M. Trivedi, Lane change intent prediction for driver assistance: on-road design and evaluation, in: Intelligent Vehicles Symposium (IV), 2011 IEEE, IEEE, 2011.

[50] A. Doshi, B. Morris, M. Trivedi, On-road prediction of driver's intent with multimodal sensory cues, IEEE Pervasive Computing 10 (3) (2011) 22−34.

[51] H. Berndt, J. Emmert, K. Dietmayer, Continuous driver intention recognition with hidden Markov models, in: Intelligent Transportation Systems, 2008. ITSC 2008. 11th International IEEE Conference on, IEEE, 2008.

[52] R. Toledo-Moreo, M.A. Zamora-Izquierdo, IMM-based lane-change prediction in highways with low-cost GPS/INS, IEEE Transactions on Intelligent Transportation Systems 10 (1) (2009) 180−185.

[53] J.C. McCall, et al., A collaborative approach for human-centered driver assistance systems, in: Intelligent Transportation Systems, 2004. Proceedings. The 7th International IEEE Conference on, IEEE, 2004.

[54] P. Kumar, et al., Learning-based approach for online lane change intention prediction, in: Intelligent Vehicles Symposium (IV), 2013 IEEE, IEEE, 2013.

[55] K. Schmidt, et al., A mathematical model for predicting lane changes using the steering wheel angle, Journal of Safety Research 49 (2014). 85-e1.

[56] F. Li, et al., Driving intention inference based on dynamic Bayesian networks, in: Practical Applications of Intelligent Systems, Springer, Berlin, Heidelberg, 2014, pp. 1109−1119.

[57] H. Hou, et al., Driver intention recognition method using continuous hidden Markov model, International Journal of Computational Intelligence Systems 4 (3) (2011) 386−393.

[58] F. Lethaus, et al., Using pattern recognition to predict driver intent, in: International Conference on Adaptive and Natural Computing Algorithms, Springer, Berlin, Heidelberg, 2011.

[59] Y.-M. Jang, R. Mallipeddi, M. Lee, Identification of human implicit visual search intention based on eye movement and pupillary analysis, User Modeling and User-Adapted Interaction 24 (4) (2014) 315−344.

[60] T. Pech, P. Lindner, G. Wanielik, Head tracking based glance area estimation for driver behaviour modelling during lane change execution, in: Intelligent Transportation Systems (ITSC), 2014 IEEE 17th International Conference on, IEEE, 2014.

[61] D. Shinar, Looks are (almost) everything: where drivers look to get information, Human Factors 50 (3) (2008) 380−384.

[62] N. Caceres, J.P. Wideberg, F.G. Benitez, Deriving origin−destination data from a mobile phone network, IET Intelligent Transport Systems 1 (1) (2007) 15−26.

[63] A. Borji, A. Lennartz, M. Pomplun, What do eyes reveal about the mind?: algorithmic inference of search targets from fixations, Neurocomputing 149 (2015) 788−799.

[64] A. Doshi, M. Trivedi, Investigating the relationships between gaze patterns, dynamic vehicle surround analysis, and driver intentions, in: Intelligent Vehicles Symposium, 2009 IEEE, IEEE, 2009.

[65] H. Zhou, M. Itoh, T. Inagaki, Toward inference of driver's lane-change intent under cognitive distraction, in: SICE Annual Conference 2010, Proceedings of IEEE, 2010.

[66] A. Doshi, M.M. Trivedi, On the roles of eye gaze and head dynamics in predicting driver's intent to change lanes, IEEE Transactions on Intelligent Transportation Systems 10 (3) (2009) 453−462.

[67] F. Lethaus, et al., A comparison of selected simple supervised learning algorithms to predict driver intent based on gaze data, Neurocomputing 121 (2013) 108−130.

[68] F. Timm, E. Barth, Accurate eye centre localisation by means of gradients, Visapp 11 (2011) 125−130.

[69] S. Wang, et al., Atypical visual saliency in autism spectrum disorder quantified through model-based eye tracking, Neuron 88 (3) (2015) 604−616.

[70] S. Martin, et al., Dynamics of driver's gaze: explorations in Behaviour Modeling and manoeuvre prediction, in: IEEE Transactions on Intelligent Vehicles, 2018.

[71] F. Lethaus, et al., Windows of driver gaze data: how early and how much for robust predictions of driver intent?, in: International Conference on Adaptive and Natural Computing Algorithms Springer, Berlin, Heidelberg, 2013.

[72] E. Ohn-Bar, M.M. Trivedi, Looking at humans in the age of self-driving and highly automated vehicles, IEEE Transactions on Intelligent Vehicles 1 (1) (2016) 90−104.

[73] A. Jain, et al., Recurrent neural networks for driver activity anticipation via sensory-fusion architecture, in: Robotics and Automation (ICRA), 2016 IEEE International Conference on, IEEE, 2016.

[74] E. Murphy-Chutorian, M.M. Trivedi, Head pose estimation in computer vision: a survey, IEEE Transactions on Pattern Analysis and Machine Intelligence 31 (4) (2009) 607−626.

[75] C. Ahlstrom, et al., Processing of eye/head-tracking data in large-scale naturalistic driving data sets, IEEE Transactions on Intelligent Transportation Systems 13 (2) (2012) 553−564.

[76] A. Tawari, S. Martin, M.M. Trivedi, Continuous head movement estimator for driver assistance: issues, algorithms, and on-road evaluations, IEEE Transactions on Intelligent Transportation Systems 15 (2) (2014) 818−830.

[77] R.K. Tiwari, S.D. Giripunje, Design approach for EEG-based human computer interaction driver monitoring system, International Journal of latest trends in Engineering and Technology 3 (4) (2014) 250−255.

[78] N. Das, E. Ohn-Bar, M.M. Trivedi, On performance evaluation of driver hand detection algorithms: challenges, dataset, and metrics, in: Intelligent Transportation

Systems (ITSC), 2015 IEEE 18th International Conference on, IEEE, 2015.

[79] L.F. Nicolas-Alonso, J. Gomez-Gil, Brain computer interfaces, a review, Sensors 12 (2) (2012) 1211–1279.

[80] X.-W. Wang, D. Nie, B.-L. Lu, Emotional state classification from EEG data using machine learning approach, Neurocomputing 129 (2014) 94–106.

[81] S. Haufe, et al., EEG potentials predict upcoming emergency brakings during simulated driving, Journal of Neural Engineering 8 (5) (2011), 056001.

[82] T. Ikenishi, T. Kamada, M. Nagai, Classification of driver steering intentions using an electroencephalogram, Journal of System Design and Dynamics 2 (6) (2008) 1274–1283.

[83] L. Bi, et al., Detecting driver normal and emergency lane-changing intentions with queuing network-based driver models, International Journal of Human-Computer Interaction 31 (2) (2015) 139–145.

[84] K. Driggs-Campbell, R. Bajcsy, Identifying modes of intent from driver behaviors in dynamic environments, in: Intelligent Transportation Systems (ITSC), 2015 IEEE 18th International Conference on, IEEE, 2015.

[85] J. Ding, et al., Driver intention recognition method based on comprehensive lane-change environment assessment, in: Intelligent Vehicles Symposium Proceedings, 2014 *IEEE*, IEEE, 2014.

[86] K. Li, et al., Lane changing intention recognition based on speech recognition models, Transportation Research Part C: Emerging Technologies 69 (2016) 497–514.

[87] D. Polling, et al., Inferring the driver's lane change intention using context-based dynamic Bayesian networks, in: Systems, Man and Cybernetics, 2005 IEEE International Conference on, vol. 1, IEEE, 2005.

[88] D. Kasper, et al., Object-oriented Bayesian networks for detection of lane change manoeuvres, IEEE Intelligent Transportation Systems Magazine 4 (3) (2012) 19–31.

[89] H.M. Mandalia, M.D.D. Salvucci, Using support vector machines for lane-change detection, in: Proceedings of the Human Factors and Ergonomics Society Annual Meeting, vol. 49, SAGE Publications, Sage CA: Los Angeles, CA, 2005. No. 22.

[90] U. Dogan, H. Edelbrunner, I. Iossifidis, Towards a driver model: preliminary study of lane change behavior, in: Intelligent Transportation Systems, 2008. ITSC 2008. 11th International IEEE Conference on, IEEE, 2008.

[91] R. Girshick, Fast r-cnn, 2015 arXiv preprint arXiv: 1504.08083.

[92] Y. Lv, et al., Traffic flow prediction with big data: a deep learning approach, IEEE Transactions on Intelligent Transportation Systems 16 (2) (2015) 865–873.

[93] A. Krizhevsky, I. Sutskever, G.E. Hinton, Imagenet classification with deep convolutional neural networks, in: Advances in Neural Information Processing Systems, 2012.

[94] A. Jain, et al., Brain4cars: car that knows before you do via sensory-fusion deep learning architecture, 2016 arXiv preprint arXiv:1601.00740.

[95] S. Hochreiter, J. Schmidhuber, Long short-term memory, Neural Computation 9 (8) (1997) 1735–1780.

[96] A. Zyner, S. Worrall, E. Nebot, A recurrent neural network solution for predicting driver intention at unsignalized intersections, in: IEEE Robotics and Automation Letters, 2018.

[97] T. Fawcett, An introduction to ROC analysis, Pattern Recognition Letters 27 (8) (2006) 861–874.

[98] J.C. McCall, et al., Lane change intent analysis using robust operators and sparse Bayesian learning, IEEE Transactions on Intelligent Transportation Systems 8 (3) (2007) 431–440.

[99] Y.-M. Jang, R. Mallipeddi, M. Lee, Driver's lane-change intent identification based on pupillary variation, in: Consumer Electronics (ICCE), 2014 IEEE International Conference on, IEEE, 2014.

[100] H. Zhou, M. Itoh, T. Inagaki, Eye movement-based inference of truck driver's intent of changing lanes, SICE Journal of Control, Measurement, and System Integration 2 (5) (2009) 291–298.

[101] M.J. Henning, et al., Modelling driver behaviour in order to infer the intention to change lanes, in: Proceedings of European Conference on Human Centred Design for Intelligent Transport Systems, vol. 113, 2008.

[102] E. Ohn-Bar, et al., Predicting driver manoeuvres by learning holistic features, in: Intelligent Vehicles Symposium Proceedings, 2014 IEEE, IEEE, 2014.

[103] G. Xu, L. Liu, Z. Song, Driver behaviour analysis based on Bayesian network and multiple classifiers, in: Intelligent Computing and Intelligent Systems (ICIS), 2010 IEEE International Conference on, vol. 3, IEEE, 2010.

[104] K. Young, M. Regan, M. Hammer, Driver distraction: a review of the literature, Distracted driving (2007) 379–405.

[105] Y. Liao, et al., Detection of driver cognitive distraction: a comparison study of stop-controlled intersection and speed-limited highway, IEEE Transactions on Intelligent Transportation Systems 17 (6) (2016) 1628–1637.

[106] J.A. Healey, R.W. Picard, Detecting stress during real-world driving tasks using physiological sensors, IEEE Transactions on Intelligent Transportation Systems 6 (2) (2005) 156–166.

[107] S.G. Klauer, et al., The Impact of Driver Inattention on Near-Crash/crash Risk: An Analysis Using the 100-car Naturalistic Driving Study Data, 2006.

[108] M.A. Regan, C. Hallett, C.P. Gordon, Driver distraction and driver inattention: definition, relationship and taxonomy, Accident Analysis & Prevention 43 (5) (2011) 1771–1781.

[109] Y.-F. Ma, et al., A generic framework of user attention model and its application in video summarization, IEEE Transactions on Multimedia 7 (5) (2005) 907–919.

[110] S. Lefèvre, D. Vasquez, C. Laugier, A survey on motion prediction and risk assessment for intelligent vehicles, ROBOMECH Journal 1 (1) (2014) 1.

[111] L. Gugerty, Situation awareness in driving, in: Handbook for Driving Simulation in Engineering, Medicine and Psychology, vol. 1, 2011.

[112] W. Yao, et al., On-road vehicle trajectory collection and scene-based lane change analysis: Part II, IEEE Transactions on Intelligent Transportation Systems 18 (1) (2017) 206−220.

[113] S. Lefèvre, C. Laugier, J. Ibañez-Guzmán, Evaluating risk at road intersections by detecting conflicting intentions, in: Intelligent Robots and Systems (IROS), 2012 IEEE/RSJ International Conference on, IEEE, 2012.

[114] S. Lefèvre, J. Ibañez-Guzmán, C. Laugier, Context-based estimation of driver intent at road intersections, in: Computational Intelligence in Vehicles and Transportation Systems (CIVTS), 2011 IEEE Symposium on, IEEE, 2011.

[115] A. Lawitzky, et al., Interactive scene prediction for automotive applications, in: Intelligent Vehicles Symposium (IV), 2013 IEEE, IEEE, 2013.

[116] M. Bahram, et al., A game-theoretic approach to replanning-aware interactive scene prediction and planning, IEEE Transactions on Vehicular Technology 65 (6) (2016) 3981−3992.

[117] M. Bahram, et al., A combined model-and learning-based framework for interaction-aware maneuver prediction, IEEE Transactions on Intelligent Transportation Systems 17 (6) (2016) 1538−1550.

[118] N. Deo, A. Rangesh, M. Mohan, Trivedi, How would surround vehicles move? A unified framework for maneuver classification and motion prediction, IEEE Transactions on Intelligent Vehicles 3 (2) (2018) 129−140.

[119] J. Li, et al., Generic probabilistic interactive situation recognition and prediction: from virtual to real, in: 2018 21st International Conference on Intelligent Transportation Systems (ITSC), IEEE, 2018.

[120] G. Buja, L. Fellow, M. Bertoluzzo, K.N. Mude, Design and experimentation of WPT charger for electric city car, IEEE Transactions on Industrial Electronics 62 (12) (2015) 7436−7447.

[121] S.K. Young, C.A. Eberhard, P.J. Moffa, Development of Performance Specifications for Collision Avoidance Systems for Lane Change, Merging and Backing Task 2 Interim Report: Functional Goals Establishment, 1995 no. February.

[122] T. Wakasugi, A study on warning timing for lane change decision aid systems based on driver's lane change maneuver, in: Proc. 19th International Technical Conference on the Enhanced Safety of Vehicles, 2005. Paper. No. 05-0290.

[123] D. Polling, M. Mulder, M.M. van Paassen, Q.P. Chu, Inferring the driver's lane change intention using context-based dynamic Bayesian networks, in: 2005 IEEE Int. Conf. Syst. Man Cybern., vol. 1, 2005, pp. 853−858.

[124] J.E. Gayko, " Lane Departure and Lane Keeping."// Eskandarian, Handbook of Intelligent Vehicles, Springer-Verlag London Ltd, 2012.

[125] C. Visvikis, et al., in: Study on Lane Departure Warning and Lane Change Assistant Systems, vol. 374, Transport Research Laboratory Project Rpt PPR, 2008.

[126] E. Portouli, et al., Long-term phase test and results, in: Final Deliverable (Del. 1.2. 4) of the AIDE IP Research Project, Commission of the European Union-DG Information Society and Media (DG INFSO)(contract no. 507674), Brussels, Belgium, 2006.

[127] R.J. Kiefer, J.M. Hankey, Lane change behavior with a side blind zone alert system, Accident Analysis & Prevention 40 (2) (2008) 683−690.

[128] A. Doshi, M. Trivedi, A comparative exploration of eye gaze and head motion cues for lane change intent prediction, in: 2008 IEEE Intelligent Vehicles Symposium, 2008, pp. 49−54.

[129] H. Zhou, M. Itoh, T. Inagaki, Influence of cognitively distracting activity on driver's eye movement during preparation of changing lanes, in: Proc. SICE Annu. Conf., 2008, pp. 866−871.

[130] P. Smith, M. Shah, N. da V. Lobo, Monitoring head/eye motion for driver alertness with one camera, in: IEEE Int. Conf. Pattern Recognit., vol. 4, 2000, pp. 636−642.

[131] C. Tran, M.M. Trivedi, 3-D posture and gesture recognition for interactivity in smart spaces, IEEE Transactions on Industrial Informatics 8 (1) (2012) 178−187.

[132] K.-R. Müller, M. Tangermann, G. Dornhege, M. Krauledat, G. Curio, B. Blankertz, Machine learning for real-time single-trial EEG-analysis: from brain−computer interfacing to mental state monitoring, Journal of Neuroscience Methods 167 (1) (2008) 82−90.

[133] T. Ikenishi, T. Kamada, Estimation of driver's steering direction about lane change maneuver at the preceding car avoidance by brain source current estimation method [C]//Systems, Man and Cybernetics (SMC), in: 2014 IEEE International Conference on, IEEE, 2014, pp. 2808−2814.

[134] J.-W. Kim, et al., Brain-computer interface for smart vehicle: detection of braking intention during simulated driving, in: Brain-Computer Interface (BCI), 2014 International Winter Workshop on, IEEE, 2014.

[135] H. Zhang, R. Chavarriaga, L. Gheorghe, et al., Inferring driver's turning direction through detection of error related brain activity[C]//Engineering in Medicine and Biology Society (EMBC), in: 2013 35th Annual International Conference of the IEEE, IEEE, 2013, pp. 2196−2199.

[136] C.H. Chuang, P.C. Lai, L.W. Ko, et al., Driver's cognitive state classification toward brain computer interface via using a generalized and supervised technology[C]//Neural Networks (IJCNN), in: The 2010 International Joint Conference on, IEEE, 2010, pp. 1−7.

# CHAPTER 3

# Road Perception in Driver Intention Inference System

## INTRODUCTION

Traffic accidents are mainly caused by human mistakes such as inattention, misbehavior, and distraction [1]. Many companies and institutes have proposed methods and techniques for the improvement of driving safety and the reduction of traffic accidents. The commonly developed technologies for road and traffic perception include obstacle detection (vehicles, pedestrians, etc.) and road detection (road driving region, lane detection, and road structure detection, etc.). As discussed in Chapter 2, traffic context can be the main stimulus for driver intention and can be used to infer driver intention accordingly. For example, detection and tracking of a leading vehicle or surrounding vehicles enable a real-time analysis of the lane change possibilities. If the situation awareness module generates low lane change possibility and high risk for lane change scenarios, then the driver may not make a lane change in the near future unless a low risk is calculated according to the traffic context. Similarly, the recognition of lane position and lane styles enables the driver intention inference system to comprehensively analyze the driver intention according to the traffic laws. A solid lane style represents that no lane change is allowed to that direction, whereas a dashed lane shows that the lane change is acceptable. Hence, traffic context and road perception can be viewed as the stimuli and knowledge base for driver intention inference.

Among the road perception techniques, road perception and lane marking detection play a vital role in helping drivers avoid mistakes. Lane detection is the foundation for many advanced driver assistance systems (ADAS), such as lane departure warning (LDW) and lane keeping assistance (LKA) [2,3]. Some successful ADAS or automotive enterprises, such as Mobileye, BMW, and Tesla, have developed their own lane detection and lane keeping products and have obtained significant achievement in both research and real-world applications. The Mobileye Series ADAS and Tesla

Autopilot products have been widely accepted by both the automotive enterprises and personal customers. Almost all the current mature lane assistance products use vision-based techniques because lane markings are painted on the road for human visual perception. The utilization of vision-based techniques detects lanes from the camera devices and prevent the driver from making unintended lane changes. Therefore accuracy and robustness are the two most important properties for lane detection systems. Lane detection systems should have the capability to be aware of unreasonable detections and should adjust the detection and tracking algorithm accordingly [4,5]. When a false alarm occurs, the ADAS should alert the driver to concentrate on the driving task. On the other hand, vehicles with high levels of automation continuously monitor their environment and should be able to deal with low-accuracy detection problems by themselves. Hence, evaluation of lane detection systems becomes even more critical with increasing automation of vehicles.

Most vision-based lane detection systems are commonly designed based on image processing techniques within similar frameworks. With the development of high-speed computing devices and advanced machine learning theories, such as deep learning, lane detection problems can be solved in a more efficient fashion using an end-to-end detection procedure. However, the critical challenge faced by lane detection systems is the demand for high reliability and the diverse working conditions. One efficient way to construct robust and accurate advanced lane detection systems is to fuse multimodal sensors and integrate lane detection systems with other object detection systems, such as detection by surrounding vehicles and road area recognition. It has been proved that lane detection performance can be improved with these multilevel integration techniques [4]. However, the highly accurate sensors such as light/laser detection and ranging (lidar/ladar) are expensive and not available in public transport.

Advanced Driver Intention Inference. https://doi.org/10.1016/B978-0-12-819113-2.00003-8

## VISION-BASED LANE DETECTION ALGORITHM

Literature reviews of lane detection algorithms and their corresponding general frameworks have been proposed in Refs. [4–6]. Bar Hillel et al. [4] concluded that road color, texture, boundaries, and lane markings are the main perception aspects for human drivers. In Ref. [5], McCall and Trivedi classified the lane detection objectives into three categories, which are LDW, driver attention awareness, and automated vehicle control system design. However, these studies paid more attention to the design of lane detection algorithms and incompletely reviewed the integration and evaluation methods. This study tries to comprehensively review the lane detection system from the perspective of algorithms, integration method, and evaluation method. First, in this section, lane detection algorithms and techniques are reviewed from the scope of conventional image processing and novel machine learning methods. In the first part of this section, basic lane detection procedures and general framework will be analyzed. The second part will concentrate on the review of commonly used conventional image processing methods. In the last part, lane detection algorithms based on machine learning and deep learning methods, especially the utilization of a convolutional neural network (CNN), will be discussed.

### General Lane Detection Procedure

Vision-based lane detection systems described in studies usually consist of three main procedures, which are image preprocessing, lane detection, and lane tracking. Among these, the lane detection process, which comprises feature extraction and model fitting, is the most important aspect of the lane detection system, as shown in Fig. 3.1. The most common procedures in the preprocessing step includes region of interest (ROI) selection, vanishing point detection, transferring the color image into a grayscale image or a different color format, noise and blur removal, inverse perspective mapping (IPM, also known as bird's-eye

view), segmentation, and edge statistics. Among these tasks, determining the ROI is usually the first step performed in most of the previous research. The main reason for focusing on ROI is to increase computation efficiency and reduce false lane detection. ROI can be roughly selected as the lower portion of an input image or dynamically determined according to the detected lanes. It can also be more efficiently determined with prior road area detection [7,8]. Details of these methods are described in the next section. Generally speaking, a carefully designed ROI will significantly improve lane detection accuracy as well as computation efficiency.

Once the input images have been preprocessed, lane features such as the colors and edge features can be extracted, and hence the lane can be detected based on these features. The Hough transform algorithm, which uses the edge pixel images, is one of the most widely used algorithms for lane detection in previous studies. However, this method is designed to detect straight lines in the beginning and is not efficient in curved lane detection. Curved lanes can often be detected based on model fitting techniques such as random sample consensus (RANSAC). RANSAC fits lane models by recursively testing the model fitting score to find the optimal model parameters. Therefore it has a strong ability to cope with outlier features. Finally, after lanes have been successfully detected, lane positions can be tracked with tracking algorithms such as Kalman filter or particle filters to refine the detection results and predict lane positions in a more efficient way.

## CONVENTIONAL IMAGE-PROCESSING-BASED ALGORITHMS

Vision-based lane detection can be roughly classified into two categories: feature based [9–19] and model based [20–29]. Feature-based methods rely on the detection of lane marking features such as lane colors, textures, and edges. For example, in Ref. [9], noisy lane edge features were detected using the Sobel

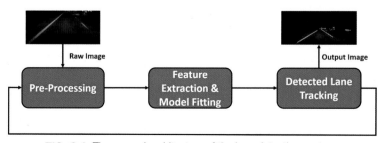

FIG. 3.1 The general architecture of the lane detection system.

operator and the road images were divided into multiple subregions along the vertical direction. Suddamalla et al. [10] detected the curves and straight lanes using pixel intensity and edge information, with lane markings being extracted with adaptive threshold techniques. To remove the camera perspective distortions from the digital images and extract real lane features, lane markings can be efficiently detected with a perspective transform. Collado et al. [11] created a bird's-eye view of the road image and proposed an adaptive lane detection and classification method based on spatial lane features and the Hough transform algorithm. A combination of IPM and clustered particle filter method based on lane features was used to estimate multiple lanes in Ref. [12]. The authors claimed that it is less robust if a strong lane model is used in the context, and they only used a weak model for particle filter tracking. Instead of using color images, lanes can also be detected using other color format images. The general idea behind the color format transform is that yellow and white lane markings can be more distinct in other color domains, so the contrast ratio is increased. In Ref. [13], lane edges were detected with an extended edge linking algorithm in the lane hypothesis stage. Lane pixels in the YUV format, edge orientation, and width of lane markings were used to select the candidate edge-link pairs in the lane verification step. In Ref. [14], lanes were recognized using an unsupervised and adaptive classifier. Color images were first converted to HSV (hue, saturation, value) format to increase the contrast. Then the binary feature image was processed using the threshold method based on the brightness values. Although in some normal cases color transform can benefit lane detection, it is not robust and has limited ability to deal with shadows and illumination variation [4].

Borkar et al. [15] proposed a layered approach to detect lanes at night. A temporal blur technique was used to reduce video noise, and binary images were generated based on an adaptive local threshold method. The lane finding in another domain (LANA) algorithm represented lane features in the frequency domain [16]. The algorithm captured the lane strength and orientation in the frequency domain, and a deformable template was used to detect the lane markings. Results showed that the LANA algorithm was robust under varying conditions. In Ref. [17], a spatiotemporal lane detection algorithm was introduced. A series of spatiotemporal images were generated by accumulating certain row pixels from the past frames and the lanes were detected using Hough transform applied on the synthesized images. In Ref. [18], a real-time lane

detection system based on field programmable gate array (FPGA) and digital signal processor (DSP) was designed based on lane gradient amplitude features and an improved Hough transform. Ozgunalp and Dahnoun [19] proposed an improved feature map for lane detection. The lane orientation histogram was first determined with edge orientations and then the feature map was improved and shifted based on the estimated lane orientation.

In general, feature-based methods have better computational efficiency and are able to accurately detect lanes when the lane markings are clear. However, as too many constraints are assumed, such as the lane colors and shapes, the drawbacks of these methods include less robustness to deal with shadows and poor visibility conditions compared with model-based methods.

Model-based methods usually assume that lanes can be described with a specific model such as a linear model, a parabolic model, or various kinds of spline models. Besides, some assumptions about the road and lanes, such as a flat ground plane, are required. Among these models, spline models were popular in previous studies, as they are flexible enough to recover any shape of the curved lanes. Wang et al. fitted lanes with different spline models [20,21]. In Ref. [20], a Catmull-Rom spline was used to model the lanes in the image. In Ref. [21], the lane model was improved to generate a B-snake model, which can model any arbitrary shape by changing the control points. In Ref. [22], a novel parallel-snake model was introduced. In Ref. [23], lane boundaries were detected based on a combination of Hough transform in near-field areas and a river-flow method in far-field areas. Finally, lanes were modeled with a B-spline model and tracked with a Kalman filter. Jung and Kelber [24] described the lanes with a linear-parabolic model and classified the lane types based on the estimated lane geometries. Aly [25] proposed a multiple lane fitting method based on the integration of Hough transform, RANSAC, and B-spline model. Initial lane positions were first roughly detected with Hough transform and then improved with RANSAC and B-spline model. Moreover, a manually labeled lane dataset called the Caltech lane dataset was introduced.

The RANSAC algorithm is the most popular way to iteratively estimate the lane model parameters. In Ref. [26], the linear lane model and RANSAC were used to detect lanes and a Kalman filter was used to refine the noisy output. Ridge features and adapted RANSAC for both straight and curved lane fitting were proposed in Refs. [27,28]. The ridge features of lane pixels, which depend on the local structures rather than on contrast, were defined as the center lines of a bright structure of a region in a grayscale image. In Refs. [29,30], the hyperbolic model and RANSAC were used for lane fitting. In Ref. [30], input images were divided into two parts known as far-field area and near-field area. In the near-field area, lanes were regarded as straight lines detected using the Hough transform algorithm. In the far-field area, lanes were assumed to be curved lines and fitted using the hyperbolic model and RANSAC.

In Ref. [31], a conditional random field method was proposed to detect lane marks in urban areas. Bounini et al. [32] introduced a lane boundary detection method for autonomous vehicles working in a simulation environment. A least squares method was used to fit the line model, and the computation cost was reduced by determining a dynamic ROI. In Ref. [33], an automated multisegment lane-switch scheme and a RANSAC lane fitting method were proposed. The RANSAC algorithm was applied to fit the lines based on the edge image. A lane-switch scheme was used to determine lane curvatures and choose the correct lane models from straight and curved models to fit the lanes. In Ref. [34], a Gabor wavelet filter was applied to estimate the orientation of each pixel and match a second-order geometric lane model. Niu et al. [35] proposed a novel curve-fitting algorithm for lane detection with a two-stage feature extraction (LDTFE) algorithm. A density-based spatial clustering of application with noise (DBSCAN) algorithm was applied to determine whether the candidate lane line segments belong to ego-lanes or not. The identified small lane line segments can be fitted with a curved model, and this method is particularly efficient for small lane segment detection tasks.

Model-based methods are more robust than feature-based methods because of the use of model-fitting techniques. The noisy measurement and the outlier pixels of lane markings usually can be ignored with the model. However, model-based methods usually entail more computational cost, as RANSAC has no upper limits on the number of iterations. Moreover, model-based methods are less easy to be implemented compared with the feature-based systems.

## MACHINE LEARNING-BASED ALGORITHMS
Despite using conventional image processing-based methods to detect lane markings, some researchers focus on detecting lane markings using novel machine learning and deep learning methods. Deep learning

techniques have been one of the hottest research areas in the past decade owing to the development of deep network theories, parallel computing techniques, and large-scale data. Many deep learning algorithms show great advantages in computer vision tasks and the detection and recognition performance increases dramatically compared with conventional approaches. CNN is one of the most popular approaches used for object recognition research. CNN provides some impressive properties such as high detection accuracy, automatic feature learning, and end-to-end recognition. Some researchers have successfully applied CNN and other deep learning techniques for lane detection. It is reported that by using the CNN model, the lane detection accuracy increased dramatically from 80% to 90% compared with traditional image processing methods [36].

Li et al. [37] proposed a lane detection system based on deep CNN and recurrent neural network (RNN). A CNN was fed with a small ROI image that was used for multiple tasks. There are two types of CNN outputs. The first is a discrete classification result indicating if the visual cues are lane markers or not. If a lane was detected, then the other output would be the continuous estimation of lane orientation and location. To recognize the global lane structures in a video sequence instead of local lane positions in a single image, RNN was used to recognize the lane structures in sequence data with its internal memory scheme. The training was based on a merged scene with three cameras each facing the front, the left side, and the rear area. Accurate detection results showed that the integrated lane detection method using CNN and RNN can work in practice. Besides, RNN can help recognize and connect lanes that are covered by vehicles or obstacles.

Gurghian [38] proposed another deep CNN method for lane marking detection using two side-facing cameras. The proposed CNN recognized the side lane positions with an end-to-end detection process. The CNN was trained with both real-world images and synthesized images and achieved a 99% high detection accuracy. To solve the low accuracy and high computational cost problem, authors in Ref. [36] proposed a novel lane marking detection method based on a point cloud map generated by a laser scanner. To improve the robustness and accuracy of the CNN result, a gradual upsampling method was introduced. The output image was in the same format as the input images to get an accurate classification result. The reported computation cost of each algorithm is 28.8 s on average, which can be used for offline high-precision road map construction.

In Ref. [39], a spiking neural network was used to extract edge images and lanes were detected based on Hough transform. This was inspired by the idea that a human neuron system produces a dense pulse response to edges while generating a sparse pulse signal to flat inputs. A similar approach can be found in Ref. [40]. The study proposed a lane detection method based on RANSAC and CNN. An eight-layer CNN including three convolution layers was used to remove the noise in edge pixels if the input images were too complex. Otherwise, RANSAC was applied to the edge image directly to fit the lane model. He et al. proposed a dual-view CNN for lane detection [41]. Two different views, which were the front view and top view of the road obtained from the same camera, were simultaneously fed into the pretrained CNN. The CNN contained two sub-CNN networks to process the two kinds of input images separately and concatenate the results eventually. Finally, an optimal global strategy taking into account lane length, width, and orientations was used to threshold the final lane markings.

Instead of using general image processing and machine learning methods, some other researchers also used evolution algorithms or heuristic algorithms to automatically search for lane boundaries. For example, Revilloud [42] proposed a novel lane detection method using a confidence map and a multiagent model inspired by human driver behavior. Similarly, an ant colony evolution algorithm for optimal lane marking search was proposed in Ref. [43]. A novel multiple-lane detection method using directional random walking was introduced in Ref. [44]. In that study, a morphology-based approach was used to extract lane mark features at the beginning. Then the directional random walk based on a Markov probability matrix was applied to link the candidate lane features. The proposed algorithm required no assumption about the road curvatures or lane shapes.

In summary, it can be stated that machine learning algorithms or intelligent algorithms increase the lane detection accuracy significantly and provide many efficient detection architectures and techniques. Although these systems usually require more computational cost and need a large amount of training data, these systems are more powerful than conventional methods. Therefore many novel, efficient, and robust lane detection methods with lower training and computation requirements are expected to be developed soon.

## INTEGRATION METHODOLOGIES FOR ROAD PERCEPTION

### Integration Methods—Introduction

Although many studies have been done to enable accurate vision-based lane detection, the robustness of the detection systems still cannot meet the real-world requirements, especially in urban areas, because of the highly random properties of the traffic and the state of roads. Therefore a reasonable way to enhance the lane detection system is to introduce redundancy algorithms, integrate with other object detection systems, or use sensor fusion methods. It is commonly agreed among automotive industries that a single sensor is not enough for vehicle perception tasks. Some companies such as Tesla, Mobileye, and Delphi developed their own intelligent on-vehicle perception system using multiple sensors such as cameras and radar (especially the millimeter-wave radar). In this section, the integration methods will be classified into three levels, which are algorithm level, system level, and sensor level, as shown in Fig. 3.2.

Specifically, algorithm level integration combines different lane detection algorithms to comprehensively determine reasonable lane positions and improve the robustness of the system. In system level integration, different object detection systems work simultaneously in real-time communication with each other. Finally, in sensor level integration, multimodal sensors are integrated. The proposed sensor fusion methods at this level are believed to improve the robustness of the lane detection system most significantly. In the following subsections, the multilevel integration techniques will be described in detail and the studies conducted within each scope will be discussed.

### Algorithm level integration

Integration of vision-based lane detection algorithms has been widely practiced in the past. Past studies have focused on two main integration architectures, which can be summarized as parallel and serial combination methods. Moreover, feature-based and model-based algorithms can also be combined. Serial combination methods are commonly seen in previous studies. Studies described in Refs. [20,21,25] demonstrate examples of methods that serially combine the Hough transform, RANSAC, and spline model-fitting methods. The other method followed in multiple studies involves applying a lane tracking system after the lane detection procedure to refine and improve the stability of the detected lanes [5,21,22,45–47].

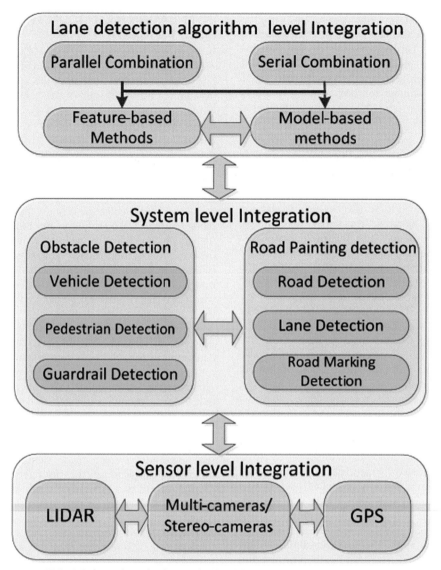

FIG. 3.2 Lane detection integration level. *Lidar*, light detection and ranging.

For lane tracking, Kalman filter and particle filter are the two most widely used tracking algorithms [4]. Shin [48] proposed a superparticle filter combining two separate particle filters for ego-lane boundary tracking. In Ref. [49], a learning-based lane detection method is proposed and tracked with a particle filter. The learning-based algorithm requires no prior road model and vehicle velocity knowledge.

Parallel combination methods can be found in Refs. [50,51]. In Ref. [50], a monocular vision-based lane detection system was proposed, which parallelly combined two independent algorithms to make a comprehensive lane detection. The first algorithm used a lane marking extractor and road shape estimation to find the potential lanes. Meanwhile, a simple feature-based detection algorithm was applied to check the candidate lanes chosen by the first algorithm. If the results from the two algorithms are compared with each other, the detection result is accepted. Douret et al. proposed three parallel integrated algorithms to pursue a robust lane detection with higher confidence [51]. Two lower-level lane detection algorithms, namely, lateral and longitudinal consistent detection methods, were processed simultaneously. Then the sampling points of the detected lanes given by these two lower-level detection algorithms were tested. If the results were close to each other, the detection was viewed as a success and the average position from the two algorithms was selected as the lane position.

Some studies also combined different lane features to construct a more accurate feature vector for lane detection. In Ref. [52], the lane detection system was based on the fusion of color and edge features. Color features were used to separate road foreground and background regions using Otsu's method, while edges were detected with a canny detector. Finally, curved lanes in the image were fitted using Lagrange interpolating polynomial. In Ref. [53], a three-feature based automatic lane detection algorithm (TFALDA) was proposed. The lane boundary was represented as a three-feature vector, which includes intensity, position, and orientation value of the lane pixels. The continuity of lanes was used as the selection criterion to choose the best current lane vector that is at a minimum distance with the previous one.

Although parallel integration methods improve the robustness of the system by introducing redundancy algorithms, the computation burden will increase correspondingly. Therefore a more efficient way is to combine algorithms in a dynamic manner and only initiate a redundancy system when necessary.

### System level integration
**Vehicle detection.** Vehicle detection is another important clue for driver lane change intention inference, which has been widely studied in the past. Integrating vehicle detection and lane detection can improve lane detection accuracy and provide a comprehensive understanding of the traffic context. Moreover, most of the current vehicle detection system is vision based, which can be easily integrated into the lane detection system. In [112], an integrated stereo camera and millimeter-wave radar are fused to find a dangerous and potentially dangerous obstacle. The cameras are used to detect the near and lateral objects and determine the ROI. The millimeter-wave radar is used for far and longitudinal dynamic objects. The cameras detect the objects based on the error vector and the millimeter-wave detects the relative dynamic objects based on the absolute speed with the host vehicle. Then the detection results from the two sensors can be fused within an obstacle ROI map. In Ref. [113], an effective nighttime vehicle detection system was proposed based on a bio-inspired image enhancement approach with weighted feature fusion technique. The

nighttime image enhancement approach was inspired by the retinal mechanism in natural visual processing. Specifically, the adaptive feedback from the horizontal cells and the center-surround antagonistic receptive fields of the bipolar cell were modeled. Then the vehicle objects were detected using the support vector machine based on the integrated features that are extracted by the CNN model, histogram of the oriented gradient, and local binary pattern. In Ref. [114], a surrounding vehicle detection system based on the FPGA-enabled panoramic camera was developed. The proposed panoramic camera-based system can perform fast image stitching in real time based on the developed deep CNN, namely, EZ-NET. The EZ-NET performs vehicle detection based on the panoramic images at a speed of 140 fps, with competitive results with the state-of-the-art detectors. Similarly in Ref. [115], detection of a leading vehicle and a turn signal during nighttime was proposed. The nighttime vehicle detection system was developed based on the Nakagami image-based method to locate the potential region. Meanwhile, a set of vehicle object proposals was generated based on a CNN-based region proposal network. Then the light regions and the proposal regions are combined to construct the ROI. The detected vehicles are tracked based on a perceptional hashing algorithm, and the light is detected by analyzing the continuous intensity variation of the vehicle box sequence. Most of the current vehicle detection systems suffer from the problem that convolutional features are scale-sensitive in the object detection task. In Ref. [116], a scale-insensitive CNN (SI-Net) for fast vehicle detection with a large variance of scales is proposed. A context-aware ROI pooling method was used to maintain the contextual information and original structure of the small-scale objects. Then a multibranch decision network was used to minimize the intraclass distance of features. Unlike most of the appearance-based vehicle detection methods that use several various sizes of sliding windows to search the vehicle region, in Ref. [117] a graph-based algorithm was proposed to precisely estimate the vehicle in a bounding box. In Ref. [118], a hard negative mining, multiscale training, and model pretraining were integrated into a YOLO-V2 detection network for pedestrian and vehicle detection. The proposed algorithm improves the detection accuracy of the different objects with an efficient computational burden.

**System level fusion.** Lane detection in the real world can be affected by surrounding vehicles and other obstacles, which may have a similar color or texture features to the lane markings in the digital images. For instance, the guardrail usually shows strong lanelike characteristics in color images and can easily result in false lane detection [54–56]. Therefore integrating the lane detection system with other onboard detection systems will enhance its accuracy. Obstacle detection and road painting detection are the two basic categories of vision-based detection techniques, as shown in Fig. 3.2. By introducing an obstacle, noise measurement or outlier pixels can be filtered. Similarly, road recognition can narrow down the searching area for lane detections and provide a reasonable result.

Lane detection algorithms usually require lane features for model fitting tasks. Nearby vehicles, especially passing vehicles, are likely to cause a false detection result due to occlusion and similar factors. With the detection of surrounding vehicles, the color, shadow, appearance, and noise generated by the vehicles ahead can be removed and a higher accuracy of lane boundaries can be achieved [30]. In Refs. [30,57–60], the lane detection result was reported to be more accurate with a front-vehicle detection system. This reduces the quantities of false-lane features and improves the model fitting accuracy. Cheng et al. proposed an integrated lane and vehicle detection system. Lane markings were detected by analyzing road and lane color features, and the system was designed so as not to be influenced by variations in illumination [57]. Those vehicles that have similar colors with the lanes were distinguished on the basis of the size, shape, and motion information.

Sivaraman and Trivedi [58] proposed a driver assistance system based on an integration of lane and vehicle tracking systems. With the tracking of nearby vehicles, the position of surrounding vehicles within the detected lanes and their lane change behaviors can be recognized. Final evaluation results showed an impressive improvement compared with the results delivered by the single lane detection algorithm. In Ref. [61], a novel lane and vehicle detection integration method called efficient lane and vehicle detection with integrated synergies (ELVIS) was proposed. The integration of vehicles and lane detection reduces the computation cost of finding the true lane positions by at least 35%. Similar results can be found in Ref. [62]. Integrated lane detection and front vehicle recognition algorithm for a forward-collision warning system was also introduced. Front vehicles were recognized with a Hough forest method. The vehicle tracking system enhanced the accuracy of the lane detection result in high-density traffic scenarios.

In terms of road painting recognition, Qin et al. [63] proposed a general framework of road marking detection and classification. Four common road markings (lanes, arrows, zebra crossing, and words) were detected and classified separately using a support vector machine. However, this system only identifies the different kinds of road marking without further context explanation of each road marking. It is believed that road marking recognition results contribute to a better understanding of ego-lanes and help decide current lane types such as a right/left turning lane [64,65]. Finally, many studies were dedicated to the integration of road detection and lane detection [4,7,66–68]. Tesla and Mobileye have been reported to use a road segmentation to refine the lane detection algorithms [69,70]. Road area is usually detected before lanes because an accurate recognition of road area increases the lane marking searching speed and provides an accurate ROI for lane detection. Besides, as the road boundaries and lanes are correlated and normally have the same direction, a road boundary orientation detection enhances the subsequent lane detection accuracy. Ma et al. [71] proposed a Bayesian framework to integrate road boundary and lane edge detection. Lane and road boundaries were modeled with a second-order model and detected using a deformable template method.

Fritsch et al. [7]. proposed a road and ego-lane detection system particularly focusing on inner-city and rural roads. The proposed road and ego-lane detection algorithm was tested in three different road conditions. Another integrated road and ego-lane detection algorithm for urban areas was proposed in Ref. [8]. Road segmentation based on an illumination invariant transform was the prior step for lane detection to reduce the lane detection time and increase the detection accuracy. The outputs of the system consisted of road region, ego-lane region and markings, local lane width, and the relative position and orientation of the vehicle.

### Sensor level integration

Sensor fusion dramatically improves the lane detection performance, as more sensors are used and perception ability is boosted. Using multiple cameras including monocular cameras, stereo cameras, or a combination of multiple cameras with different fields of view is the most common way to enhance the lane detection system [46,55,72]. In Ref. [72], a dense vanishing point detection method for lane detection using the stereo camera was proposed. The combination of global dense vanishing point detection and the stereo camera makes the system very robust to various road conditions and multiple-lane scenarios. Bertozzi and Broggi [55] proposed a generic obstacle and lane detection (GOLD) system to detect obstacles and lanes based on a stereo camera and IPM image. The system was tested on the road for more than 3000 km and it showed robustness under exposure to shadow, illumination, and road variation. Three wide-field cameras and one telelens camera were combined and sampled at the frequency of 14 Hz. Raw images were converted to HSV format and IPM was performed. An around-view monitoring (AVM) system with four fish-eye cameras and one monocular front-looking camera is used for lane detection and vehicle localization. The benefit of using the AVM system is that a whole picture of the top view of the vehicle can be generated, which contains the front, surrounding, and rear views of the vehicle in one single image.

Instead of using only camera devices, lane detection system can also be realized by combining cameras with GPS and radar [75–81]. An integration system based on vision and radar was proposed in Ref. [71]. Radar was particularly used for road boundary detection in ill-illuminated conditions. Jung et al. proposed an adaptive ROI-based lane detection method aimed at designing an integrated adaptive cruise control (ACC) and LKA system [75]. Range data from ACC was used to determine a dynamic ROI and improve the accuracy of the monocular vision-based lane detection system. The lane detection system was designed using a conventional method, which includes edge distribution function, steerable filter, model fitting, and tracking. If nearby vehicles were detected with the range sensor, all the edge pixels were eliminated to enhance lane detection. Final results show that recognition of nearby vehicles based on the range data improves lane detection accuracy and simplifies the detection algorithm.

Cui et al. [76] proposed an autonomous vehicle positioning system based on GPS and vision system. Prior information such as road shape was first extracted from GPS and then used to refine the lane detection system. The proposed method was extensively evaluated and found to be robust in varying road conditions. Jiang et al. [77] proposed an integrated lane detection system in a structured highway scenario. Road curvatures were determined using GPS and digital maps in the beginning. Then two-lane detection modules designed for straight lanes and curved lanes were selected accordingly. Schreiber et al. [82] introduced a lane marking-based localization system. Lane markings and curbs were detected with a stereo camera and vehicle localization was performed with the integration of a global navigation satellite system, high accuracy map, and stereo vision system. The integrated

localization system achieved accuracy up to a few centimeters in rural areas.

An integrated LDW system using GPS, an inertial sensor, high-accuracy map, and vision system was introduced in Ref. [83]. Vision-based LDW was easily affected by various road conditions and weather. A sensor fusion scheme increases the stability of the lane detection system and makes the system more reliable. Moreover, a vision-based lane detection system and an accurate digital map help reduce the position errors from GPS, which lead to a more accurate vehicle localization and lane keeping.

Lidar was another widely used sensor and was the primary sensor used in most autonomous vehicles in the DARPA challenge [84,85] because of its high accuracy and robust sensing ability. Lane markings are on-road paintings that have higher reflective properties than the road surface in the three-dimensional (3D) points cloud map given by lidar. Therefore lidar can detect lane markings according to those high reflectance points on the road. Lidar uses multiple channel laser lights to scan surrounding surfaces and build 3D images. Therefore lidar and vision integrated lane detection systems can be more accurate and robust to shadows and illumination change than vision-based systems [86]. Shin et al. [87] proposed a lane detection system using a camera and lidar. The algorithm consists of ground road extraction, lane detection with multimodal data, and lane information combination. The proposed method shows a high detection accuracy performance (up to 90% accuracy) in real-world experiments. Although camera and lidar-based methods can cope with curved lanes, shadow, and illumination issues, it requires a complex cocalibration of the multimodal sensors. Amaradi et al. [88] proposed a lane-following and obstacle detection system using a camera and lidar. Lanes are first detected with Hough transform. Lidar was used to detect obstacles and measure the distance between the ego-vehicle and front obstacles to plan an obstacle-free driving area. In Ref. [56], a fusion system of multiple cameras and lidar was proposed to detect lane markings in urban areas. The test vehicle was reported as the only vehicle that used a vision-based lane detection algorithm in the final stage of the DARPA urban challenge. The system detects multiple lanes followed by the estimation and tracking of the center lines. Lidar and cameras were first calibrated to detect road paint and curbs. Lidar was used to reduce the false-positive detection rate by detecting obstacles and drivable road area.

According to the implementation angle and surveying distances, the laser scanner device can efficiently identify the lane marking. Lane detection using this laser reflection method has also been widely applied [79,89−93]. Li et al. [79] proposed a drivable region and lane detection system based on lidar and vision fusion at the feature level. The test-bed vehicle uses two cameras mounted at different angles and three laser scanners. The algorithm detects the optimal drivable region using multimodal sensors. The system was able to work under both structured and unstructured roads without any prior terrain knowledge. A laser-camera system for lane detection was introduced in Ref. [90]. The two-dimensional (2D) laser reflectivity map was generated on the roof of the vehicle. Instead of using constrained rule-based methods to detect lanes on the reflectivity map, a DBSCAN algorithm was applied to automatically determine the lane positions and the number of lanes in the field according to the 2D map. In Ref. [92], an integration system with a laser scanner and s stereo cameras was proposed. The system achieved an accurate driving area detection result even in a desert. However, in some unstructured road or dirty road, the signals from the laser scanner may carry more noise than the frame signals from the camera. Therefore a signal filter for the laser scanner and sensor fusion usually needed for integrated systems.

In this section, some sensors that are relevant to lane detection task are reviewed. Other sensors including vehicle dynamic signals, such as the velocity, longitudinal/lateral acceleration, and yaw angle, and inertial measurement unit are also commonly used in the construction of a complete vehicle perception system. Although lidar-based lane detection system can be more precise than other systems, the cost is still too high for public transport. Therefore studies [78] tend to fuse sensors such as GPS, digital map, and cameras, which are already available in commercial vehicles, to design a robust lane detection and driver assisting system.

## EVALUATION METHODOLOGIES FOR VISION-BASED ROAD PERCEPTION SYSTEMS

Most of the previous lane detection studies used visual verification to evaluate the system performance because of the lack of ground data, and only a few research studies proposed quantitative performance analysis and evaluation. In addition, lane detection evaluation is a complex task because the detection methods can vary across hardware and algorithms. There are still no common metrics that can be used to comprehensively evaluate each aspect of lane detection algorithms. An

accurate lane detection system in one place is not guaranteed to be accurate in another place because the road and lane situation in different countries or areas differ significantly. Some detection algorithms may even show significantly different detection results in days and nights. It is also not fair to say that a monocular vision-based system is not as good as a system with vision and lidar fusion and use a complex synergistic algorithm because the system cost is higher.

Therefore the performance evaluation of lane detection systems is necessary, and it should be noted that the best index for the lane detection performance is the driving safety issues and how robust the system is to the environment change. In this section, the evaluation methodologies used in studies are divided into offline evaluation and online evaluation, where online evaluation can be viewed as a process of calculating the detection confidence in real time. The main evaluation architecture is shown in Fig. 3.3. As mentioned earlier, a common vision-based lane detection system can be roughly separated into three parts, which are preprocessing, lane detection, and tracking. Accordingly, evaluation can be applied to all the three parts and the performance of these modules can be assessed separately. In the following section, influencing factors that affect the performance of a lane detection system will be summarized first. Then the offline and online evaluation methods used in past studies and other literature are described. Finally, the evaluation metrics will be discussed.

### Influential Factors for Lane Detection Systems

Vision-based lane detection systems studied in previous studies differed in terms of hardware, algorithms, and application situations. Some focus on highway implementation, while some systems were tested in urban areas. An accurate highway-oriented lane detection system is not guaranteed to be accurate in urban road areas, as more disturbance and dense traffic will be observed in such areas. Therefore it is impossible to use one single evaluation method or metric to assess all the existing systems. Some important factors that can affect the performance of the lane detection system are listed in Table 3.1. Fair evaluation and comparison of lane detection systems should take these factors and the system working environment into consideration. As different lane detection algorithms are designed and tested for different places, different road and lane factors in different places will affect the detection performance. Moreover, the data recording device, the camera, or other vision hardware are other aspects

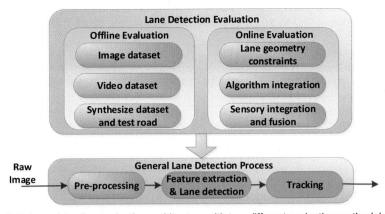

FIG. 3.3 Lane detection evaluation architecture with two different evaluation methodologies.

| TABLE 3.1 Factors That Influence Lane Detection Systems. | |
| --- | --- |
| Lane and Road Factors | Crosswalk Stop Lane, Lane Color, Lane style, Road Curvature, Poor-Quality Lane Markings, complex Road Texture |
| Hardware factors | Camera types, camera calibration, camera mounting position, other sensors |
| Traffic factors | Road curbs, guardrail, surrounding vehicles, shadow, illumination issues, vibration |
| Weather factors | Cloudy, snowy, rainy, foggy |

that can significantly influence lane detection systems. For example, the lane detection systems may have different resolution and field of view with different cameras, which will influence the detection accuracy. Lastly, some traffic and weather factors can also lead to a different lane detection performance.

As shown in Table 3.1, many factors can cause a less accurate detection result and make the performance vary with other systems. For example, some lane detection systems were tested under a complex traffic context, which had more disturbances such as crosswalks or poor-quality lane markings, while some other systems were tested in standard highway environments with few influencing factors. Therefore an ideal way is to use a common platform for algorithm evaluation, which is barely possible in real life. Also a mature

evaluation system should take as many influential factors as possible into account and comprehensively assess the performance of the system. One potential solution for these problems is using parallel vision architecture, which will be discussed in the next section.

In the following section, the methodologies and metrics that can be used to propose a reasonable performance evaluation system are described.

### Offline Evaluation

Offline evaluation is commonly used in previous studies. After the framework of a lane detection system has been determined, system performance is first evaluated offline using still images or video sequences. There are some public datasets such as KITTI Road and Caltech Road [7,25] that are available on the internet. KITTI Road dataset consists of 289 training images and 290 testing images separated into three categories. The road and ego-lane areas were labeled in the dataset. The evaluation is usually done using receiver operating characteristic (ROC) curves to illustrate the pixel-level true and false detection rates. Caltech Road dataset contains 1224 labeled individual frames captured in four different road situations. Both these datasets focus on evaluating road and lane detection performance in urban areas. The main drawbacks of image-based evaluation methods are that they are less reflective of real traffic environments and the datasets contain limited annotated test images.

On the other hand, video datasets depict much richer information and enable the reflection of real-life traffic situations. However, it normally requires more human resources to label ground-truth lanes. To deal with this problem, Borkar et al. [94] proposed a semiautomatic method to label lane pixels in video sequences. They used the time-sliced (TS) images and the interpolation method to reduce the labeling workload. The TS images were constructed by selecting the same rows from each video frame and rearranging these row pixels according to the frame order. Two or more TS images were required, and the accuracy of ground-truth lanes was directly proportional to the number of images. The lane labeling tasks are converted to point labeling in the TS images. After the labeled ground-truth points were selected from each TS image, the interpolated ground-truth lanes can be recovered into the video sequence accordingly. The authors significantly reduced the ground-truth labeling workload by converting lane labeling into a few point labeling tasks. This method was further improved in Ref. [49] by using the so-called modified min-between-max thresholding algorithm (M2BMT) applied to both the TSs and spatial stripes of the video frames.

Despite manually annotated ground truth, some researchers use the synthesis method to generate lane images with known position and curvature parameters in simulators [28,56]. López et al. [28] used a MATLAB simulator to generate video sequences and ground-truth lanes. Lane frames were created with known lane parameters and positions. This method was able to generate arbitrary road and lane models with an arbitrary number of video frames. Using a simulator to generate lane ground truth is an efficient way to assess the lane detection system under ideal road conditions. However, there are few driving simulators that can completely simulate real-world traffic context at this moment. Therefore the detection performance still has to be tested with real-world lane images or videos after evaluation using simulators. Another way is to test the system on real-world testing tracks to assess the lane detection system compared to the accurate lane position ground truth provided by GPS and high-precision maps [78].

### Online Evaluation

The online evaluation system combines road and lane geometry information and integrates with other sensors to generate detection confidence. Lane geometry constraints are reliable metrics for online evaluation. Once the camera is calibrated and mounted on the vehicle, road and lane geometric characteristics, such as the ego-lane width, can be determined. In Ref. [95], a real-time lane evaluation method was proposed based on the width measurement of the detected lanes. The detected lanes were verified based on three criteria, which are the slopes and intercept of the straight lane

model, the predetermined road width, and the position of the vanishing point. The distribution of lane model parameters was analyzed, and a lookup table was created to determine the correctness of the detection. Once the detected lane width exceeds the threshold, reestimation is proposed with respect to the lane width constraints.

In Ref. [5], the authors used a world-coordinate measurement error instead of using errors in image coordinates to assess the detection accuracy. A roadside down-facing camera was used to directly record lane information, generate ground truth, and estimate vehicle position within the lanes. In Refs. [50 and 51], real-time confidence was calculated based on the similarity measurement of the results given by different detection algorithms. The evaluation module calculates if the detected lane positions from different algorithms are within a certain distance. If similar results are obtained, then the detection results are averaged and high detection confidence is reported. However, this method requires performing two algorithms simultaneously at each step, which increases the computation burden.

In Ref. [56], vision- and lidar-based algorithms were combined to build a confidence probability network. The traveling distance was adopted to determine the lane detection confidence. The system was said to have high estimation confidence at certain meters in front of the vehicle if the vehicle can travel safely at that distance. Other online evaluation methods such as estimating the offsets between the estimated center line and lane boundaries were also used in previous studies. Instead of using a single sensor, vision-based lane detection results can be evaluated with other sensors such as GPS, lidar, and highly accurate road models [56,76]. A vanishing point lane detection algorithm was introduced in Ref. [96]. The vanishing point of lane segments was first detected according to a probabilistic voting method. Then the vanishing point, along with the line orientation threshold, was used to determine correct lane segments. To further reduce the false detection rate, a real-time interframe similarity model for evaluation of lane location consistency was adopted. This real-time evaluation idea was also under the assumption that lane geometry properties do not change significantly within a short period of continuous frames.

### Evaluation Metrics

Existing studies mainly use visual evaluation or simple detection rates as evaluation metrics because there are still no common performance metrics to evaluate lane detection performance. Li et al. [97] designed a complete testing scheme for intelligent vehicles mainly focusing on the whole vehicle performance rather than on just the lane detection system. In Ref. [20], five major requirements for a lane detection system were given: shadow insensitivity, suitable for unpainted roads, handling of curved roads, meeting lane parallel constraints, and reliability measurement. Kluge [98] introduced feature-level metrics that measure the gradient orientation of the edge pixels and angular deviation entropy. The proposed metrics evaluate edge points and required road curvatures and vanishing point information.

Veit et al. [99] proposed another feature level evaluation based on a hand-labeled dataset exceeding 100 images. Six different lane feature extraction algorithms were compared. The authors concluded that the lane feature extraction, which combines photometric and geometric features, will achieve the best result. McCall and Trivedi [100] examined the most important evaluation metrics to assess the lane detection system. They concluded that it is not appropriate to view the system as a whole and to use detection rates as the metrics. Instead, three different metrics, which include the standard deviation of error, mean absolute error, and standard deviation of the error in the rate of change were used.

Satzoda and Trivedi [101] introduced five metrics to measure different properties of lane detection systems and to examine the trade-off between accuracy and computational efficiency. The five metrics consist of the measurement of lane feature accuracy, ego-vehicle localization, lane position deviation, computation efficiency and accuracy, and cumulative deviation in time. Among these metrics, cumulative deviation in time helps determine the maximum amount of safety time and can be used to evaluate if the proposed system meets the critical response time of ADAS. However, all these metrics pay more attention to the detection accuracy assessment and do not consider the robustness.

In summary, a lane detection system can be evaluated separately from the preprocessing, lane detection algorithms, and tracking aspects. Evaluation metrics are not only limited to measuring the error between detected lanes and ground-truth lanes but can also be extended to assess the lane prediction horizon, the shadow sensitivity, the computational efficiency, etc. The specific evaluation metrics for a system should be determined based on the real-world application requirements. There are three basic properties of a lane detection system, which are the accuracy, robustness, and efficiency. The primary objective of the lane detection algorithm is to meet the real-time safety requirement with acceptable accuracy and at a low computational cost. Accuracy metrics measure if the algorithm can detect lanes with a small error for both straight and curved lanes. Lane detection accuracy issues have been widely studied in the past and many metrics can be found in the literature. However, the robustness issues of the detection system are still not sufficiently studied. Urban road images are usually used to assess the robustness of the system, as more challenges will be encountered in such situations.

Some representative lane detection studies are illustrated in Table 3.2, in which the "Preprocessing" column records the image processing methods used in the literature. The "Integration" column describes the integration methods used in the study, which may contain different levels of integration. Frame images and visual assessment in the "Evaluation" column indicate that the proposed algorithm was only evaluated with still images and visual assessment method without any comparison with ground-truth information. As shown in previous studies, a robust and accurate lane detection system usually combines detection and tracking algorithms. Besides, the most advanced lane detection systems integrate with other object detection systems or sensors to generate a more comprehensive detection network.

## DISCUSSION

In this section, the current limitations of the vision-based lane detection algorithm, integration, and evaluation are analyzed based on the context of the abovementioned sections. Next, the framework of parallel vision-based lane detection system, which is regarded as a possible efficient way to solve the generalization and evaluation problems for lane algorithm design, will be discussed.

### Current Limitations and Challenges

Lane detection systems have been widely studied and successfully implemented in some commercial ADAS products in the past decade. A large volume of literature can be found, which uses vision-based algorithms due to the low cost of camera devices and extensive background knowledge of image processing. Although vision-based lane detection system suffers from illumination variation, shadows, and bad weathers, it is still widely adopted and will continue to dominate the future ADAS markets. The main objective of the lane detection system is to design an accurate and robust

**TABLE 3.2**
**Summary of Various Previous Lane Detection Systems.**

| Ref. | Pre-processing | Lane Detection | Tracking | Integration | Evaluation | Comments |
|---|---|---|---|---|---|---|
| [12] | IPM | Lane marking clustering | Particle filter | Lidar and CCD camera | Frame images and visual assessment | Avoid strong assumption to lane geometry and use weak tracking models |
| [5] | IPM, steerable filters, adaptive template | Statistical and motion-based outlier removal | Kalman filter | Cameras, Laser ranger, GPS, CAN | Quantitative analysis using evaluation metrics | Rich experiments and metrics applied to test the VioLET system |
| [35] | Temporal blur, IPM, adaptive Threshold | RANSAC | Kalman filter | Camera | Quantitative analysis and visual assessment | The proposed ALD 2.0 is used for efficient Video ground truth labeling |
| [102] | EDF | Hough Transform | None | Camera | Frame images and visual assessment | Road are divided into near field and far field with straight and curve model |
| [56] | Road detection, centerline estimation | RANSAC | Route Network Description File | Lidar and cameras | Confidence and centerline evaluation | Obstacle detection and free road area is determined before lane detection |
| [21] | Vanishing point detection, Canny edge dctector | Control point detection | None | Camera | Frame images and visual assessment | The proposed B-snake model is robust to shadow and illumination variation |
| [28] | Ridge feature | RANSAC | None | Camera | Quantitative analysis | Synthesized lane ground truth data are generated with known geometry parameters |
| [25] | IPM, Gaussian kernel filter | Hough Transform, RANSAC | None | Camera | Quantitative analysis with public Caltech dataset | The proposed method is robust to shadow and curves but can be influence by crosswalks and road painting |
| [75] | Layered ROI, steerable filter | Hough Transform | Kalman filter | Radar and camera fusion | Visual assessment and correct detect rate metrics | Adaptive ROI created with range data makes lane detection robust to nearby vehicles and other road markings |

*(continued)*

| TABLE 3.2 Summary of Various Previous Lane Detection Systems.—cont'd | | | | | |
|---|---|---|---|---|---|
| **Ref.** | **Pre-processing** | **Lane Detection** | **Tracking** | **Integration** | **Evaluation** | **Comments** |

| Ref. | Pre-processing | Lane Detection | Tracking | Integration | Evaluation | Comments |
|---|---|---|---|---|---|---|
| [103] | IPM, 2nd and 4th steerable filter | RANSAC | None | Camera, | Performed on KITTI dataset using correct and false positive rate | Detection algorithm is robust to shadow, integrate with optical flow for lane departure aware |
| [37] | ROI, IPM | CNN, RNN | None | Surrounding cameras | Quantitative analysis with ROC curve | Proposed RNN use long-short-term-memory can capture lane spatial structures over a period of time in the video sequences |
| [38] | ROI, artificial image generating | CNN | None | Two lateral cameras facing down the road | Pixel level distance evaluation | End-to-End lane recognition procedure and able to apply in real time |
| [26] | IPM, temporal blur | RANSAC | Kalman filter | Camera | Visual assessment and correct detect rate metrics | Lane detection algorithm is designed mainly focus on night vision |
| [56] | YCbCr colour space transform, vanishing point detection | Lane turning point detection | None | Lane and vehicle integration using single camera | Frame images and visual assessment | Vehicles that have same colour with lanes are distinguished with shape, size, and motion information |
| [59] | YIQ colour space transform, vanishing point detection | Fan-scanning line detection | None | Lane and front vehicle integration using single camera | visual assessment and correct detection rate | The highway lane departure warning and front collision system is built with straight lane model |
| [88] | Median filter, ground plane extraction | Lane segmentation | None | Camera and Lidar integration | visual assessment and correct detection rate | Lane position detected with vision and Lidar is fused with a voting scheme |

*(continued)*

**TABLE 3.2**
**Summary of Various Previous Lane Detection Systems.—cont'd**

| Ref. | Pre-processing | Lane Detection | Tracking | Integration | Evaluation | Comments |
|---|---|---|---|---|---|---|
| [55] | IPM, adaptive threshold | Morphological filters | None | Stereo camera for lane and obstacle detection | Frame images and visual assessment | Lanes are detected mainly with colour features which may be less robust to illumination change |
| [58] | IPM, steerable filter | RANSAC | Kalman filter | Lane and nearby vehicles integration using single camera | Evaluate using hand label frames with multiple metrics | Lane detection is robust in heavy traffic situation with improved surrounding vehicle detection and localization |
| [76] | IPM | Template matching | None | Camera, IMU and GPS fusion | Evaluate using hand label frames with mean absolute error (MAE) metrics | Lanes detected with camera is cross validate with road geometry knowledge given by road map and GPS to improve detection accuracy |
| [51] | Lane marking texture extraction | Scanning line | Kalman filter | Camera | Frame images and visual assessment | Two low level detections is combined with results similarity comparison |
| [78] | Dynamic thresholding, Canny edge detector | Hough Transform and least square model fitting | Kalman filter | Camera, IMU, Lidar, and GPS fusion | Spatial and slope criterion for real time assessment and MAE with ground truth position | A robust redundant lane detection and lateral offset measurement is proposed based on the detection given by camera and Lidar |
| [79] | Prewitt vertical gradient, adaptive threshold | probabilistic Hough Transform | None | IMU, GPS, Lidar and cameras fusion | visual assessment and correct detection rate | Lane marking detection is performed only after road and optimal drivable area is detected based on sensor fusion |

*CAN*, Controller Area Network; *CCD*, charge-coupled device; *CNN*, convolutional neural network; *EDF*, edge distribution function; *IMU*, inertial measurement unit; *IPM*, inverse perspective mapping; *MAE*, mean absolute error; *RNN*, recurrent neural network; *ROC*, receiver operating characteristic; *ROI*, region of interest.

detection algorithm. Accuracy issues were the main concerns of previous studies and many novel methods that are based on machine learning and deep learning methods are designed to construct a more precise system. However, robustness issues are the key aspects that determine if a system can be applied in real life. The huge challenge to future vision-based systems is to maintain a stable and reliable lane measurement under heavy traffic and adverse weather conditions.

Considering this problem, one efficient method is to use integration and fusion techniques. It has been proved that a single vision-based lane detection system has its limitations to deal with the varying road and traffic situation. Therefore it is necessary to prepare a backup system that can enrich the functionality of ADAS. Basically, a redundancy system can be constructed in three ways based on algorithm, system, and sensor level integration. Algorithm integration is a choice with the lowest cost and easiest to be applied. A system level integration combines the lane detection system with other perception systems such as road and surrounding vehicle detection to improve the accuracy and robustness of the system. However, the two integration methods still rely on camera vision systems and have their inevitable limitations. Sensor level integration, on the other hand, is the most reliable way to detect lanes under different situations.

Another challenging task in lane detection systems is to design an evaluation system that can verify the system performance. Nowadays, a common problem is the lack of public benchmarks and datasets due to the difficulty of labeling lanes as the ground truth. Besides, there are no standard evaluation metrics that can be used to comprehensively assess the system performance with respect to both accuracy and robustness properties. Online confidence evaluation is another important task for the lane detection system. For ADAS and lower-level automated vehicles, the driver should be alerted once low detection confidence occurs. In terms of autonomous vehicles, it is also important to let the vehicle understand how it does in the lane detection task, which can be viewed as a self-aware and diagnostic process.

## Applying the Parallel Theory to Road Perception Systems

Considering the issues, a novel parallel framework for lane detection system design will be proposed in this section. The parallel lane detection framework is expected to be an efficient tool to assess both the robustness and the evaluation issues for the lane detection system.

A parallel system is the product of advanced control systems and computer simulation systems. It was introduced by Fei-Yue Wang and developed to control and manage complex systems [104−106]. The parallel theory is an efficient tool that can compensate for the hard modeling and evaluating issue for the complex systems. The main objective of the parallel system is to connect the real-world system with one or multiple artificial virtual systems that are in the cyberspace. The constructed virtual systems will have similar characteristics as the real-world complex system but not the same. Here, parallel refers to a parallel interaction between the real-world system and its corresponding virtual counterparts. By connecting these systems together and analyzing and comparing their behaviors, the parallel system will be able to predict the future status of both the real-world systems and the artificial ones. According to the response and behaviors of the virtual system, the parallel system will automatically adjust the parameters of the real-world model to control and manage the real-world complex system such that an efficient solution will be applied.

The construction of the parallel system requires the ACP theory as background knowledge. ACP is short for Artificial Society, Computational experiments, and Parallel execution, which are the three major components of the parallel system. The complex system is first modeled using a holistic approach, whereas the real-world system is represented using an artificial system. After this step, the virtual system in the cyberspace becomes another solution domain of the complex system, which contributes to the potential complete solution along with the natural system in the physical space. It is hard to say that one solution will satisfy all the real-world challenges. An effective solution for the control of complex system should have the ability to deal with various situations occurring in the future. However, the limited testing scenarios in the real world cannot guarantee the potential solution being comprehensively tested. Therefore the computation experiment module will execute a large number of virtual experiments according to the constructed artificial system in the last step. Finally, considering there are normally no unique solution for the complex system, the parallel execution provides an effective fashion to validate and evaluate various solutions. The parallel execution module will update the local optimal solution online to the real-world system, which is found in the cyberspace for better control and management [107].

A parallel vision architecture based on the ACP theory has been summarized and introduced to the computer vision society [108]. The parallel vision theory offers an efficient way to deal with the detection and evaluation problems of the vision-based object detection systems. Similarly, the general ideal of the ACP theory within the parallel vision scope is to achieve perception and understanding of the complex real-world environment according to the combination of virtual realities and the real-world information. In terms of lane detection task, the first artificial society module can be used to construct a virtual traffic environment and various road scenes using computer graphics and virtual reality techniques. Next, in the computation experiment module, the unlimited labeled traffic scene images and the limited real-world driving images can be combined to train powerful lane detectors using the machine learning and deep learning methods. This process also contains two subprocedures, namely, learning and training, testing and evaluating. The large-scale dataset will benefit the model training task, after that, a large amount of near-real data will sufficiently facilitate the model evaluation. Finally, in the parallel execution process, the lane detection model can be trained and evaluated in a parallel scheme in both the real world and the virtual environment. The lane detector can be optimized online according to its performance in the two parallel worlds.

In addition, the application of ACP parallel vision system will efficiently solve the generalization and evaluation problems because of the utilization of the large-scale near-real synthesis images. To improve the generalization of the lane detection system, the detectors can be tested on virtual environments that have high similarity with the real world. The performance can also be sufficiently evaluated from the accuracy and robustness perspectives. Various computation experiments and model testing procedures can be continuously executed. In the computational experiments, the cutting-edge deep learning and reinforcement learning techniques can be applied to improve the accuracy and generalization of the system without considering the lack of labeled data. Meanwhile, some deep learning models such as the generative adversarial network (GAN) can be used to generate near-real road scene images that can reflect the real-world road characteristics such as illumination, occlusion, and poor visualization. In addition, in the virtual computational world, GAN can be trained to discriminate whether the lane markings exist in the input image.

Fig. 3.4 shows a simplified architecture of the ACP-based lane detection system. The road and lane images are in parallel collected from the real world and the artificial cyberspace. The real-world data is then used as a guide for generating near-real artificial traffic scenes, which are automatically labeled. Both the real-world data and the synthesized data are fed into the data-driven computational level. Machine learning and deep learning methods are powerful tools in this level. For various driving scenarios occurring in both real world and the parallel virtual world, the model training

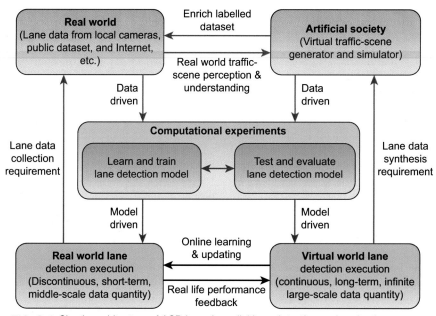

FIG. 3.4 Simple architecture of ACP-based parallel lane detection and evaluation system.

process will try to come up with the most satisfying model. After that, the lane detection model will be exhaustively evaluated and validated in the cyberspace world according to large-scale labeled data. Once a well-trained and evaluated lane detection model is constructed, the model can be applied in parallel to both real-world environment and virtual world for real-time lane detection evaluation. Owing to the safety, human resource limitation, and energy consumption, the number of experiments in real world is limited, which may not be enough to deal with all the challenges from the road [109,110]. In contrary, the experiments in the parallel virtual world are safer and economical to be applied; moreover, the virtual world can simulate much more situations that less possibly occur in the real world. However, by using an online learning technique, the experience from the continuous learning and testing module in the virtual world will improve the real-world performance.

Some previous literature has partially applied the parallel vision theory into the construction of the lane detection system [28,56]. These studies try to simulate the lane detection model within the simulation environment and process the lane detection model with the first two steps of ACP architecture. However, to construct an actual parallel system, the ACP architecture should be treated. The final parallel execution step of ACP theory is the core of the parallel system. This step will update the real-world model online and adjust the corresponding model parameters according to the testing results in the parallel worlds. This step is also the core step, which guarantees that the learned lane detection model can be satisfied in various real-world driving scenarios. Despite applying the parallel theory into the design of intelligent transport and vehicles, it has been widely used in some other domains. For example, DeepMind uses multiple processors to train their AlphaGo based on deep reinforcement learning methods [111]. The idea behind the reinforcement learning, in this case, is to construct a parallel virtual world for the virtual go player to do exercise. In summary, the parallel theory is drawing increasing attention from researchers. The utilization of parallel vision techniques in the future is expected to become another efficient way to solve the generalization and evaluation problems for lane detection algorithms. The ACP-based parallel lane detection system not only will assist in building an accurate model that is well tested and assessed but also will enable the intelligent vehicles to carefully adjust their detection strategies in real time. Also as there are too many different lane detection methodologies that are hardly evaluated uniformly, a public virtual simulation platform can be used to compare these algorithms in the future. Those algorithms that achieve satisfactory performance in the parallel virtual worlds can then be implemented in the real world.

## CONCLUSION

In this study, vision-based lane detection systems are reviewed from three aspects, namely, algorithms, integration methods, and evaluation methods. Existing algorithms are summarized into two categories, which are conventional image processing based and novel machine learning (deep learning) based. Next, previous integration methods of the lane detection system are divided into three levels, which are algorithm level, system level, and sensor level. In the algorithm level, multiple lane detection and tracking algorithms are combined in a serial or parallel manner. System level integration combines vision-based lane detection with other road marking or obstacle detection systems. Sensor fusion enhances the vehicle perception system most significantly by fusion of multimodal sensors. Finally, lane detection evaluation issues are analyzed from different aspects. Evaluation methods are divided into offline performance assessment and online real-time confidence evaluation.

As mentioned earlier, although the vision-based lane detection system has been widely studied in the past two decades, it is hard to say that research in this area has advanced. In fact, there are still many critical studies that need to be done, such as efficient low-cost system integration and the evaluation system design, especially the construction of a parallel lane detection system. An increasing amount of advanced object detection algorithms and architecture has been developed to optimize the lane detection systems. The continuous studies and the application of these techniques will further benefit ADAS and the automated driving industry. The ACP-based parallel lane detection approach holds significant potentials for future implementation.

## REFERENCES

[1] E. Bellis, J. Page, National Motor Vehicle Crash Causation Survey (NMVCCS) SAS Analytical User's Manual. No. HS-811 053, 2008.

[2] J.E. Gayko, Lane departure and lane keeping, in: Handbook of Intelligent Vehicles, Springer London, 2012, pp. 689−708.

[3] C. Visvikis, T.L. Smith, M. Pitcher, et al., Study on Lane Departure Warning and Lane Change Assistant Systems, Transport Research Laboratory Project Report PPR 374, 2008.

[4] A. Bar Hillel, et al., Recent progress in road and lane detection: a survey, Machine Vision and Applications (2014) 1−19.

[5] J.C. McCall, M.M. Trivedi, Video-based lane estimation and tracking for driver assistance: survey, system, and evaluation, IEEE Transactions on Intelligent Transportation Systems 7 (1) (2006) 20−37.

[6] S. Yenikaya, G. Yenikaya, E. Düven, Keeping the vehicle on the road: a survey on on-road lane detection systems, ACM Computing Surveys (CSUR) 46 (1) (2013) 2.

[7] J. Fritsch, K. Tobias, A. Geiger, A new performance measure and evaluation benchmark for road detection algorithms, in: Intelligent Transportation Systems-(ITSC), 2013 16th International IEEE Conference on, IEEE, 2013.

[8] M. Beyeler, F. Mirus, V. Alexander, Vision-based robust road lane detection in urban environments, in: Robotics and Automation (ICRA), 2014 IEEE International Conference on, IEEE, 2014.

[9] D.-J. Kang, M.-H. Jung, Road lane segmentation using dynamic programming for active safety vehicles, Pattern Recognition Letters 24 (16) (2003) 3177−3185.

[10] U. Suddamalla, et al., A novel algorithm of lane detection addressing varied scenarios of curved and dashed lanemarks, in: Image Processing Theory, Tools and Applications (IPTA), 2015 International Conference on, IEEE, 2015.

[11] J.M. Collado, et al., Adaptative road lanes detection and classification, in: International Conference on Advanced Concepts for Intelligent Vision Systems, Springer Berlin Heidelberg, 2006.

[12] S. Sehestedt, et al., Robust lane detection in urban environments, in: Intelligent Robots and Systems, 2007. IROS 2007. IEEE/RSJ International Conference on, IEEE, 2007.

[13] Q. Lin, Y. Han, H. Hahn, Real-time lane departure detection based on extended edge-linking algorithm, in: Computer Research and Development, 2010 Second International Conference on, IEEE, 2010.

[14] A.F. Cela, et al., Lanes detection based on unsupervised and adaptive classifier, in: Computational Intelligence, Communication Systems and Networks (CICSyN), 2013 Fifth International Conference on, IEEE, 2013.

[15] A. Borkar, et al., A layered approach to robust lane detection at night, in: Computational Intelligence in Vehicles and Vehicular Systems, 2009. CIVVS'09. IEEE Workshop on, IEEE, 2009.

[16] C. Kreucher, S. Lakshmanan, LANA: a lane extraction algorithm that uses frequency domain features, IEEE Transactions on Robotics and Automation 15 (2) (1999) 343−350.

[17] S. Jung, J. Youn, S. Sull, Efficient lane detection based on spatiotemporal images, IEEE Transactions on Intelligent Transportation Systems 17 (1) (2016) 289−295.

[18] J. Xiao, S. Li, B. Sun, A real-time system for lane detection based on FPGA and DSP, Sensing and Imaging 17 (1) (2016) 1−13.

[19] U. Ozgunalp, D. Naim, Lane detection based on improved feature map and efficient region of interest extraction, in: Signal and Information Processing (GlobalSIP), 2015 IEEE Global Conference on, IEEE, 2015.

[20] Y. Wang, D. Shen, E.K. Teoh, Lane detection using spline model, Pattern Recognition Letters 21 (8) (2000) 677−689.

[21] Y. Wang, E.K. Teoh, D. Shen, Lane detection and tracking using B-Snake, Image and Vision Computing 22 (4) (2004) 269−280.

[22] X. Li, et al., Lane detection and tracking using a parallel-snake approach, Journal of Intelligent and Robotic Systems 77 (2015) 3−4, 597.

[23] K.H. Lim, P.S. Kah, L.-M. Ang, River flow lane detection and Kalman filtering-based B-spline lane tracking, International Journal of Vehicular Technology (2012) 2012.

[24] C.R. Jung, C.R. Kelber, An improved linear-parabolic model for lane following and curve detection, in: Computer Graphics and Image Processing, 2005. SIBGRAPI 2005. 18th Brazilian Symposium on, IEEE, 2005.

[25] M. Aly, Real time detection of lane markers in urban streets, in: Intelligent Vehicles Symposium, 2008 IEEE, IEEE, 2008.

[26] A. Borkar, M. Hayes, M.T. Smith, Robust lane detection and tracking with ransac and Kalman filter, in: Image Processing (ICIP), 2009 16th IEEE International Conference on, IEEE, 2009.

[27] A. Lopez, et al., Detection of lane markings based on ridgeness and RANSAC, in: Intelligent Transportation Systems, 2005. Proceedings. 2005 IEEE, IEEE, 2005.

[28] A. López, et al., Robust lane markings detection and road geometry computation, International Journal of Automotive Technology 11 (3) (2010) 395−407.

[29] Q. Chen, H. Wang, A real-time lane detection algorithm based on a hyperbola-pair model, in: Intelligent Vehicles Symposium, 2006 IEEE, IEEE, 2006.

[30] H. Tan, et al., Improved river flow and random sample consensus for curve lane detection, Advances in Mechanical Engineering 7 (7) (2015), 1687814015593866.

[31] J. Hur, S.-N. Kang, S.-W. Seo, Multi-lane detection in urban driving environments using conditional random fields, in: Intelligent Vehicles Symposium (IV), 2013 IEEE, IEEE, 2013.

[32] F. Bounini, et al., Autonomous vehicle and real time road lanes detection and tracking, in: Vehicle Power and Propulsion Conference (VPPC), 2015 IEEE, IEEE, 2015.

[33] D. Wu, R. Zhao, Z. Wei, A multi-segment lane-switch algorithm for efficient real-time lane detection, in: Information and Automation (ICIA), 2014 IEEE International Conference on, IEEE, 2014.

[34] S. Zhou, et al., A novel lane detection based on geometrical model and gabor filter, in: Intelligent Vehicles Symposium (IV), 2010 IEEE, IEEE, 2010.

[35] J.W. Niu, J. Lu, M.L. Xu, P. Lv, X.K. Zhao, Robust lane detection using two-stage feature extraction with curve fitting, Pattern Recogn 59 (2016) 225−233.

[36] B. He, et al., Lane marking detection based on Convolution Neural Network from point clouds, in: Intelligent Transportation Systems (ITSC), 2016 IEEE 19th International Conference on, IEEE, 2016.

[37] Jun Li, et al., Deep neural network for structural prediction and lane detection in traffic scene, IEEE Transactions on Neural Networks and Learning Systems 28 (3) (2016) 690−703.

[38] A. Gurghian, et al., DeepLanes: end-to-end lane position estimation using deep neural networks, in: Proceedings of the IEEE Conference on Computer Vision and Pattern Recognition Workshops, 2016.

[39] X. Li, et al., Lane detection based on spiking neural network and Hough transform, in: Image and Signal Processing (CISP), 2015 8th International Congress on, IEEE, 2015.

[40] J. Kim, et al., Fast learning method for convolutional neural networks using extreme learning machine and its application to lane detection, in: Neural Networks, 2016.

[41] B. He, et al., Accurate and robust lane detection based on dual-view convolutional neutral network, in: Intelligent Vehicles Symposium (IV), 2016 IEEE, IEEE, 2016.

[42] M. Revilloud, D. Gruyer, M.-C. Rahal, A new multi-agent approach for lane detection and tracking, in: Robotics and Automation (ICRA), 2016 IEEE International Conference on, IEEE, 2016.

[43] M. Bertozzi, et al., An Evolutionary Approach to Lane Markings Detection in Road Environments, Atti del 6, 2002, pp. 627−636.

[44] L.-W. Tsai, et al., Lane detection using directional random walks, in: Intelligent Vehicles Symposium, 2008 IEEE, IEEE, 2008.

[45] L. Bai, Y. Wang, Road tracking using particle filters with partition sampling and auxiliary variables, Computer Vision and Image Understanding 115 (10) (2011) 1463−1471.

[46] R. Danescu, S. Nedevschi, Probabilistic lane tracking in difficult road scenarios using stereovision, IEEE Transactions on Intelligent Transportation Systems 10 (2) (2009) 272−282.

[47] Z.W. Kim, Robust lane detection and tracking in challenging scenarios, IEEE Transactions on Intelligent Transportation Systems 9 (1) (2008) 16−26.

[48] B.-S. Shin, J. Tao, R. Klette, A superparticle filter for lane detection, Pattern Recognition 48 (11) (2015) 3333−3345.

[49] D. Apurba, S.S. Murthy, U. Suddamalla, Enhanced algorithm of automated ground truth generation and validation for lane detection system by M2BMT, IEEE Transactions on Intelligent Transportation Systems 18 (4) (2016) 996−1005.

[50] R. Labayrade, S.S. Leng, A. Didier, A reliable road lane detector approach combining two vision-based algorithms, in: Intelligent Transportation Systems, 2004. Proceedings. The 7th International IEEE Conference on, IEEE, 2004.

[51] R. Labayrade, et al., A reliable and robust lane detection system based on the parallel use of three algorithms for driving safety assistance, IEICE Transactions on Information and Systems 89 (7) (2006) 2092−2100.

[52] D.C. Hernández, D. Seo, K.-H. Jo, Robust lane marking detection based on multi-feature fusion, in: Human System Interactions (HSI), 2016 9th International Conference on, IEEE, 2016.

[53] Y.U. Yim, S.-Y. Oh, Three-feature based automatic lane detection algorithm (TFALDA) for autonomous driving, IEEE Transactions on Intelligent Transportation Systems 4 (4) (2003) 219−225.

[54] M. Felisa, P. Zani, Robust monocular lane detection in urban environments, in: Intelligent Vehicles Symposium (IV), 2010 IEEE, IEEE, 2010.

[55] M. Bertozzi, A. Broggi, GOLD: a parallel real-time stereo vision system for generic obstacle and lane detection, IEEE Transactions on Image Processing 7 (1) (1998) 62−81.

[56] A.S. Huang, et al., Finding multiple lanes in urban road networks with vision and lidar, Autonomous Robots 26 (2) (2009) 103−122.

[57] H.-Y. Cheng, et al., Lane detection with moving vehicles in the traffic scenes, IEEE Transactions on Intelligent Transportation Systems 7 (4) (2006) 571−582.

[58] S. Sivaraman, M.M. Trivedi, Integrated lane and vehicle detection, localization, and tracking: a synergistic approach, IEEE Transactions on Intelligent Transportation Systems 14 (2) (2013) 906−917.

[59] C.-F. Wu, C.-J. Lin, C.-Y. Lee, Applying a functional neurofuzzy network to real-time lane detection and front-vehicle distance measurement, IEEE Transactions on Systems, Man, and Cybernetics, Part C (Applications and Reviews) 42 (4) (2012) 577−589.

[60] S.-S. Huang, et al., On-board vision system for lane recognition and front-vehicle detection to enhance driver's awareness, in: Robotics and Automation, 2004. Proceedings. ICRA'04. 2004 IEEE International Conference on, vol. 3, IEEE, 2004.

[61] R.K. Satzoda, M.M. Trivedi, Efficient lane and vehicle detection with integrated synergies (ELVIS), in: Computer Vision and Pattern Recognition Workshops (CVPRW), 2014 IEEE Conference on, IEEE, 2014.

[62] H. Kim, et al., Integration of vehicle and lane detection for forward collision warning system, in: Consumer Electronics-Berlin (ICCE-Berlin), 2016 IEEE 6th International Conference on, IEEE, 2016.

[63] B. Qin, et al., A general framework for road marking detection and analysis, in: Intelligent Transportation Systems-(ITSC), 2013 16th International IEEE Conference on, IEEE, 2013.

[64] A. Kheyrollahi, T.P. Breckon, Automatic real-time road marking recognition using a feature driven approach, Machine Vision and Applications 23 (1) (2012) 123−133.

[65] J. Greenhalgh, M. Mirmehdi, Detection and recognition of painted road surface markings, ICPRAM 1 (2015).

[66] G.L. Oliveira, W. Burgard, T. Brox, Efficient deep models for monocular road segmentation, in: Intelligent Robots and Systems (IROS), 2016 IEEE/RSJ International Conference on, IEEE, 2016.

[67] H. Kong, J.-Y. Audibert, J. Ponce, Vanishing point detection for road detection, in: Computer Vision and Pattern Recognition, 2009. CVPR 2009. IEEE Conference on, IEEE, 2009.

[68] D. Levi, N. Garnett, E. Fetaya, I. Herzlyia, StixelNet: A deep convolutional network for obstacle detection and road segmentation, British Machine Vision Conference (2015) 109−111.

[69] G.P. Stein, Y. Gdalyahu, S. Amnon, Stereo-assist: top-down stereo for driver assistance systems, in: Intelligent Vehicles Symposium (IV), 2010 IEEE, IEEE, 2010.

[70] E. Raphael, et al., Development of a camera-based forward collision alert system, SAE International Journal of Passenger Cars-Mechanical Systems 4 (2011) 467−478, 2011-01-0579.

[71] B. Ma, S. Lakahmanan, A. Hero, Road and lane edge detection with multisensor fusion methods, in: Image Processing, 1999. ICIP 99. Proceedings. 1999 International Conference on, vol. 2, IEEE, 1999.

[72] U. Ozgunalp, et al., Multiple lane detection algorithm based on novel dense vanishing point estimation, IEEE Transactions on Intelligent Transportation Systems 18 (3) (2017) 621−632.

[73] C. Lipski, et al., A fast and robust approach to lane marking detection and lane tracking, in: Image Analysis and Interpretation, 2008. SSIAI 2008. IEEE Southwest Symposium on, IEEE, 2008.

[74] D. Kim, et al., Lane-level localization using an AVM camera for an automated driving vehicle in urban environments, IEEE/ASME Transactions on Mechatronics 22 (1) (2017) 280−290.

[75] H.G. Jung, et al., Sensor fusion-based lane detection for LKS+ ACC system, International Journal of Automotive Technology 10 (2) (2009) 219−228.

[76] D. Cui, J. Xue, N. Zheng, Real-time global localization of robotic cars in lane level via lane marking detection and shape registration, IEEE Transactions on Intelligent Transportation Systems 17 (4) (2016) 1039−1050.

[77] Y. Jiang, F. Gao, G. Xu, Computer vision-based multiple-lane detection on straight road and in a curve, in: Image Analysis and Signal Processing (IASP), 2010 International Conference on, IEEE, 2010.

[78] C. Rose, et al., An integrated vehicle navigation system utilizing lane-detection and lateral position estimation systems in difficult environments for GPS, IEEE Transactions on Intelligent Transportation Systems 15 (6) (2014) 2615−2629.

[79] Q. Li, et al., A sensor-fusion drivable-region and lane-detection system for autonomous vehicle navigation in challenging road scenarios, IEEE Transactions on Vehicular Technology 63 (2) (2014) 540−555.

[80] S. Kammel, B. Pitzer, Lidar-based lane marker detection and mapping, in: Intelligent Vehicles Symposium, 2008 IEEE, IEEE, 2008.

[81] M. Manz, et al., Detection and tracking of road networks in rural terrain by fusing vision and LIDAR, in: Intelligent Robots and Systems (IROS), 2011 IEEE/RSJ International Conference on, IEEE, 2011.

[82] M. Schreiber, C. Knöppel, U. Franke, Laneloc: lane marking based localization using highly accurate maps, in: Intelligent Vehicles Symposium (IV), 2013 IEEE, IEEE, 2013.

[83] J.M. Clanton, D.M. Bevly, A.S. Hodel, A low-cost solution for an integrated multisensor lane departure warning system, IEEE Transactions on Intelligent Transportation Systems 10 (1) (2009) 47−59.

[84] M. Montemerlo, et al., Junior: the stanford entry in the urban challenge, Journal of Field Robotics 25 (9) (2008) 569−597.

[85] M. Buehler, K. Iagnemma, S. Singh (Eds.), The DARPA Urban Challenge: Autonomous Vehicles in City Traffic, vol. 56, Springer, 2009.

[86] P. Lindner, et al., Multi-channel lidar processing for lane detection and estimation, in: Intelligent Transportation Systems, 2009. ITSC'09. 12th International IEEE Conference on, IEEE, 2009.

[87] S. Shin, I. Shim, I.S. Kweon, Combinatorial approach for lane detection using image and LIDAR reflectance, in: Ubiquitous Robots and Ambient Intelligence (URAI), 2015 12th International Conference on, IEEE, 2015.

[88] P. Amaradi, et al., Lane following and obstacle detection techniques in autonomous driving vehicles, in: Electro Information Technology (EIT), 2016 IEEE International Conference on, IEEE, 2016.

[89] K. Dietmayer, et al., Roadway detection and lane detection using multilayer laserscanner, in: Advanced Microsystems for Automotive Applications 2005, Springer Berlin Heidelberg, 2005, pp. 197−213.

[90] D.C. Hernandez, V.-D. Hoang, K.-H. Jo, Lane surface identification based on reflectance using laser range finder, in: System Integration (SII), 2014 IEEE/SICE International Symposium on, IEEE, 2014.

[91] J. Sparbert, K. Dietmayer, D. Streller, Lane detection and street type classification using laser range images, in: Intelligent Transportation Systems, 2001. Proceedings. 2001 IEEE, IEEE, 2001.

[92] A. Broggi, et al., A laserscanner-vision fusion system implemented on the terramax autonomous vehicle, in: Intelligent Robots and Systems, 2006 IEEE/RSJ International Conference on, IEEE, 2006.

[93] H. Zhao, et al., A laser-scanner-based approach toward driving safety and traffic data collection, IEEE Transactions on Intelligent Transportation Systems 10 (3) (2009) 534−546.

[94] A. Borkar, M. Hayes, M.T. Smith, A novel lane detection system with efficient ground truth generation, IEEE Transactions on Intelligent Transportation Systems 13 (1) (2012) 365−374.

[95] C.-W. Lin, H.-Y. Wang, D.-C. Tseng, A robust lane detection and verification method for intelligent vehicles, in: Intelligent Information Technology Application, 2009.

IITA 2009. Third International Symposium on, vol. 1, IEEE, 2009.

[96] Ju Han Yoo, et al., A robust lane detection method based on vanishing point estimation using the relevance of line segments, IEEE Transactions on Intelligent Transportation Systems 18 (12) (2017) 3254–3266.

[97] L. Li, et al., Intelligence testing for autonomous vehicles: a new approach, IEEE Transactions on Intelligent Vehicles 1 (2) (2016) 158166.

[98] K.C. Kluge, Performance evaluation of vision-based lane sensing: some preliminary tools, metrics, and results, in: Intelligent Transportation System, 1997. ITSC'97., IEEE Conference on, IEEE, 1997.

[99] T. Veit, et al., Evaluation of road marking feature extraction, in: Intelligent Transportation Systems, 2008. ITSC 2008. 11th International IEEE Conference on, IEEE, 2008.

[100] J.C. McCall, M.M. Trivedi, Performance evaluation of a vision based lane tracker designed for driver assistance systems, in: Intelligent Vehicles Symposium, 2005. Proceedings. IEEE, IEEE, 2005.

[101] R.K. Satzoda, M.M. Trivedi, On performance evaluation metrics for lane estimation, in: Pattern Recognition (ICPR), 2014 22nd International Conference on, IEEE, 2014.

[102] C.R. Jung, C.R. Kelber, A robust linear-parabolic model for lane following, in: Computer Graphics and Image Processing, 2004. Proceedings. 17th Brazilian Symposium on, IEEE, 2004.

[103] M. Haloi, D. Babu Jayagopi, A robust lane detection and departure warning system, in: Intelligent Vehicles Symposium (IV), 2015 IEEE, IEEE, 2015.

[104] F.Y. Wang, Parallel system methods for management and control of complex systems, Control Decision 19 (5) (2004), 485–489, 514.

[105] F.Y. Wang, Parallel control and management for intelligent transportation systems: concepts, architectures, and applications, IEEE Transactions on Intelligent Transportation Systems 11 (3) (Sep. 2010) 630–638.

[106] F. Y. Wang, "Artificial societies, computational experiments, and parallel systems: a discussion on computational theory of complex social economic systems," Complex Systems Complexity Science, vol. 1, no. 4, pp.25-35, Oct.

[107] L. Li, Y.L. Lin, D.P. Cao, N.N. Zheng, F.Y. Wang, Parallel learning-a new framework for machine learning, Acta Automatica Sinica 43 (1) (2017) 1–18.

[108] K.F. Wang, C. Gou, N.N. Zheng, J.M. Rehg, F.Y. Wang, Parallel vision for perception and understanding of complex scenes: methods, framework, and perspectives, Artificial Intelligence Review 48 (3) (Oct. 2017) 298–328.

[109] F.Y. Wang, N.N. Zheng, D.P. Cao, C.M. Martinez, L. Li, T. Liu, Parallel driving in CPSS: a unified approach for transport automation and vehicle intelligence, IEEE/CAA Journal of Automatica Sinica 4 (4) (2017) 577–587.

[110] C. Lv, Y. H. Liu, X. S. Hu, H. Y. Guo, D. P. Cao, and F. Y. Wang, "Simultaneous observation of hybrid states for cyber-physical systems: a case study of electric vehicle powertrain," IEEE Trans. Cybern., 2018, to be Published, doi: 10.1109/TCYB.2017.2738003.

[111] D. Silver, et al., Mastering the game of Go with deep neural networks and tree search, Nature 529 (7587) (2016) 484–489.

[112] W. Song, et al., Real-time obstacles detection and status classification for collision warning in a vehicle active safety system, IEEE Transactions on Intelligent Transportation Systems 19 (3) (2018) 758–773.

[113] H. Kuang, et al., Nighttime vehicle detection based on bio-inspired image enhancement and weighted score-level feature fusion, IEEE Transactions on Intelligent Transportation Systems 18 (99) (2016) 1–10.

[114] L. Chen, et al., Surrounding vehicle detection using an FPGA panoramic camera and deep CNNs, IEEE Transactions on Intelligent Transportation Systems (2019).

[115] L. Chen, et al., Turn signal detection during nighttime by CNN detector and perceptual hashing tracking, IEEE Transactions on Intelligent Transportation Systems (2017) 1–12.

[116] X. Hu, et al., SINet: a scale-insensitive convolutional neural network for fast vehicle detection, IEEE Transactions on Intelligent Transportation Systems (2018) 99.

[117] X. Yuan, S. Su, H. Chen, A graph-based vehicle proposal location and detection algorithm, IEEE Transactions on Intelligent Transportation Systems (2017) 1–8.

[118] Z. Yang, J. Li, H. Li, Real-time pedestrian and vehicle detection for autonomous driving, in: 2018 IEEE Intelligent Vehicles Symposium (IV), IEEE, 2018, pp. 179–184.

# CHAPTER 4

# Design of Integrated Road Perception and Lane Detection System for Driver Intention Inference

## ROAD DETECTION

### Introduction

Road detection is one of the primary tasks for autonomous vehicles. However, it can be challenging to detect the road or the drivable region with only color cameras on the unstructured road. In this study, a low-cost and low-resolution camera–Lidar fusion-based deep segmentation network is proposed to detect the front road region. The fusion network can capture both the color features from the image and spatial features from the point cloud, which can be significantly effective in the unstructured road condition. A deep segmentation convolutional neural network is designed to process both the RGB image and the calibrated Lidar point cloud. The Lidar coordinate system was first transformed into the camera coordinate, and the 3D point cloud information is transformed to the 2D color image, which maintained the depth features and textures of the road. A low-cost, low-resolution solution is proposed by rescaling the original high-resolution images into a low-resolution format to increase the real-time inference speed. A cross-fusion segmentation network is trained to process the two different inputs simultaneously. To evaluate the efficient model optimization methodologies, several different criteria and learning rate adjustment methods are evaluated. The models are trained and tested with the KITTI public dataset. Results indicate that low-cost cross-fusion network can provide a reasonable road detection with an exponential learning rate adjustment.

Road detection is one of the preliminary tasks for intelligent and automated driving vehicles and has been richly studied in the past 2 decades [1–3]. Road area or drivable region detection can assist the path planning and motion control strategies of the intelligent vehicle [4–6]. In addition, it can enhance the lane detection accuracy and lead to a more comprehensive understanding of current road context and structures [7,8]. Road detection can be roughly divided into two categories, which are for structured road and unstructured off-road environments [9]. Conventional road-detection algorithms for formal roads usually rely on the computer vision-based methods, such as RGB cameras and stereo cameras [10]. Cameras are the most popular computer vision sensors for both academic research and industrial application due to the low cost, high frame rate, and high precision. The color and textures feature of the road context can be precisely recorded by the color image, which enables accurate object detection and segmentation with current deep convolutional neural networks [11,12]. Therefore for the structured road, the onboard multicamera system usually can provide a reasonable detection accuracy for road regions, vehicles, and pedestrians [3,10].

Although cameras are widely used owing to their low cost and easy-to-implement characters, it suffers from illumination, weather, different terrains, and occlusion issues [10]. The passive detection fashion for the cameras cannot provide robust long-range sensing. In addition, as most of the cameras map 3D context into the 2D images, the distortion will affect the object recovery from 2D to the real world. For example, as shown in Fig. 4.1, the color image can reflect the color and edge features for the unstructured off-road environment, and the collapse in the middle of the road is severe to be processed with 2D images only. Therefore for the unstructured road region detection, a more robust and accurate manner is to combine the active sensing such as Lidar systems and the passive sensing method.

Lidar has advantages in illumination variation and tough weather conditions. A combined approach with Lidar and cameras proposes an opportunity for deep CNN to learn different characteristics from the two types of sensors. It has been shown that by integrating the two sensors, road detection can be more accurate in the unstructured situation. In Ref. [13], the authors proposed three different approaches for data fusion,

Advanced Driver Intention Inference. https://doi.org/10.1016/B978-0-12-819113-2.00004-X

FIG. 4.1 Illustration of difficulties for the unstructured road region detection.

which are early fusion, late fusion, and cross fusion. Cross fusion is proved to be the most efficient and accurate method among the three typical approaches. In this study, to further exploit the characteristics of the cross-fusion method, a lightweight version is developed based on the SegNet [14]. In addition, several model learning strategies are evaluated for the light version of the dynamic fusion network. This work intends to propose an efficient method for the model training of the Lidar−camera fusion-based segmentation networks.

The remainder of this part is organized as follows. The Related Works part introduces the related works of on-road detection. In the Data Processing part, the experimental design and data processing is proposed. In the Model Construction part, the construction of the fused segmentation network and its corresponding training and testing process are discussed. The road detection results are illustrated in the Experimental Results part.

## Related Works

Vehicle onboard road detection is an open topic that has been widely studied in the past. According to Ref. [9], existing road-detection methodologies can be divided into the unimodal-based approach and multimodal approaches. The unimodal approach mainly relies on single sensor systems such as cameras and Lidar. On the contrary, the multimodal approaches are based on sensor and data fusion.

The multimodal road detection can be traced back to 2 decades ago, where Broggi et al. developed the GOLD system for intelligent vehicles [15], which detected the lanes and obstacles with image binarization and classification. In Ref. [16], a general road-detection system was proposed based on the computer vision method. Given a single image with either a structured or unstructured road, the system will first estimate the vanishing point of the central part of the road and detect the road region with the estimated vanishing point and adaptive soft voting-based image segmentation. In Ref. [17], an illumination invariance road detection model was proposed based on the color images. The illumination invariance angle of the camera was calibrated and determined in the first. Then, a road class-likelihood classifier was used to process the illumination-invariance space features and estimate the road area via pixel-level classification. In Ref. [18], an online-learning road area classification based on the support vector machine (SVM) was developed. The road samples were selected and featured with different extractors. Then, a binary structured SVM was trained to classify the road boundary, which can be further updated online to ensure the model's adaptability.

Recent advances in deep learning methods enable more accurate road detection. In Ref. [19], a direct mapping between the CNN filters and classes was proposed at the expansion side of the segmentation network, which makes the model 20 times faster than conventional deep segmentation networks. In Ref. [20], the authors developed an "s-FCN-loc" model that takes the RGB images, semantic contours, and location priors simultaneously to segment the road. The s-FCN-loc had two streams to process the RGB images and contour maps separately. In Ref. [21], to maintain the real-time inference, a network-in-network (NiN) model was proposed to take the large contextual window while keeping the fast model inference. In Ref. [22], the road detection problem was transformed into a single-scale problem represented by a top-view point cloud. A fast and straightforward fully convolutional neural network (FCN) was adopted for pixel-wise semantic segmentation.

In Ref. [23], a fused road detection system was developed for desert terrain. A laser range finder, a pose estimation system, and a color camera were combined to estimate the front drivable region. The system explored a nearby patch of drivable surface based on the laser and vehicle pose estimation system. Then, the far range appearance model was developed based on the surface batch and color images. In Ref. [24], a joint learning scheme for road detection based on Lidar and camera fusion was proposed. The authors developed three different inputs for the YOLO CNN, which were the upsampled version of the sparse Lidar point cloud, the high-resolution reflectance map, and RGB images. It was shown that the dense depth map and dense reflectance map could significantly increase the model performance. In Ref. [25], a hybrid conditional random field framework was applied for pixel-wise road detection and Lidar–camera fusion. The point cloud of the Lidar was treated as random variables and used for pixel label inference via minimizing a hybrid energy function. The boost decision tree model was used to predict the unary potentials of both pixels and cloud points.

Similar studies can be found in Refs. [9,13]. In Ref. [9], an ENet framework was proposed to fuse the Lidar point cloud features and RGB images. The multimodal network takes the 2D images and 3D point cloud roughness and porosity as inputs. Results on the seasoned unstructured road dataset demonstrate that the fusion network overweight the single-mode or single-input networks. In Ref. [13], three fusion strategies, namely early fusion, late fusion, and cross fusion, were evaluated. The cross-fusion method that applies trainable parameters between two networks generates the most accurate detection on the standard and challenging road datasets. In Ref. [26], a road driving region detection system was proposed based on the high definition of 3D LiDAR data and the continuity of road in image representation. The efficient representation of LiDAR data and the 2D inverse depth map are projected onto the image plane. The intermediate representations of road scenes can be obtained by extracting the vertical and horizontal histograms of the normalized inverse depth. A row and column scanning strategy in the approximate road region is proposed to refine the detected road area and accurately find the road area.

## Data Processing
### KITTI road dataset
The KITTI dataset was collected around the Karlsruhe area in Germany. The vehicle platform was equipped with two PointGray grayscale cameras, two PointGray color scale cameras, four Edmund Optics lens, one Velodyne HDL-64E laser scanner, and one OTXS RT3000 inertial and GPS navigation system [27]. The sensors are well calibrated and synchronized. The timestamps of the Velodyne Lidar were used as a reference. The camera and Lidar coordinates were defined as follows:

$$\text{camera } [x, y, z] \rightarrow [\text{right, down, forward}] \quad (4.1)$$

$$\text{Lidar } [x, y, z] \rightarrow [\text{forward, left, up}] \quad (4.2)$$

The detailed coordinate of the sensors can be found in Ref. [27].

The dataset used in this study is the KITTI road dataset, which contains 289 annotated samples for model training and 290 unlabeled samples for testing. Each example includes the images from the left of the stereo camera, the right images, gray images, and point cloud

images. The left images are annotated with ground truth values. There are three different road contexts within the dataset, which are urban marked, urban multiple marked, and urban unmarked. The proportion of data split is shown in Table 4.1.

### Lidar camera calibration
The point cloud files in the KITTI dataset were collected with a Velodyne HDL64E rotating 3D Lidar with a 10-Hz scanning rate and 64 beams, which can generate about 120 thousand points. Each point is represented by a four-dimensional vector, which is the spatial coordinate given by the Lidar along with the $X, Y,$ and $Z$ directions and an additional reflectance value of that point. To fuse the point clouds and images together, the Lidar and camera must be collaborated. Specifically, to calibrate the Lidar and camera into the same coordinate, the camera coordinate is used as domain coordinate. Given the point cloud representation $p = [x, y, z, 1]^T$, the point can be transformed into the 2D image plane with the Lidar-to-camera transformation matrix $T$, rectification matrix $R$, and the camera after the rectification projection matrix $P$. The $[u, v]$ coordinates within the image plane can be then calculated as follows:

$$[u, v, 1]^T = \lambda PRTp \quad (4.3)$$

where $\lambda$ is a scaling factor that is determined with (4.3). $P \in R^{3 \times 4}$, $R \in R^{4 \times 4}$ which is expanded by appending a fourth zero-row and column with $R(4, 4) = 1$.

$$T = \begin{pmatrix} R_{\text{velo}}^{\text{cam}} & t_{\text{velo}}^{\text{cam}} \\ 0 & 1 \end{pmatrix} \quad (4.4)$$

where $R_{\text{velo}}^{\text{cam}} \in R^{3 \times 3}$ is the rotation matrix from Lidar to the camera, and $t_{\text{velo}}^{\text{cam}} \in R^{1 \times 3}$ is the translation vector from Lidar to the camera.

In this part, only the points cloud within the image plane will be retained, while those points that are located out of the image plane will be discarded. Figs. 4.2 and 4.3 indicate the calibrated results of the Lidar point clouds and RGB color images.

After the point clouds are transformed into the image domain, it can be used for data fusion. The $[X, Y, Z]$ spatial features are mapped into the image

**TABLE 4.1**
**Categories and Data Size of KITTI Road Dataset.**

| Categories | UM | UMM | UU | Total |
|---|---|---|---|---|
| Training | 95 | 96 | 98 | 289 |
| Testing | 96 | 94 | 100 | 290 |

FIG. 4.2 Lidar camera calibration within the image plane.

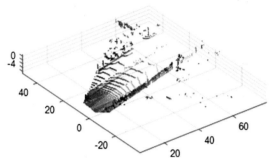

FIG. 4.3 Lidar camera calibration with image color information on the 3D point clouds.

plane, which can be used to generate the $H \times W \times 3$ Lidar images. The usage of RGB and Lidar images for model training is discussed in the next section.

### Model construction
#### Fusion network architectures
In this study, the road detection model is proposed based on SegNet architecture [14]. SegNet segmentation engine consists of encoder layers, decoder layers, and a pixel-wise classification layer. The first few encoder layers SegNet is topologically identical to the VGG16 network, which is responsible for extracting semantic features from the raw images. The encoder network consists of 13 convolutional layers that have the same structures as the VGG16. Each encoder layer has its corresponding decoder layers. The decoder part is designed to transform the low-resolution encoder features into the full-resolution feature map so that the classification layer can make pixel-level identities. The max-pooling indices from the downsampling process are remembered and used to upsample the feature map through the decoder layers.

In this part, the cross-fusion method was chosen to train the deep SegNet model jointly. However, some difference exists between the proposed cross-fusion model and the models shown in Ref. [13]. First, we adopted a SegNet model as our base model in this study, which can make a faster training and inference

than the FCN model. Second, despite applying trainable parameters after each convolutional layer, the cross-parameters are added block by block which is to follow the general structure of SegNet or VGG16. By doing this, the function of each block can be maintained and will be more efficient when inheriting the pretrained model parameters. Last, the number of the cross-parameters can be reduced, and low-resolution inputs are used so that the model can be trained with fewer data and the inference speed would increase.

There are two separate SegNet models that were used to take the RGB images and converted Lidar images, respectively. A detailed model structure is shown in Fig. 4.4. As shown in Fig. 4.4, the cross-parameters are applied after each block to connect the two models dynamically and enable the joint training processing. The models are jointly connected with the following topology.

$$F_{cam, \ N} = Conv\_cam_N \left( F_{cam, \ N-1} + b_{n-1} F_{Lid, \ N-1} \right) \quad (4.5)$$

$$F_{Lid, \ N} = Conv\_lid_N \left( F_{Lid, \ N-1} + a_{n-1} F_{cam, \ N-1} \right) \quad (4.6)$$

where $F_{cam, \ N}$ and $F_{Lid, \ N}$ are the feature maps of the $N$th block of camera SegNet and Lidar SegNet, respectively. The $Conv\_cam$ and $Conv\_lid$ represent the convolutional process of the $N$th block. $a_{n-1}$ and $b_{n-1}$ are the scalar cross-parameters of each model. The two models are connected block by block before the final pixel-wise classification layer. The high-representative features from the two networks are simply summed linearly.

$$F_{tot} = a_9 \ F_{cam,9} + b_9 F_{lid,9} \quad (4.7)$$

where $F_{tot}$ is the final high-level representative map of the road semantic, $a_9$ and $b_9$ are the final layer cross-parameters, $F_{cam,9}$ and $F_{lid,9}$ are the final feature maps from the camera network and Lidar network, respectively.

#### Model optimization
This study focused on a low-resolution solution for the cross-fusion model construction. The computing devices on the vehicles usually have a specific

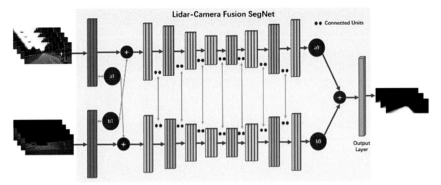

FIG. 4.4 Lidar–camera fusion SegNet framework. The upper SegNet takes the color images as input, and the bottom SegNet takes the X, Y, and Z channels of the point clouds as inputs. The two models are cross-fused block by block (shown in the red circles of the image). The last layer takes the final summed feature vectors and outputs the predicted road segmentation.

requirement on the size, cost, and performance. A road detection system usually can be used to assist other object detection systems such as lane detection, vehicle, and pedestrian detection. A road detection system should not occupy too many computing resources. Therefore in this part, a low-resolution model, which is trained with rescaled images of the Lidar and camera, is proposed. The original image size of the KITTI dataset is $384 \times 1248 \times 3$, which is in a high-resolution format that can require plenty of GPU resources and make it challenging to make inference in real time. Accordingly, we rescale the Lidar and image inputs three times less than the original format and adopted the $100 \times 300 \times 3$ resolutions. This format can speed up the training and testing process significantly. However, the low-resolution model inputs will reduce the model performance on the road detection tasks. Hence, a trade-off between the real-time inference and model accuracy must be determined.

### Experimental Results

In this part, the experimental results for the cross-fusion network-based road detection are analyzed. Then we compared the model performance with different optimization strategies and different fusion methods with the low-resolution inputs.

The road detection results on the three different urban streets are shown in Table 4.2. The network achieved the best results on the urban marked road, followed by the urban multiple marked roads. The model is evaluated based on the six benchmarks given in Ref. [28], which are pixel-wise maximum F-measure (Maxx), average precision (AvgPrec), precision (PRE), recall (REC), false-positive rate (FPR), and false-negative rate (FNR). As manually labels are not provided for the test dataset, the model is evaluated based on a subset of the training data, where 16 μm cases, 16 μmm cases, and 17 μμ cases are randomly selected from the training set, respectively.

The cross-fusion network is also compared with the early fusion and late fusion methods. The learning rate is fixed with exponential dropping in these cases. The comparison between different fusion methods on the low-resolution solution is shown in Table 4.3. As indicated in Table 4.3, the cross-fusion way provides the most accurate inference on the road region. However, the early fusion manner that concatenates the Lidar and camera inputs into the form of $100 \times 300 \times 6$ leads to a better prediction than the late fusion. The cross-fusion network with an exponential learning rate is compared with the constant learning rate (CLR) and multistep learning rate (MLR). As shown in Table 4.3, the exponential learning rate leads to the most accurate prediction accuracy and the network performance with MLR is slightly lower than the exponential methods.

A visualization of the proposed road detection network is shown in Fig. 4.5. As given in Fig. 4.5, most of the road region can be correctly identified by the cross-fusion network. However, the boundary information is not as accurate as of the high-resolution studies in Ref. [13]. We believe a further road boundary detection algorithm can increase the detection rate and provide a more realistic estimation on the road region.

### Conclusion

In this study, a low-resolution solution for road detection is proposed based on the dynamic fusion of Lidar and camera input. The proposed method uses

**TABLE 4.2**
**Road Detection Results on the Different Urban Streets Based on the Cross-Fusion Network.**

|     | MaxF | AvgPrec | PRE | REC | FPR | FNR |
|-----|------|---------|-----|-----|-----|-----|
| UM | 86.20 | 86.82 | 85.64 | 86.77 | 3.13 | 13.22 |
| UMM | 83.38 | 85.71 | 79.45 | 87.71 | 7.85 | 12.89 |
| UU | 81.17 | 84.08 | 78.63 | 83.87 | 3.82 | 16.12 |
| **Ave** | 83.58 | 85.54 | 81.24 | 86.11 | 4.93 | 14.07 |

**TABLE 4.3**
**Road DDetection Results Based on Different Fusion Methods and Learning Strategies.**

|     | MaxF | AvgPrec | PRE | REC | FPR | FNR |
|-----|------|---------|-----|-----|-----|-----|
| Early | 82.68 | 80.50 | 77.55 | 88.82 | 6.49 | 11.26 |
| Late | 81.93 | 82.58 | 78.70 | 85.71 | 6.03 | 14.27 |
| CLR | 67.87 | 50.38 | 52.54 | 98.28 | 20.61 | 1.71 |
| MLR | 73.86 | 57.96 | 60.91 | 94.92 | 13.96 | 5.06 |
| **Cross** | 83.58 | 85.54 | 81.24 | 86.11 | 4.93 | 14.07 |

low-resolution data for joint model training and testing, which can increase the real-time model inference speed with a reasonable identification rate. The fusion method is critical to both structured road and unstructured road context. By evaluating different fusion strategies and learning rate adjustment method, it can be found that the cross-fusion way is still the most efficient manner for the low-resolution case, and the exponential learning rate strategy generates the most accurate model performance. Although the accuracy of the low-resolution road detection model is no better than some of the existing methods, the computation cost in terms of the speed and financial cost is significantly lower, which can provide a reasonable real-time prediction for multiobstacle detection tasks on different kinds of road. Future works will concentrate on investigating more efficient and accurate fusion networks, which are efficient in the unstructured road context.

## LANE DETECTION

Regarding the lane detection module, first, the steerable filter and Hough transform are used as a primary detection algorithm. A secondary algorithm, which combines the Gaussian mixture model (GMM) for unsupervised lane segmentation and random sample consensus (RANSAC) for lane model fitting, is used to detect lanes when the primary algorithm encounters low detection confidence. Then, a Kalman filter is applied to track and smooth the detected lane parameters. To detect the color and line style of ego lane and evaluate the lane detection system in real time, a sampling and voting technique is proposed, and the corresponding drivable area is determined. By combining the sampling and voting system with prior lane geometry knowledge, the evaluation system can efficiently recognize false detections. The proposed real-time lane detection and evaluation systems use a front-facing monocular camera. The system can work robustly in various complex situations (e.g., shadows, night, and lane missing scenarios). The proposed lane detection system is evaluated with both Caltech lanes dataset and Cranfield local dataset, and test results and comparisons are discussed.

### Introduction
Traffic accidents are mainly caused by human operational mistakes, such as inattention, misbehavior, and distraction [29]. A large number of companies and institutes have participated in improving driving safety and reducing traffic accidents. Among all those techniques, road perception and lanes detection are of great significance. Lane detection is the foundation of many

FIG. 4.5 Illustration of low-cost and low-resolution solutions for road detection. The detected road region is shown in green.

advanced driver assistance system (ADAS) products such as lane departure warning and lane-keeping assistance [30–34]. As lane markings are painted on road for human visual perception and the cost of camera devices is acceptable to public transport, thus most of ADAS products adopt vision-based techniques to detect front lanes. Apart from the accuracy and robustness improvement of the lane detection system, another important task is performance evaluation. From the functional perspective, the lane detection system should be aware of inaccurate detections and properly adjust the detection and tracking algorithms and parameters accordingly [33,34]. For ADAS products, when encountering low detection confidence, it should alert the driver and request the driver to concentrate on the driving task. On the other hand, different from ADAS that

allows the driver to intervene the vehicle control, highly automated vehicles are responsible for traffic monitoring all the time and must deal with low-accurate detection by themselves. This makes the self-evaluation of the lane detection system more crucial in high-level automated vehicles.

System integration is an efficient policy for the construction of robust lane detection systems [35]. In this section, the lane detection algorithms, integration, and evaluation methods investigated in the literature are reviewed first, and an integrated robust lane detection system combining two distinct detection methods is proposed. The primary algorithm is built with steerable filter and Hough transforms with the assumption that lanes are straight in most situations in real life. To further improve the system performance, a second algorithm, which is based on GMM lane segmentation and RANSAC, is designed to detect both straight and curve lanes when the first algorithms fail. Then, the lane position parameters are tracked with the Kalman filter. To evaluate the lane detection performance in real time, an evaluation scheme based on prior lane geometry information, such as lane width and lane orientation range, is applied. Next, another evaluation method based on the proposed sampling and voting technique is used to further evaluate the system performance. Instead of detecting all candidate lanes in the image, our work focusses on ego-lane detection and determines whether the left and right areas of ego lane are drivable or not on the basis of the lane-type recognition. The lane types that contain lane color and the line style (e.g., solid, dashed, and double solid) are determined with the lane sampling algorithm. Finally, the integrated lane detection system is tested with a local dataset of Cranfield and a public Caltech lanes dataset. According to the test results, the system achieves steady and robust performance on both datasets. The proposed approach contributes to a robust lane detection with an algorithm-level integration and online evaluation method.

## Popular lane detection techniques

The survey of lane detection algorithms and the general framework of the detection systems have been proposed in Refs. [34–37]. In Ref. [36], McCall and Trivedi defined three-lane detection objectives, which are lane departure warning, driver attention awareness, and automated vehicle control system design. Hillel and Lerner et al. [34] concluded that road color, texture, boundaries, and lane markings are the main perception aspects for human drivers, and lane detection system is one of the major perception tasks of ADAS.

Previous research usually divided lane detection algorithms into two categories, namely the feature-based [38–42] and model-based ones [43–47]. Feature-based methods rely on the detection of lane-marking features such as color, texture, and edges. Then, based on these features, the Hough transform is commonly used to detect straight lanes in the image. Model-based methods, on the other hand, usually assume that lanes can be described with a specific model such as linear model, parabola model, and various kinds of spline models. Then, a lane model can be fitted using RANSAC or other model-fitting methods.

One challenge of the vision-based lane detection system is that its robustness usually cannot meet real-life requirements due to the highly random properties of the traffic and road conditions. Therefore to improve the accuracy and stability of ADAS, the lane detection system is usually integrated with other object detection systems. Previous integration methods can be summarized into three levels, which are algorithm level, system level, and sensor level. Algorithm-level integration has been widely studied in the literature. There are two basic integration architectures, which are the parallel [48,49] and serial [44–47,50] combination types. The parallel-type integration processes multiple algorithms simultaneously, while the serial type usually combines algorithms in sequence. System integration increases the accuracy of lane detection and reduces the false detection rate by combining other detection systems to remove the nonlane pixels. For example, lane detection can be integrated with vehicle and road detections to enhance the overall detection accuracy [28,51,52]. The last integration method is sensory fusion. The integration at the sensor level improves the detection accuracy most significantly as more sensors are used so that the sensing ability of the vehicle is increased. Despite using the monocular camera, multiple cameras, stereo cameras, Lidar, and GPS are also widely used to enhance the lane detection system [7,53–55].

The construction of lane detection evaluation systems is another challenging task. Evaluation of the lane detection system is very complex because of detection methods distinct from hardware to algorithms. Besides, road conditions in different scenarios differ significantly. An accurate lane detection system in a certain area is not guaranteed to be accurate in another area, and the performance can also vary from day tonight. Previous lane detection system usually uses a visual evaluation method to assess the final detection performance because of lacking public dataset and difficulties in labeling ground truth lanes [56]. Popular evaluation methods can be roughly divided into the offline evaluation and online evaluation. The offline evaluation has been widely used in the literature and public datasets, such as those studies conducted in KITTI road and Caltech road [47,54]. Online confidence evaluation often combines road and lane geometry properties and other sensors to evaluate the detection performance and provide a confidence measurement. In Ref. [57], a real-time lane evaluation method was proposed based on the width of the detected lanes. In Ref. [36], a side-facing camera was used to directly sample road lane information and generate ground truth.

In the following, a robust lane detection system with efficient algorithm integration is designed. Two different algorithms are integrated and compensate for each other. Then, drivable area recognition and online system evaluation are proposed based on a lane sampling and voting scheme. The detailed methodology used in this work with test results will be introduced in the next sections.

## Lane Detection System Setup

As aforementioned, lane detection evaluation usually suffers from data variation, which means that there are no existing standards and metrics for lane detection performance evaluation. In this work, a fair performance comparison will be made between the proposed algorithms and the previous Caltech lane detection system [47]. Therefore the open-access Caltech road dataset will be adopted in this study. Meanwhile, a local dataset named Cranfield local road dataset will be used for further evaluation of our system. The Cranfield local road dataset contains the video captured in the motorway of M1 from Cranfield to Northampton. Videos are captured at the frequency of 30 Hz using a CMOS camera mounted in the middle of the dashboard. One common passenger car was used as an experiment vehicle to collect all the data during day and night. All the images are of VGA resolutions (640 ×480). Two datasets are used in this work. In this experiment, the algorithms are implemented in an Intel Core i7 2.5 GHz computer using MATLAB 2016a. Currently, the system works with a frequency of 25 fps in the laboratory environment.

## Algorithm-Level Integrated Lane Detection

In this section, the lane detection algorithms used in this work are introduced. In the first part, the primary lane detection method using Gaussian steerable filter (SF) and hough transform (HT) is described. In the second part, the proposed secondary detection algorithm based on GMM and RANSAC is introduced, which is initiated when the primary algorithm fails or is detected to have low detection confidence according to the lane sampling and voting method. In part C, a Kalman filter is introduced to track the detected lane positions. Lastly, in part D, the sampling and voting scheme for lane-type recognition and lane evaluation is described.

### Lane detection using Sobel filter and Hough transform method

In this work, a primary lane detection algorithm based on the combination of steerable filter and Hough transform is built. The detection process follows a basic lane detection procedure, which consists of image capture and preprocessing edge extraction, and lane detection.

**Image processing.** Three tasks are proposed in this step. The captured images are first cropped with a predetermined ROI, as shown in Fig. 4.1 A. Then, decorrelation stretching is used to enhance the color contrast between lane makings and road surfaces. Lastly, the image is converted to a grayscale image.

**Edge extraction.** A second-order Gaussian steerable filter is used to extract edge information. The steerable filter is able to detect edges in any orientation [58]. The idea behind is that the directional derivative operator of Gaussian function G shown in Eq. (4.8) is steerable.

$$G(x,y) = e^{-\frac{x^2+y^2}{\sigma^2}} \tag{4.8}$$

where $x$ and $y$ are Cartesian coordinates, $\sigma$ is the variance of Gaussian kernel.

The steerable filter is constructed with a set of base filters. Due to the separable properties, it is more flexible and faster than other edge detectors. The second-order Gaussian steerable filter used in this work for edge extraction is defined as follows:

$$G_{xx}(x,y) = \frac{\partial}{\partial x^2}e^{\frac{-(x^2+y^2)}{\sigma^2}} = \left(\frac{4x^2}{\sigma^4} - \frac{2}{\sigma^2}\right)e^{\frac{-(x^2+y^2)}{\sigma^2}} \tag{4.9}$$

$$G_{xy}(x,y) = \frac{\partial}{\partial x}\frac{\partial}{\partial y}e^{-\frac{(x^2+y^2)}{\sigma^2}} = \frac{4xy}{\sigma^2}e^{-\frac{(x^2+y^2)}{\sigma^2}} \tag{4.10}$$

$$G_{yy}(x,y) = \frac{\partial^2}{\partial y^2}e^{-\frac{(x^2+y^2)}{\sigma^2}} = \left(\frac{4y^2}{\sigma^2} - \frac{2}{\sigma^2}\right)e^{-\frac{(x^2+y^2)}{\sigma^2}} \tag{4.11}$$

The combination of these base filters makes steerable filter able to calculate the filter response at arbitrary orientations, and the combination method is shown as follows [58].

$$G_2^\theta = \cos^2(\theta)G_{xx} + \sin^2(\theta)G_{yy} - 2\sin(\theta)\cos(\theta)G_{x,y} \tag{4.12}$$

where $\theta$ represents the orientation angle.

In this work, two steerable filters, which detect left and right lane edge features, are applied individually in Fig. 4.6C and D represent the lane features given by steerable filter with two different orientation responses.

**Hough transform.** Hough transform line detection has been successfully used for lane detection in previous studies and has been proved efficiency in lane detection. It was first designed to recognize lines [59] and then improved to detect arbitrary shapes [60]. It is an efficient line detection method, which can be applied to binary images.

To apply standard Hough transform, lanes are assumed to be straight, which is true in most situations. In terms of curve lanes, lanes in the near field of the

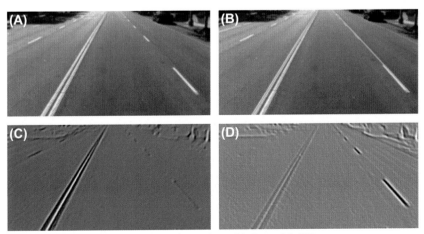

FIG. 4.6 **Lane Detection with Steerable Filter.**

image can be viewed as a straight lane and can be detected with Hough transform. Therefore the Hough transform is used to detect lanes in the primary algorithm for fast and rough lane detection as no model fitting is required. The Hough transform converts models in Cartesian coordinate into Polar coordinate, such that the straight-line model in Eq. (4.13) can be transformed into Eq. (4.14) or Eq. (4.15).

$$y = k \cdot x + b \tag{4.13}$$

$$\rho = x \cos \theta + y \sin \theta \tag{4.14}$$

$$y = -\frac{\cos \theta}{\sin \theta} x + \frac{\rho}{\sin \theta} \tag{4.15}$$

where $\theta \in [0°, 180°]$ and $\rho \in R$, $\theta$ is the line orientation in Polar space, and $\rho$ is the distance between the original point and the line in Polar space. $k$ and $b$ are the slopes of the and the Y-intercept line in Cartesian space, respectively.

The process of the primary lane detection algorithm is illustrated in Table 4.4.

Lines in the original image are transferred to single points in the Polar coordinate, and this point can be represented as a sinusoid line in Hough space. An accumulator array that contains the votes of the number of sinusoid Hough lines that pass through a point is used. The points with most Hough lines intersect are determined as a line in the edge map Fig. 4.6B shows the result of lane detection given by SF and HT. To make a fair illustration and comparison, all the illustration images in this part are selected from Caltech public dataset [47].

## Lane detection using Gaussian mixture model and RANSAC method

GMM and RANSAC are combined as the secondary lane detection system when the primary algorithm fails, or low detection confidence is returned. Hough transform-based method has been proved efficient for lane detection in a vast amount of the previous literature. However, it is difficult to process curve lanes. Meanwhile, in the case to increase the diversity of the lane detection algorithms, different image segmentation and model fitting-based method, which integrates the GMM and RANSAC, will be used as a secondary algorithm to enlarge the robustness of the system. The Hough transform-based algorithm can be viewed as a feature-based method, while the RANSAC method belongs to the model-based algorithm. Therefore by dynamically integrating the feature-based algorithm and model-based algorithm, the lane detection system will have the advantages given from these two different algorithms. In this part, GMM is first used to segment lane markings, and RANSAC is then applied to fit both straight and curve lane models based on the features given by GMM. A slightly smaller ROI than the one used in the first part is used to increase the detection accuracy and reduce the false detection rate of lane pixels.

**GMM-based feature extraction.** GMM is an unsupervised machine learning method, which can be used for data clustering and data mining [61]. In this part, GMM is used to cluster the image into different parts, such as roads, lanes, and shadows. In GMM, the

---

**TABLE 4.4**
**The First Lane Detection Procedure.**

**Lane Detection Algorithm Based on Hough Transform**

**Input**: Raw image
**Outputs**: Straight lane parameters
**Process**:
1. Crop the raw image with predetermined ROI.
2. Enhance color contrast using decorrelation stretching.
3. Convert RGB image to grayscale image.
4. Perform second-order steerable filter to find lane edge map.
5. Using Hough line transform to detect left and right lanes.
6. **if** slopes and intercept meet lane geometry constraint
7. Lanes are found.
8. **else**
9. Lane detection error.
10. **end if**
11. Return output value.

distribution of the input data is a mixture of a set of Gaussian distribution as shown in the following:

$$p(x) = \sum_{k=1}^{K} \pi_k N\left(x \middle| \mu_k, \sum\nolimits_k\right) \qquad (4.16)$$

where $x$ represents data point, $K$ is the number of components, and $\mu$ and $\sum$ are the mean and covariance parameter of multivariate Gaussian function. Each component follows a multivariate normal distribution format that can be described as

$$N\left(x \middle| \mu, \sum\right) = \frac{1}{(2\pi)^{d/2} \sum^{1/2}} exp\left(-\frac{1}{2}(x-\mu)^T \sum\nolimits^{-1}(x-\mu)\right)$$
$$(4.17)$$

where $\pi_k$ is the weight for each Gaussian component and their sum equals to one, and $0 \leq \pi_k \leq 1$.

The feature vector of GMM is a four-element vector that contains the RGB values and the Sobel magnitude value to minimize the nonedge noise.

$$x(k) = [R(u,v), G(u,v), B(u,v), S(u,v)] \qquad (4.18)$$

where $x$ is the feature vector of a cropped ROI image, $k$ is the index of $x$. $(u,v)$ is image coordinate.

The GMM-based feature extraction performance is examined by comparing GMM with Otsu's segmentation method and examine GMM with the different number of cluster centers. First, the road segmentation result is compared with Otsu's method [62]. Otsu's algorithm segments the image by searching an optimal threshold in the grayscale image and has been widely used in many studies. When it is difficult to segment the lane and road as shown in Fig. 4.7A, Otsu's segmentation fails to extract the lane pixels, as shown in Fig. 4.7B. However, GMM can efficiently remove the road texture noise and extract the true lane pixels, as illustrated in Fig. 4.7C.

A GMM with two cluster centers sometimes clusters nonlane pixels into the lane group, such as the bright area shown in the top-left corner in Fig. 4.7C. However, the effect of these lane outlier pixels can be removed by the following RANSAC algorithm. In addition, another powerful property of GMM-based lane segmentation is that shadows can be removed by controlling the number of cluster centers. By increasing the number of clustering centers, GMM will generate a more accurate segmentation result as shown in Fig. 4.8. The subplots (C) and (E) are the segmentation results obtained with three and four cluster centers. Graphs (D) and (F) are the corresponding binary images that only contain lane and nonlane pixels.

FIG. 4.7 **A Segmentation Comparison Between Otsu's Method and Gmm in a Complex Texture Road.**

**RANSAC model fitting.** RANSAC algorithm is widely used in computer vision systems due to the good robustness property. RANSAC iteratively estimates the parameters of the given model by dividing data into outlier and inlier sets. A sample subset containing the minimal points to describe the model is chosen iteratively. Then, all the data are tested with the fitted model, and data items that are consistent with the model within a certain threshold are selected as inliers. RANSAC keeps fitting the given model unless the maximum iteration number is reached or the best model that has most inliers is found.

RANSAC algorithm is used to fit a Hyperbola-pair lane model in this work. Lane models are divided into the near-field and far-field part. Near-field model is a common straight-line model as shown in Eq. (4.19), and a hyperbola model, which is given by Eq. (4.19) [43], is used to represent lanes in far-field part:

$$u = \frac{C}{v-h} + k \times (v-h) + vp \qquad (4.19)$$

where $(u,v)$ is the coordinate in the image plane, $C$ is the road curvature in far-field, $k$ is the slope of the straight line in near filed, $h$ is the $y$-coordinate of road horizon, and $vp$ denotes the estimated vanishing point of straight lines (Fig. 4.9).

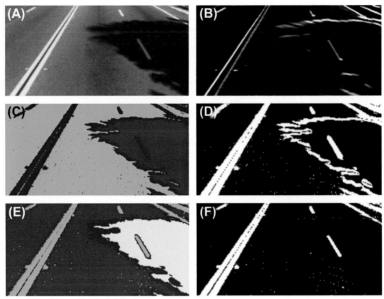

FIG. 4.8 **Results of Lane Segmentation Using Gmm on Shadow Road.** (**A**) is raw RGB image, (**B**) is the magnitude image calculated with Sobel operator, (**C**) and (**E**) are GMM segmentation result with 3 and 4 clustering centers while (**D**) and (**F**) are the binary value images of (**C**) and (**E**), respectively. As shown in this graph, the more clusters used, the more accurate lane extraction can be achieved.

FIG. 4.9 **Caltech Road Lane Detection Result Based on the Hyperbola-Pair Model.** *Red line* represents the estimated horizon line in the image, *blue line* is the estimated straight line for both right and left line, the intersection of *blue lines* and *red horizon line* is regarded as the lane vanishing point, and *green lines* are the estimated lanes for both left and right lanes following road curvature. The left and right lane models are fitted using RANSAC, separately. Fig. 4.8 shows an example of curve lane detection results based on GMM and RANSAC lane detection. Table 4.5 illustrates the lane detection procedure using the proposed GMM and RANSAC algorithms.

### Lane tracking with Kalman filter

Lanes are tracked with Kalman filter after a successful detection to smooth and improve the stability of the detection results. Kalman filter is used for many applications such as noise removal, generating nonobservable states, and object navigation and tracking. It provides the optimal state estimation for the linear systems. Kalman filter involves two steps, prediction and update. The prediction step uses previous states to estimate the current states, along with their uncertainties. The update step uses the current measurements to correct the estimation results given by the prediction step using a weighted average that gives more weights to the estimates with high certainty. The prediction and update steps run in a recursive way and estimate the optimal states at each time step.

The Kalman filter can be simply described as the following process.

First, based on the linear dynamic model, the Kalman filter model assumes that the true states at time $t$ have the following relationship with the states at $t-1$:

$$x_k = F_k x_{k-1} + B_k u_k + w_k \qquad (4.20)$$

where $F_k$ is the state transition model, $B_k$ is the control-input model, and $w_k$ is the process noise with zero

---

**TABLE 4.5**
**Second Lane Model-Fitting Procedure.**

**GMM and RANSAC Lane Model-Fitting Algorithm**

**Input:** ROI image
**Outputs:** Lane curvatures $C_l, C_r$, straight-line models of left and right lanes
**Process:**
1. Construct feature vectors based on the ROI image.
2. Fed into the GMM clustering module.
3. Extracted lane features from GMM result and binarization.
4. Fit near-field straight-line model and calculate slopes of both left and right lanes if no slopes exist.
5. **if** slopes and intercept meet lane geometry constraint
6. Fit far-field curve line model using the slopes.
7. **else**
8. Lane model-fitting error.
9. **end if**
10. Return output value.

---

mean multivariate normal distribution $N$, with covariance $Q_k$.

The observation $z_k$ at each time step is described as follows:

$$z_k = H_k x_k + v_k \qquad (4.21)$$

where $H_k$ represents the observation model and $v_k$ is the Gaussian noise model that exists in the measurement with the covariance of $R_k$.

The two-stage process can be described as follows.

At predict step, the state and error covariance is estimated,

$$\widehat{x}_{k|k-1} = F_k \widehat{x}_{k-1|k-1} B_k u_k \qquad (4.22)$$

$$P_{k|k-1} = F_k P_{k-1|k-1} F_k^T + Q_k \qquad (4.23)$$

In the update stage, the system output will be corrected according to the estimated *prior* state as well as the measurements.

$$\widetilde{y}_k = z_k - H_k \widehat{x}_{k|k-1} \qquad (4.24)$$

$$S_k = R_k + H_k P_{k|k-1} + H_k^T \qquad (4.25)$$

$$K_k = P_{k|k-1} H_k^T S_k^{-1} \qquad (4.26)$$

$$\widehat{x}_{k|k} = \widehat{x}_{k|k-1} + K_k \widetilde{y}_k \qquad (4.27)$$

$$P_{k|k} = (I - K_k H_k) P_{k|k-1} (I - K_k H_k)^T + K_k R_k K_k^T \qquad (4.28)$$

$$\widetilde{y}_{k|k} = z_k - H_k \widehat{x}_{k|k} \qquad (4.29)$$

where $\widetilde{y}_k$ is the innovation prefit residual, $S_k$ is the innovation covariance, $K_k$ represents the Kalman gain, and

$\widehat{x}_{k|k}$ and $P_{k|k}$ are the estimated posterior state and covariance, respectively. $\widetilde{y}_{k|k}$ is the measurement postfit residual.

In this part, the Kalman filter is used to track and smooth the detected lane parameters. The lane detection model is selected as a simple constant velocity model, which ignores the variation of the vehicle velocity. The discrete state-space model for the Kalman filter is described as follows:

$$x[t] = Fx[t-1] + w(t-1) \qquad (4.30)$$

$$y[t] = Hx[t] + v(t) \qquad (4.31)$$

The selection of the covariance matrices did not rely on complex covariance matrix estimation algorithms such as the autocovariance least-square technique; instead, they are selected merely based on manual selection and the performance are simplify verified according to the visual evaluation on the public dataset and Cranfield local dataset. Kalman filter tracks the lane positions with a linearized system dynamic model and Gaussian noise to estimate the optimal states at each time step. In this work, the left and right lanes are tracked separately while using the same system model. For the straight lane case, the detected line slope and intercept are tracked. For curve lane case, extra curvature parameters $C_l, C_r$ and their variation is added. State vector $x(t)$ and output vector $y(t)$ of the straight lane model is shown as follows:

$$X = \left[ k(t), \dot{k}(t), b(t), \dot{b}(t) \right]^T \qquad (4.32)$$

$$Y = [k(t), b(t)]^T \qquad (4.33)$$

FIG. 4.10 **Illustration of Lane Sampling and Voting.** **(A)** The traffic image in the YCbCr color space, **(B)** The sampling points on both lanes in the original image. **(C)** and **(D)** are the sampling lines generated based on the sampling points in original and edge images, respectively.

where $t$ denotes for $t$, $k$, and $b$ are the slope and intercept, $\dot{k}$ and $\dot{b}$ are the derivatives. The state transition matrix $F$ and the output matrix $H$ are defined as follows:

$$F = \begin{bmatrix} 1 & \Delta T & 0 & 0 \\ 0 & 1 & 0 & 0 \\ 0 & 0 & 1 & \Delta T \\ 0 & 0 & 0 & 1 \end{bmatrix} \quad (4.34)$$

$$H = \begin{bmatrix} 1 & 0 & 0 & 0 \\ 0 & 0 & 1 & 0 \end{bmatrix} \quad (4.35)$$

### Lane sampling and voting for lane recognition

In this part, a lane-type recognition method based on lane sampling and voting (LSV) is proposed. Lane type contains two properties that are the color and line style of the lanes. Three lane styles, namely the solid, double solid, and dashed ones, and two-lane colors, that is, yellow and white, are recognized. A set of sampling points with the same interval is selected according to the detected lanes. The two recognition steps are explained in detail as follows.

**Lane color detection.** To detect lane colors of both the left and right lanes, the RGB images are first converted into the YCbCr color space to increase the contrast between yellow and white color. In addition, converting raw images to YCbCr space improves the detection efficiency, as the two colors can be efficiently distinguished only with the Cb channel of the YCbCr images. The subgraphs **(A)** and **(B)** in Fig. 4.10

illustrate the image in YCbCr space and the sampling points on the lanes. Fig. 4.11. shows the value of the sampling points in the Cb space. A clear boundary between the yellow lane and the white lanes in the Cb image can be found. Then, lane color is determined according to the voting results of the sampling points. If the values of most sampling points are below the threshold, the lane is determined to be yellow. Otherwise, the color is determined to be white.

**Lane-type detection.** After the lane sampling points have been determined, the line style of left and right

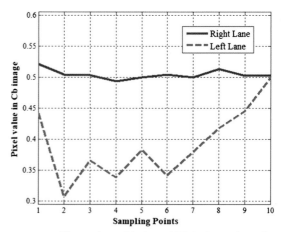

FIG. 4.11 **Illustration of Left and Right Lane Sampling Points in the Cb Image.**

lanes can be recognized by extending the sampling points into short sampling lines. The line width of each sampling line is chosen as about three times the original one to cover the double lanes. Besides, the width of the sampling lines can control the robustness of the style recognition algorithm. Sometimes the detected and tracked lanes will drift or become inaccurate, due to the disturbances on road, such as a nearby vehicle appears. A slightly longer sampling line can still cover and recognize the line style at this moment. The scanning is proposed on the edge image given by the Sobel edge detector. The lane style is determined by counting the rising edge along each sampling line. Fig. 4.10C and D illustrate the sampling lines in the raw and edge image. Fig. 4.12 shows the value distribution of the sampling line points on the edge image. As shown in Fig. 4.12, the double-solid left lane is expected to have four maximums. The dashed right lane consists of two parts, which are a short single-solid line segment and no line segment, respectively. The single-solid line part is expected to have two maximums while the empty segment should have zero values along the corresponding sampling line. The line style determination logic based on the LSV method is shown in Table 4.6.

## Lane Algorithms Integration and Evaluation

In this section, the approach used to integrate the proposed algorithms is described in detail. As mentioned in Section II, algorithm integration methods in the literature can be divided into parallel and serial methods. In this work, the two existing detection algorithms are combined in a dynamic manner to create a robust lane detection and evaluation system.

### Integration of lane detection algorithms

The main processing procedure follows a serial sequence. The GMM-based model-fitting algorithm is applied as a redundancy algorithm to increase the robustness of lane detection in a dynamic way. Instead of running multiple algorithms simultaneously, the GMM-based detection system will only be activated when the first algorithm reports low detection confidence. The real-time evaluation system combines prior lane geometry knowledge and the proposed LSV method. Fig. 4.13 illustrates the architecture of the proposed integration and evaluation system.

Hough line detection is used as the primary lane detection method, and GMM and RANSAC method is used as a redundancy one that is only activated in certain situations. There are two situations when the secondary lane detection system needs to be activated. The first case is that Hough line detection continuously fails to meet the road geometry constraints in a period. To reduce the impact of noisy measurement, the secondary lane detection system is initiated, and the detection results are tracked with the Kalman Filter. The other condition to activate the second algorithm is when the LSV system detects low confidence in current measurement. At this moment, the detection results given by the two-lane detection systems are evaluated together. If straight-line parameters are within a vicinity, then the detection result is adopted; otherwise, the low detection confidence is reported to the user, and current tracked results are maintained. The HT-based lane detection system will be reinitiated if the estimated curvatures are small enough.

### Integration of lane detection algorithms

The online evaluation system combines the prior lane geometry constraints and the lane sampling method. In this study, as inverse perspective mapping is not used in this work, lanes cannot be assumed to be parallel. However, the left and right lanes still meet some geometry constraints. For example, normally the detected slope of the left and right lanes with the straight-line model are opposite to each other and their summation is around zero. In this work, the prior lane position knowledge is used as the first evaluation metric. The prior knowledge contains the reasonable line slopes and the intercepts of both lanes. These parameters can be determined after mounting the front-face camera in the vehicle. If the camera does not change its focal length or other parameters automatically, the adjacent

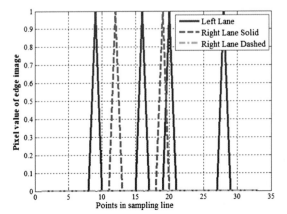

**FIG. 4.12 Sampling Point Values Along the Sampling Lines.**

**TABLE 4.6**
**Lane-Type Recognition Procedure.**

**Lane Points Sampling and Voting for Lane Recognition**

**Input**: ROI image, lanes position
**Outputs**: Lane line style and color
**Process**:
1.  Convert RGB image into YCbCr color space.
2.  Edge extraction and binarization of RGB image.
3.  Lane points sampling based on lanes position.
4.  **for** lines equal to the left lane and right lane
5.     **if** more points value in Cb image larger than the threshold
6.        Lane color is determined as white.
7.     **else**
8.        Lane color belongs to yellow.
9.     **end if**
10. **end for**
11. Generate scanning lines based on sampling points.
12. **for** lines equal to the left lane and right lane
13.    **if** the most minimum value of sampling lines equals to zero
14.       Lane style belongs to the dashed line.
15.    **else if** most scanning lines values less than four
16.       Lane style belongs to a solid line.
17.    **else if** most scanning lines values larger or equals to four
18.       Lane style belongs to double line
19.    **else if** all scanning lines values equal to zero
20.       Detection fails.
21.    **end if**
22. **end for**
23. Return output value.

lanes are expected to appear in a certain range of locations. When the detected line parameters locate in a reasonable interval (i.e., $\left|\frac{1}{3}k,\ 3k\right|$ and $\left|\frac{1}{3}b,3b\right|$ in this work), the detection results can be seen as reasonable. The reasonable interval is not necessary to be very restricted, because the results will be further filtered with a Kalman filter. When the primary lane detection continuously fails within a period, the secondary algorithm will be activated. A similar limitation is also applied to the second algorithm. Another assumption is that the reasonable road curvature should be less than 90 degrees and the two lanes' curvatures are in the same direction. Fail detection is marked when the primary detection system continuously fails, and the secondary algorithm returns unreasonable results.

The second evaluation method is based on the proposed LSV method. A set of points will be sampled along each lane and acts as virtual sensors. As mentioned in Section III part D, LSV is used to recognize the lane types. In this part, this method is used to evaluate the lane detection system. The idea is that all the

points should locate on the real lanes ideally. If the tracked lanes drift due to false detections or road curvature changes, the sampling points will no longer locate on the lanes anymore.

In this part, both the sampling points and the extended scanning lines are used to evaluate the detection result. The distance between two parallel points on the adjacent lanes is normally within a certain range. This distance is used to meet the ego-lane constant width constraints. When the distance is out of range, low confidence is reported. As mentioned before, scanning lines proposed on the edge image are used to detect the lanes. If detected lanes drift a little from the real position, the scanning is still able to find the peak values. However, if the detected results are far away from the real position, there will be no peak values along the line and the values of each point in the line equal to zero, as the bottom green dashed line shown in Fig. 4.12. At this time, low confidence in the current detection will be reported, and the secondary algorithm will be initiated. Fig. 4.14 shows the tracked lane

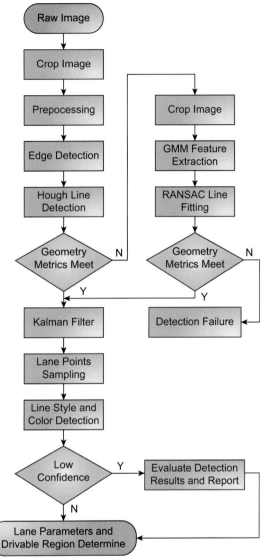

FIG.   4.13 **Integrated   Lane   Detection   System Architecture.**

FIG.   4.14 **Compensation   of   the   Secondary   Lane Detection Algorithm.**

the red lines maintain a correct prediction of the road curve. Low confidence is reported if a certain number or all the points vote. The threshold and the number of sampling points can be adjusted to control the sensitivity of the evaluation system. When the evaluation system reports low confidence, the detection results given by different algorithms will be compared. If the result given by the secondary algorithm meets the lane geometry constraints, the detection results will be updated, and tracking will be kept until the road can be modeled as a straight line.

### Experimental Results

To make a fair comparison and evaluation, in this section, our algorithm is compared with the one reported in Ref. [44] using Caltech public dataset. There are four clips of sequences in the dataset with 1224 manual labeled in total.

The proposed algorithms are evaluated using the ground truth lanes in the dataset and the same lane similarity measurement method as described in Ref. [44]. Lane detection system based on HT only and the integrated method based on HT and GMM are evaluated separately. Different from previous research, road curbs are regarded as a lane boundary in this work. This is a reasonable assumption, which has been discussed in Ref. [44]. In addition, it can be found in the dataset that sometimes lanes are not labeled when the lanes are close to the road curb. Therefore the detected lanes using the proposed algorithm are with more numbers than that of the ground truth as shown in Table 4.7. For example, the quantity of our detected lanes in clip two is significantly larger than the given number in Table 4.7. The reason is that some lanes are not given a

positions given by Kalman filter (the green lines) drifted when facing curve road, and the sampling points fail to detect the lanes, especially the left lane. Thus GMM-RANSAC is activated to detect lanes by fitting a curve lane model, as the red lines shown in Fig. 4.14.

The green dashed line is a false detection given by the HT detection and tracking system, and the red solid line shows the detection results given by the secondary GMM based lane detection algorithm. Although road curb in the right affects the accuracy of the right lane,

**TABLE 4.7**
**Detection Result Using the Algorithm in Ref. [44].**

| Clip | Frames | Total | Detected | Correct Rate (%) | False Pos. Rate (%) |
|---|---|---|---|---|---|
| 1 | 250 | 466 | 467 | 97.12 | 3.00 |
| 2 | 406 | 472 | 631 | 96.16 | 38.38 |
| 3 | 336 | 639 | 645 | 96.70 | 4.72 |
| 4 | 232 | 452 | 440 | 95.13 | 2.21 |
| Total | 1224 | 2026 | 2183 | 96.34 | 11.57 |

ground truth value because they are divided into road curbs.

In this work, lane tracking after the detection process to keep tracking the lane position and minimize the influence of noise detection. This is another reason that our system detects more lanes than the previous method. By using the tracking algorithm, the lanes are tracked when there are no detected lanes, such as at intersections.

Table 4.8 shows the performance of the proposed integrated algorithm. The results given by the primary algorithm with tracking achieve comparable performance with previous research. When applying the redundancy algorithm, the correct rate increases, which shows that the secondary algorithm can compensate for the primary algorithm. The general performance of the proposed algorithm achieves a high precision in both the Caltech road dataset and the local one. The main issue of the result is that a high value of false-positive detection rate occurs in the second clip. As explained before, this is because some lanes, which are close to the curb, are not labeled and curbs are not regarded as a lane in the dataset. However, curb recognition is as important as lane detection. Therefore this high value of false positive is not a big problem in real-world applications.

*Experimental results*
Some result samples of the proposed integrated lane detection system are shown in Figs. 4.15 and 4.16. Fig. 4.15 shows some successful detections according to Caltech and Cranfield datasets. Caltech dataset was captured in the urban area, while the Cranfield dataset was captured in both urban and highway roads. To make a fair illustration of the proposed system, we majorly adopt Caltech images for verification. As shown in Fig. 4.16 the integrated lane detection system is robust to shadows, surrounding vehicles, curvy roads, intersections, and other road markings. The system

can successfully detect lanes with large shadows on the road and is robust to nonlane road markings. Besides, with the redundancy algorithm and tracking algorithm, the system can successfully detect lanes with poor visualization, for example when the camera is facing toward the sun. However, this monocular vision-based algorithm still has its limitations under some conditions.

Fig. 4.16 shows some samples when the algorithm cannot find the lanes correctly. The first case shows the drawback of the curvy road detection algorithm. Hyperbola model is a second-order model, which is suitable for most curvy lane conditions. However, it cannot give a precise estimation of very complex lanes comparted with higher-order models. There always exists a trade-off between computational efficiency and accuracy. Most curvy lanes in the real world, especially on the highway, can be modeled with a hyperbola model for faster lane model fitting purposes. Therefore in this work, the second-order hyperbola model is chosen to fit curvy lanes. The second case shows that lane detection would fail when there are no clear lane markings because of the reflection of light or shadows. However, in these situations, it is also very difficult to distinguish the lane markings even with human eyes. Therefore both of the two detection algorithms fail to find the lane. The third case shows that no lane marking is next to the road boundary and the detection and tracking system choose the guardrail as the lane boundary. Guardrail and lane marking distinction are one of the most common and difficult tasks because guardrail has similar properties with lanes from the perspective of computer vision. In this case, the distance is beyond the predetermined range and the secondary algorithm is imitated to find the lane boundaries. The last two case shows that no lane marking is in the middle of the road and the detection system finds two wider lanes. As there is no clear ego-lane marking, the secondary algorithm cannot find a correct lane position.

**TABLE 4.8**
**The Detection Result of Proposed Methods.**

| | | | HT ONLY | | HT + GMM | |
|---|---|---|---|---|---|---|
| Clip | Total | Detected | Correct Rate (%) | False Pos. Rate (%) | Correct Rate (%) | False Pos. Rate (%) |
| 1 | 466 | 492 | 94.85 | 15.45 | 98.50 | 7.98 |
| 2 | 472 | 806 | 92.16 | 74.2 | 96.19 | 70.40 |
| 3 | 639 | 669 | 93.15 | 9.54 | 97.18 | 7.51 |
| 4 | 452 | 459 | 94.47 | 7.08 | 98.01 | 3.54 |
| Total | 2026 | 2434 | 94.37 | 25.77 | 97.58 | 22.56 |

## Discussion

The proposed two algorithms in this work are expected to enrich the system diversity. Robust straight lane detection and tracking system are used as the primary algorithm under the assumption that in most situations lanes can describe with straight-line models. This is reasonable in real life as curvy roads do not appear frequently in most of the situations. Besides, in the near field part of curves, lanes can be modeled as straight ones. Moreover, this method is computational efficiency and can be easily implemented in the on-board control system. The GMM used in the secondary detection algorithm has been successfully applied to many image segmentation tasks. As shown in Section III part B, GMM can distinguish the road, shadow, and lane markings accurately. Features extracted by GMM carry more information, which not only gives accurate estimations of lane markings but is also able to segment the road and shadows. Another property of GMM and RANSAC model-fitting method is that it can find lane marks with very low image resolution due to the robustness characteristics. The proposed algorithm is also tested with images in the format of $60 \times 80$ pixels, and the algorithm is still able to find lanes in most situations. However, color-based image segmentation can be affected by those objects that have similar color distribution. In that case, the extracted lane mark pixels can be noisy. Therefore a smaller ROI image is used in this work to reduce the noise influence.

Lane-type recognition is proposed based on the sampling points and lines. Currently, the lane type is estimated frame by frame, which increases the computation burden. In real-world applications, lane type is not necessary to be detected at each step. A more efficient way is to combine a lane departure or lane change aware system and only initiate lane-type recognition after a lane-change maneuver.

Previous lane detection evaluation method often uses visual check without quantity measurement. This is mainly because ground truth labeling is a tough work that costs many human resources. Another reason that makes the lane detection system difficult to be evaluated that is no common metric exists in the literature. This gives rise to the utilization of different testing images and ROI. Therefore an adaptive lane detection system, which can work under various application situations without or with only a few modifications, is expected to be developed. For example, an automatic lane marking recognition can be combined with a road area detection system. There are many approaches to achieve an adaptive lane detection system, and with such a system, lane detection will be much easier to be evaluated and implemented.

## Conclusions

In this work, a novel real-time integration and evaluation architecture toward a robust lane detection system is proposed. Robust lane detection is achieved by increasing the diversity of detection algorithms. Two distinct lane detection algorithms based on Hough transform and GMM-RANSAC model-fitting methods are integrated in an efficient way. The detected lane positions are further tracked and smoothed with the Kalman filter. Instead of detecting all the lane markings on the road, the boundaries of ego lanes with corresponding lane type and drivable area recognition are constructed. This will reduce the computation cost compared with the methods that detect all the candidate lanes in the image. Finally, a real-time lane detection evaluation method is constructed based on the combination of prior lane geometry information and LSV scheme. The contribution of this part is as follows:

GMM-based lane feature extraction is proposed and analyzed in detail. The results show that with more

**FIG. 4.15** **Detection Samples Show the Robustness and Accuracy Under Different Conditions.**

**FIG. 4.16** **Inaccurate and False Detection Samples Showing The Influence of Curbs, No or Poor Lane Marks, Lane Changing and Model Weakness.**

cluster centers, GMM can classify road, lane marks, shadow, and other objects accurately.

A novel efficient integration method to combine two-lane detection algorithms is designed. Different from previous parallel architecture, the second algorithm is only activated when the primary algorithm continuously fails or low detection confidence occurs.

A novel online evaluation system based on prior lane geometry constraints and sampling and voting scheme is proposed. The system continuously evaluates the lane detection result and tries to keep the detection as accurate as possible.

The lane detection and evaluation system designed in this work uses a monocular camera and algorithm-level integration to achieve an accurate and robust detection performance in the real world. However, still some limitation exists in both hardware and software

aspects because the camera cannot deal with all the kinds of serious situations. Future works will combine lane detection systems with other object detection system as mentioned in Section II to increase the robustness of the lane detection system. Besides, more works are expected to be done toward an efficient real-time lane detection evaluation system.

## REFERENCES

[1] M. Mokhtarzade, M.J.V. Zoej, Road detection from high-resolution satellite images using artificial neural networks, International Journal of Applied Earth Observation and Geoinformation 9 (1) (2007) 32–40.

[2] J.M. Alvarez, T. Gevers, A.M. Lopez, 3D scene priors for road detection, in: 2010 IEEE Computer Society Conference on Computer Vision and Pattern Recognition, IEEE, 2010.

[3] Y. Xing, et al., Advances in vision-based lane detection: algorithms, integration, assessment, and perspectives on ACP-based parallel vision, IEEE/CAA Journal of Automatica Sinica 5 (3) (2018) 645−661.

[4] R. Labayrade, D. Jerome, D. Aubert, A multi-model lane detector that handles road singularities, in: 2006 IEEE Intelligent Transportation Systems Conference, IEEE, 2006.

[5] A.V. Nefian, G.R. Bradski, Detection of drivable corridors for off-road autonomous navigation, in: 2006 International Conference on Image Processing, IEEE, 2006.

[6] D. Song, et al., Vision-based motion planning for an autonomous motorcycle on ill-structured roads, Autonomous Robots 23 (3) (2007) 197−212.

[7] Q. Li, et al., A sensor-fusion drivable-region and lane-detection system for autonomous vehicle navigation in challenging road scenarios, IEEE Transactions on Vehicular Technology 63 (2) (2014) 540−555.

[8] Y. Xing, et al., Dynamic integration and online evaluation of vision-based lane detection algorithms, IET Intelligent Transport Systems 13 (1) (2018) 55−62.

[9] D.-K. Kim, et al., Season-invariant semantic segmentation with a deep multimodal network, in: Field and Service Robotics, Springer, Cham, 2018.

[10] A.B. Hillel, et al., Recent progress in road and lane detection: a survey, Machine Vision and Applications 25 (3) (2014) 727−745.

[11] L. Zhang, et al., Road crack detection using deep convolutional neural network, in: 2016 IEEE International Conference on Image Processing (ICIP), IEEE, 2016.

[12] Z. Zhang, Q. Liu, Y. Wang, Road extraction by deep residual u-net, IEEE Geoscience and Remote Sensing Letters 15 (5) (2018) 749−753.

[13] L. Caltagirone, et al., LIDAR−camera fusion for road detection using fully convolutional neural networks, Robotics and Autonomous Systems 111 (2019) 125−131.

[14] V. Badrinarayanan, A. Kendall, R. Cipolla, Segnet: a deep convolutional encoder-decoder architecture for image segmentation, IEEE Transactions on Pattern Analysis and Machine Intelligence 39 (12) (2017) 2481−2495.

[15] M. Bertozzi, A. Broggi, GOLD: a parallel real-time stereo vision system for generic obstacle and lane detection, IEEE Transactions on Image Processing 7 (1) (1998) 62−81.

[16] H. Kong, J.-Y. Audibert, J. Ponce, General road detection from a single image, IEEE Transactions on Image Processing 19 (8) (2010) 2211−2220.

[17] J.M.Á. Alvarez, A.M. Lopez, Road detection based on illuminant invariance, IEEE Transactions on Intelligent Transportation Systems 12 (1) (2011) 184−193.

[18] Y. Yuan, Z. Jiang, Q. Wang, Video-based road detection via online structural learning, Neurocomputing 168 (2015) 336−347.

[19] G.L. Oliveira, W. Burgard, T. Brox, Efficient deep models for monocular road segmentation, in: 2016 IEEE/RSJ International Conference on Intelligent Robots and Systems (IROS), IEEE, 2016.

[20] Q. Wang, J. Gao, Y. Yuan, Embedding structured contour and location prior in siamesed fully convolutional networks for road detection, IEEE Transactions on Intelligent Transportation Systems 19 (1) (2018) 230−241.

[21] C.C.T. Mendes, V. Frémont, D.F. Wolf, Exploiting fully convolutional neural networks for fast road detection, in: 2016 IEEE International Conference on Robotics and Automation (ICRA), IEEE, 2016.

[22] L. Caltagirone, et al., Fast LIDAR-based road detection using fully convolutional neural networks, in: 2017 IEEE Intelligent Vehicles Symposium (IV), IEEE, 2017.

[23] H. Dahlkamp, et al., Self-supervised monocular road detection in desert terrain, Robotics: Science and Systems 38 (2006).

[24] A. Asvadi, et al., Multimodal vehicle detection: fusing 3d-lidar and color camera data, Pattern Recognition Letters 115 (2018) 20−29.

[25] L. Xiao, et al., Hybrid conditional random field based camera-lidar fusion for road detection, Information Sciences 432 (2018) 543−558.

[26] S. Gu, et al., Histograms of the normalized inverse depth and line scanning for urban road detection, IEEE Transactions on Intelligent Transportation Systems (2018) 1−11.

[27] A. Geiger, et al., Vision meets robotics: the KITTI dataset, The International Journal of Robotics Research 32 (11) (2013) 1231−1237.

[28] J. Fritsch, K. Tobias, A. Geiger, A new performance measure and evaluation benchmark for road detection algorithms, in: 16th International IEEE Conference on Intelligent Transportation Systems (ITSC 2013), IEEE, 2013.

[29] National Highway Traffic Safety Administration, National Motor Vehicle Crash Causation Survey: Report to Congress, National Highway Traffic Safety Administration. "National motor vehicle crash causation survey: Report to congress." National Highway Traffic Safety Administration Technical Report DOT HS 811 059, 2008.

[30] C.M. Martinez, et al., Energy management in plug-in hybrid electric vehicles: recent progress and a connected vehicles perspective. IEEE Transactions on Vehicular Technology 66 (6) (2016) 4534−4549.

[31] Y.S. Son, et al., Robust multirate control scheme with predictive virtual lanes for lane-keeping system of autonomous highway driving, IEEE Transactions on Vehicular Technology 64 (8) (2015) 3378−3391.

[32] S. Joerer, et al., A vehicular networking perspective on estimating vehicle collision probability at intersections, IEEE Transactions on Vehicular Technology 63 (4) (2014) 1802−1812.

[33] T. Qu, H. Chen, D. Cao, H. Guo, B. Gao, Switching-based stochastic model predictive control approach for modeling driver steering skill, IEEE Transactions on Intelligent Transportation Systems 16 (1) (2015) 365−375.

[34] C. Lv, J. Zhang, Y. Li, Extended-Kalman-filter-based regenerative and friction blended braking control for electric vehicle equipped with axle motor considering damping and elastic properties of electric powertrain, Vehicle System Dynamics 52 (11) (2014) 1372−1388.

[35] A. Bar Hillel, et al., Recent progress in road and lane detection: a survey, Machine Vision and Applications (2014) 1−19.

[36] J.C. McCall, M.M. Trivedi, Video-based lane estimation and tracking for driver assistance: survey, system, and evaluation, IEEE Transactions on Intelligent Transportation Systems 7 (1) (2006) 20−37.

[37] S. Yenikaya, G. Yenikaya, E. Düven, Keeping the vehicle on the road: a survey on on-road lane detection systems, ACM Computing Surveys (CSUR) 46 (1) (2013) 2.

[38] J. Xiao, S. Li, B. Sun, A real-time system for lane detection based on FPGA and DSP, Sensing and Imaging 17 (1) (2016) 1−13.

[39] U. Suddamalla, et al., A novel algorithm of lane detection addressing varied scenarios of curved and dashed lanemarks, in: Image Processing Theory, Tools and Applications (IPTA), 2015 International Conference on, IEEE, 2015.

[40] M. Braga de Paula, C.R. Jung, Automatic detection and classification of road Lane markings using onboard vehicular cameras, IEEE Transactions on Intelligent Transportation Systems 16 (6) (2015) 3160−3169.

[41] A.F. Cela, et al., Lanes detection based on unsupervised and adaptive classifier, in: Computational Intelligence, Communication Systems and Networks (CICSyN), 2013 Fifth International Conference on, IEEE, 2013.

[42] F. Bounini, et al., Autonomous vehicle and real time road lanes detection and tracking, in: Vehicle Power and Propulsion Conference (VPPC), 2015 IEEE, IEEE, 2015.

[43] H. Tan, et al., Improved river flow and random sample consensus for curve lane detection, Advances in Mechanical Engineering 7 (7) (2015), 1687814015593866.

[44] Y. Wang, E.K. Teoh, D. Shen, Lane detection and tracking using B-Snake, Image and Vision Computing 22 (4) (2004) 269−280.

[45] K.H. Lim, P.S. Kah, L.-M. Ang, River flow lane detection and Kalman filtering-based B-spline lane tracking, International Journal of Vehicular Technology (2012).

[46] J. Li, et al., Deep neural network for structural prediction and lane detection in traffic scene, IEEE transactions on neural networks and learning systems 28 (3) (2016) 690−703.

[47] M. Aly, Real time detection of lane markers in urban streets, in: Intelligent Vehicles Symposium, 2008 IEEE, IEEE, 2008.

[48] B.-S. Shin, J. Tao, R. Klette, A superparticle filter for lane detection, Pattern Recognition 48 (11) (2015) 3333−3345.

[49] R. Labayrade, et al., A reliable and robust lane detection system based on the parallel use of three algorithms for driving safety assistance, IEICE Transactions on Info and Systems 89 (7) (2006) 2092−2100.

[50] A. Borkar, M. Hayes, M.T. Smith, A novel lane detection system with efficient ground truth generation, IEEE Transactions on Intelligent Transportation Systems 13 (1) (2012) 365−374.

[51] S. Sivaraman, M.M. Trivedi, Integrated lane and vehicle detection, localization, and tracking: a synergistic approach, IEEE Transactions on Intelligent Transportation Systems 14 (2) (2013) 906−917.

[52] V.D. Nguyen, et al., A fast evolutionary algorithm for real-time vehicle detection, IEEE Transactions on Vehicular Technology 62 (6) (2013) 2453−2468.

[53] C. Rose, et al., An integrated vehicle navigation system utilizing lane-detection and lateral position estimation systems in difficult environments for GPS, IEEE Transactions on Intelligent Transportation Systems 15 (6) (2014) 2615−2629.

[54] R. Danescu, S. Nedevschi, Probabilistic lane tracking in difficult road scenarios using stereovision, IEEE Transactions on Intelligent Transportation Systems 10 (2) (2009) 272−282.

[55] D. Cui, J. Xue, N. Zheng, Real-time global localization of robotic cars in lane level via lane marking detection and shape registration, IEEE Transactions on Intelligent Transportation Systems 17 (4) (2016) 1039−1050.

[56] R.K. Satzoda, M.M. Trivedi, On performance evaluation metrics for lane estimation, in: Pattern Recognition (ICPR), 2014 22nd International Conference on, IEEE, 2014.

[57] C.-W. Lin, H.-Y. Wang, D.-C. Tseng, A robust lane detection and verification method for intelligent vehicles, in: Intelligent Information Technology Application, 2009. IITA 2009. Third International Symposium on, vol. 1, IEEE, 2009.

[58] W.T. Freeman, E.H. Adelson, The design and use of steerable filters, IEEE Transactions on Pattern Analysis and Machine Intelligence 13 (9) (1991) 891−906.

[59] H.P. VC, Method and Means for Recognizing Complex Patterns, U.S. Patent No. 3,069,654. 18 Dec., 1962.

[60] D.H. Ballard, Generalizing the Hough transform to detect arbitrary shapes, Pattern Recognition 13 (2) (1981) 111−122.

[61] L. Xu, M.I. Jordan, On convergence properties of the EM algorithm for Gaussian mixtures, Neural Computation 8 (1) (1996) 129−151.

[62] N. Otsu, A threshold selection method from gray-level histograms, Automatica 11 (1975) 23−27, 285−296.

# CHAPTER 5

# Driver Behavior Recognition in Driver Intention Inference Systems

## INTRODUCTION

Driver behavior is the most important factor for on-road driving safety [1−6]. As humans are the major users of roads, their driving behaviors influence traffic safety and efficiency. More than 90% of traffic accidents for light vehicles in the United States were reported to be caused by driver errors such as misbehavior and inadvertent errors, which are similar to other countries worldwide. It was also mentioned in Refs. [7−11] that traffic accidents could be reduced by 10%−20% by correctly recognizing driver behaviors. Therefore it is critical to have a clear perspective of driver behavior and the tasks being performed.

Human drivers have been extensively studied since the 1970s. The study of human drivers is a massive project in many aspects. Most of the existing research lies in the scopes of driver behaviors, driver attention, and intention, driver drowsiness and fatigue, driver cognitive and neural muscles, etc. All of these studies have a common objective, which is to gain a better understanding of driver status from either a psychological or physiological aspect to assist in driving tasks and increase driving safety [12−14].

Understanding human drivers are necessary both for conventional vehicles and for automated vehicles. In the United States and China, accidents have occurred when a Tesla driver trusted or solely relied on the autopilot system while driving. For lower-level automated vehicles, especially for level two and level three automated vehicles (based on the automation definition in SAE standard J3016), human drivers need to sit in the driver seat and are responsible for the safety issues. In these vehicles, the driver is allowed to perform secondary tasks for entertainment; however, due to the partially automated limitation, the driver has to take control of emergencies. Therefore the monitoring of human drivers and determining whether they can return to the driving task is more important than in conventional vehicles.

In this part, a driver monitoring system is designed to detect driving and secondary tasks in real time.

Specifically, the recognition model is designed to identify seven tasks performed by different drivers. There are four tasks considered as normal driving tasks: normal driving (front looking), right mirror checking, left mirror checking, and rear mirror checking. Meanwhile, according to Ref. [13], the three most common secondary tasks in automated vehicles are selected, which are using a video device, answering a mobile phone, and texting using a mobile phone. The mirror-checking behaviors are also critically important to the driver's intention inference as these behaviors are the outer reflection of the mental states. Hence, precisely estimating these behaviors is the fundamental of driver intention inference. To identify the driver postures, multimodal data are collected using a Kinect consumer RGB-D camera including the head rotation and body joint positions. The main objective of this study is to design a real-time driver behavior model that does not require any history information for the recognition of normal driving and secondary driving tasks. Additionally, the importance of driver posture features to the identification of driving tasks is evaluated.

In this section, the research scope is narrowed to the range of driving task recognition toward a normal driving and secondary task monitoring system for lower-level and middle-level automated vehicles. According to previous studies, driver behavior can be classified into intended and nonintended behaviors [15,44]. The intended behavior of the driver is the extension of the driver's mental thought, which can be used to infer the mental state and intent of the driver. In contrast, nonintended behaviors are usually caused by distractions due to outside and inside disturbances. Driver behavior has been widely studied in previous literature. General driver behaviors include the study of driver head pose [15,16], eye-gaze dynamics [17,32,59] hand motions and gestures [18], body movement [19,20], and foot dynamics [21]. This behavior information has been successfully used to estimate driver fatigue, driver distraction, driver attention, etc. In this study, driver head and upper body information detected using a

Advanced Driver Intention Inference. https://doi.org/10.1016/B978-0-12-819113-2.00005-1

Kinect will be evaluated for normal driving and distraction identification.

When drivers are performing secondary tasks while driving, they are regarded as being distracted and many studies use the duration of eye-off-road to detect whether a driver is distracted by the secondary tasks. Therefore the most common features for driver distraction detection are head pose and eye-gaze information. Along with the driver behavior, information on the vehicle such as vehicle speed, heading, and acceleration are important features for evaluating the level of driver distraction. In Ref. [22], an integration method combining the driver's hand, head, and eye for driver activity recognition was proposed. Rezaei and Klette introduced an intelligent driver assistance system to prevent rear-end crashes based on driver monitoring and front vehicle detection [23]. The head pose was estimated based on the proposed face appearance model and 3D head model mapping. In Ref. [24], a driver drowsiness alert system was proposed according to the driver head and eye dynamics. The driver head pose was estimated based on an Euler angle comparison between a single head region image and a 3D head model with known rotations. In Ref. [27], the authors analyzed the relationship between head pose and eye gaze. A strong correlation was found between the head and gaze direction. The study showed that during natural driving, the participants tend to have less head rotation but more gaze searching to maintain safe driving.

In Ref. [44], a comprehensive in-vehicle perception system for driver surveillance and assistance was proposed. Multimodal sensors were fused to integrate the major driver's physical cues and traffic situations. In Ref. [28], driver acceleration profiles for a car following scenario on a highway were generated using recurrent neural networks. Specifically, an long short-term memory (LSTM) recurrent network was adapted as it can automatically learn the spatial and temporal features of the naturalistic driving data. In Ref. [29], an LSTM-based recurrent neural network was proposed to detect driver distraction behaviors based on the simulated controller area network (CAN) bus signals. Thirty participants performed eight typical secondary tasks independently and the distraction levels were classified into binary, three levels, and six levels. In Ref. [30], the authors claimed that applying eye tracking is much more difficult in real vehicles than in the simulator. Therefore gaze estimation was not adopted and only driving information through the CAN bus was used for driver visual searching distraction detection. In Ref. [31], a driving behavior model for teenage drivers was studied. Different machine-learning methods were evaluated based on the driving data, which were collected with a driving simulator. The authors reported that instead of predicting driving behavior (steer, throttle, and brake) directly, more accurate results can be achieved using context-based prediction and indirect prediction methods.

Despite the driver and driving behaviors, other studies have used physiological sensors to identify driver distraction and other abnormal statuses. According to the study in Ref. [34], driver monitoring systems for drowsiness and distraction detection can be classified into visual-based and nonvisual-based methods. Visual-based methods monitor driver head pose, eye movement and blinking, yawning, and facial expression. In contrast, nonvisual-based systems detect driver status with physiological sensors such as EEG, ECG, and EOG, along with the vehicle CAN bus signals. However, the effects of hand, arm, and body on the recognition of driver status were not discussed. Similarly, a stress detection system for drivers was studied in Ref. [33]. A specific type of continuous recurrent neural network named cellular neural network was used for the binary classification task. In Ref. [45], an EEG-based driver secondary task recognition system was proposed. Seven driving activities, which were normal driving, phone conversation, texting, question, spelling, listening to energetic, and calming music, were evaluated. A 16-channel EEG device was used, and 31 features were extracted from each channel, which contains short-time Fourier Transform, band power, discrete wavelet-mean, discrete wavelet power, discrete wavelet standard deviation, Katz fractal dimension, Higuchi fractal dimension, autoregressive parameters, and entropy of the discrete wavelet transform coefficients. Five classifiers were used to detect the secondary tasks performing states, which were support vector machine (SVM), random forest (RF), naïve Bayes (NB), decision trees, and nearest neighbor. The final results show that the driver distraction state can be detected based on the EEG features with 98.99 accuracies.

In Ref. [46], a Wi-Fi sensor-based distracted arm and head movement detection were proposed. The fluctuations in channel state information were captured and analyzed. The K-nearest neighbor method was used to classify the driver state, and final results achieved an average of 94.5% accuracy on the arm and head movement. In Ref. [47], a driver's attention and behavior monitoring robot, namely NAMIDA, was developed to quickly draw a driver's attention on the road and reduce the response time of the driver. In Ref. [48], the driver workload level was classified into a normal and overload level based on the EEG signal. Then, considering EEG signals are difficult to be acquired in the normal driving scenarios, the relationship between the EEG patterns and the vehicle driving information, such as engine revolutions per minute, vehicle speed, lane changes, and turns, is used as the input signal for driver workload estimation. In Ref. [49], a combined head poses estimation module, and a fuzzy logic module for driving dangerous inference was proposed based on a dataset collected with a driving simulator. The convolutional neural network was used to recognize normal driving, texting, and answering phone behaviors. In Ref. [50], a physiological measurement for the driver to estimate the driving performance was proposed. The normal drive and load driver (text back words) are classified based on several methods. The physiological measurements, such as palm electrodermal activity, heart rate, breathing rate, and perinasal perspiration, are analyzed. Based on the data collected from 68 participants, it was found that the physiological signals are more informative and important to reflect driver status than the biographic features such as gender and age.

The Kinect sensor, a low-cost range camera, has been successfully applied to human and driver behavior detection because it was first made available by Microsoft in 2012. Kinect was first designed for indoor motion sensing, and it provides a color image, depth image, and infrared image. In Ref. [35], the general architecture for human activity recognition was proposed using Kinect. The human activities were viewed as the spatiotemporal evolution of body postures. The estimated postures are classified using SVMs, and finally, the hidden Markov model (HMM) was used to model the activities as a time sequence of the different estimated postures. In Ref. [36], a Kinect-based wearable face recognition system for people with low vision or blindness was proposed. The color and depth images were simultaneously captured to identify the face and generate the 3D location for the user. In Ref. [26], a seven-point skeleton-based driver upper body tracking system using Kinect depth images was applied. The proposed system is efficient for detecting driver merging and turning behaviors according to the detected body pose and arm motion. The system can also be used to analyze and compare the driving maneuver styles of different drivers.

In this study, Kinect is adopted as the driver monitoring sensor to identify normal driving and secondary tasks. Similar research can be found in Ref. [25], where driver mirror-checking behavior during normal driving and performing secondary tasks were analyzed. The authors reported that mirror-checking behavior is one of the most important driving perception processes and reflects the attention level of the driver. In addition, mirror-checking behaviors are highly detectable maneuvers and can achieve 95% detection accuracy using machine learning. However, that work only studied the binary classification scenarios without reporting the recognition accuracy for each task. Additionally, that study did not analyze the impact of body postures to the recognition of complex driving behaviors.

## Machine-Learning Methods for Human Activities Recognition—A Case Study on Activity Recognition

A human activity recognition example is selected as a case study for human activity recognition in this part. Human and driver activities usually can be treated as a multiclass classification task. Several human activities can be recognized based on combined machine-learning methods and representative features. The raw human activity data used in this example are obtained from http://archive.ics.uci.edu/ml/datasets/Human+Activity+Recognition+Using+Smartphones. This time-series dataset is multivariate that can be used to classify multiple classes in human activity. The data are public and free to download. To recognize human actions, SVM and HMM are trained separately, and their performance evaluated in statistic table and receiver operating characteristics (ROC) form is given.

### Initial data processing

The raw human activity data were obtained from the University of California Irvine (UCI) machine-learning website. The data were collected from 30 participants within 19−48 years old. The participants were required to wear a smartphone with embedded inertial sensors. There are 10,299 instances in total, and 70% of them are randomly chosen as train data while the rest 30% are test data. Each participant performed six behaviors, which were walking, walking upstairs, walking downstairs, sitting, standing, and laying. The experiment was also video recorded and manually labeled. The classification task, therefore, is to train the classifier model to recognize the six human activities based on train data and test its performance with the test dataset.

### Data dimension reduction

After the preprocessing step, the raw dataset was expanded to 561 elements feature vector for one instance. For this training dataset that has more than 7000 instances, the computation cost is very expensive. This kind of large dataset is neither easy to analyze nor helpful to train the classification models. Therefore an extra step, namely, dimension reduction is taken before the training data are fed into the model.

According to Ref. [51], high dimensional data are not always necessary. It can be replaced by a much lower-dimensional dataset that can keep most of the key information of the original data. This lower dimension is called the intrinsic dimension (ID), which means the dataset with this intrinsic dimension is large enough to stand for the original data. According to the statistical study [52], the performance of the classifier depends on ID; therefore, an appropriate estimation of ID will lead to improved classification performance.

To estimate the data ID and reduce the dimension, an open-source data dimension reduction toolbox is used. The MATLAB toolbox is designed by Laurens van der Maaten from the Delft University of Technology, which is free to use and modify without commercial objects. A detail description of this toolbox can be found in https://lvdmaaten.github.io/drtoolbox/. This toolbox offers 34 kinds of popular dimension reduction methods and 6 kinds of ID estimation methods such as eigenvalue-based maximum likelihood estimation. In this part, the ID of trained data is estimated based on the maximum likelihood estimation. The ID of the training dataset is estimated for 17. Then, based on this number, a principal component analysis (PCA) method dimension reduction method is applied. The original instance vector with 561 elements is then transformed into a 17 dimension vector.

PCA is a linear unsupervised data reduction method, which makes it become one of the most popular dimension reduction tools. It does not need class information of the original data that make it unsupervised. In terms of linear, the main idea of PCA is a linear projection that projects the high dimensional data into a lower dimension and tries to keep the variance between different groups as large as possible to keep as much as original data characters. Thus PCA can be viewed as a reduction method that loses a few unnecessary data information. In mathematical terms [51], PCA tries to find a linear mapping $M$, which maximizes the cost function:

$$J = \mathrm{trace}\left(M^T \mathrm{cov}(X)M\right) \qquad (5.1)$$

where $\mathrm{cov}(X)$ is the sample covariance matrix of data X. It can be shown that the linear mapping is governed by the $d$ principle eigenvectors, which means the high dimension data can be represented by a lower $d$ dimension dataset. Hence, PCA solves the Eigen problem like

$$\mathrm{cov}(X)M = \lambda M \qquad (5.2)$$

where $\lambda$ are eigenvalues to solve the Eigen problems. The lower-dimensional data $y_i$ with respect to the original $x_i$ are then computed by mapping them onto the linear basis $M$:

$$Y = XM \qquad (5.3)$$

After the human activity datasets are processed through filters and dimension reduction by PCA, it is available to be used to train the classifiers. Here, two kinds of machine-learning methods known as SVM and HMM are applied and the results of these two classifiers will be compared also.

### Support vector machine method

This part mainly focuses on the application procedure and classification performance of SVM. Another open-source MATLAB toolbox, namely LIBSVM, is adopted here. LIBSVM was a toolbox for SVM that supports multiple computer languages such as C, MATLAB, and Python. LIBSVM is an integrated software for support vector classification, regression, and distribution estimation. It also well supports multiple class classification problems. Detailed information of SVM can be found at https://www.csie.ntu.edu.tw/~cjlin/libsvm/. User is allowed to select SVM types, kernel functions, and relative parameters to design their own SVM model. Particularly, recently, LIBSVM supports four kinds of kernel functions that are linear, polynomial, radial basis function, and sigmoid tan function.

In this part, a radial basis function kernel is selected to build the SVM model. According to the data format, the classification task can be viewed as training a classifier that is able to recognize the six types of certain human activities. The performance of the SVM model can achieve 89.20% accuracy. It should be noticed that this performance is achieved based on the data after dimension reduction. Better performance can be obtained by using optimistic algorithms such as genetic algorithm and particle swarm optimization to select the best model parameters; however, this requires extra training time. Besides, the performance here is good enough, so no further processing is necessary.

The model is tested with the test data group. An ROC graph, as shown in Fig. 5.1, is plotted based on a binary classification perspective of view that is also known as one class versus all other classes. Table 5.1 represents a comparison of the six classifiers' performance.

### Hidden Markov model method

HMM is a powerful tool in dealing with time sequence data. It has been widely used in the past for time-series modeling and has achieved great success in speech

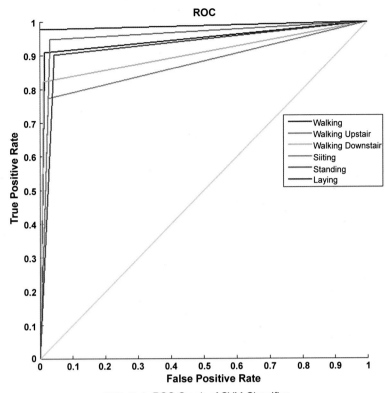

FIG. 5.1 ROC Graph of SVM Classifier.

TABLE 5.1
Comparison of Six Activities SVM Classifiers.

| Predict / True | C1 | C2 | C3 | C4 | C5 | C6 | Total Number | Accuracy |
|---|---|---|---|---|---|---|---|---|
| C1 | 451 | 19 | 26 | | 0 | 0 | 496 | 0.909 |
| C2 | 21 | 447 | 3 | 0 | 0 | 0 | 471 | 0.949 |
| C3 | 16 | 58 | 346 | 0 | 0 | 0 | 420 | 0.824 |
| C4 | 0 | 2 | 0 | 380 | 109 | 0 | 491 | 0.773 |
| C5 | 0 | 0 | 0 | 52 | 480 | 0 | 532 | 0.902 |
| C6 | 0 | 0 | 0 | 12 | 0 | 525 | 537 | 0.977 |
| | | | | | | Total Accuracy | | 0.892 |

recognition, image recognition, and fault diagnosis area [53–55]. It is a kind of dynamic Bayesian network that can lead to a good performance for pattern inference problems. To infer human activities, HMM is used as an activity recognizer. The process of applying HMM in this work is similar to its usage in isolated word recognition problems. From the HMM point of view, each activity is constructed by a sequence of states.

The object of the recognizer is to affect a mapping between an observed feature vector and the state's sequence of action. For each activity, each state in the state sequence at a time is responsible for an observation. To recognize which action is happening, different action models are built in the first. Each model represents a certain kind of activity. Second, according to the observation vector, the probability for each model that generates this vector is calculated. The model that gives the maximum probability will be selected, and the action that is undertaking is determined; the activity inference process is shown in Fig. 5.2.

According to the output distribution of hidden state, HMM can be classified into discrete HMM (DHMM) and continuous HMM (CHMM). From the definition, it is clear to see that the output of DHMM is discrete values while the output for CHMM is continuous. Here, the data from the wearable sensor are all continuous; therefore, CHMM will be used in this work. The output of CHMM can be various; however, the most popular ways to model the output distribution is assumed that it meets Gaussian distribution. This is a very simple assumption and not always give an acceptable result. Instead of using a single multivariate Gaussian distribution, a structure called Gaussian

mixture model that is a linear combination of multiple Gaussian functions is proved to be much powerful.

The Output Distribution of Hidden Markov Model Becomes

$$b_j(x) = p(x|s_j) = \sum_{m=1}^{M} c_{jm} N(x; \mu^{jm}, \Sigma^{jm}) \tag{5.4}$$

where $N(x; \mu^{jm}, \Sigma^{jm})$ is the Gaussian function with mean $\mu$ and covariance $\Sigma$ (Fig. 5.3).

Considering Bayes' Theorem

$$p(j|x) = \frac{p(x|j)p(j)}{p(x)} = \frac{p(x|j)p(j)}{\sum_{j=1}^{M} p(x|j)p(j)} \tag{5.5}$$

The posterior probability $p(j|x)$ gives the probability that the component $j$ was responsible for generating data point $x$ and it is called the component occupation probability. When the component density becomes Gaussian, it makes the Gaussian mixture model.

In GMM, it is not possible to compute $p(x|j)$ and $P(j)$, when combined GMM with HMM, this problem still exists. Therefore the way to train GMM-HMM is by using a two-step method called expectation-maximization (EM), which in essence, is a recursive process that optimizes model parameters until

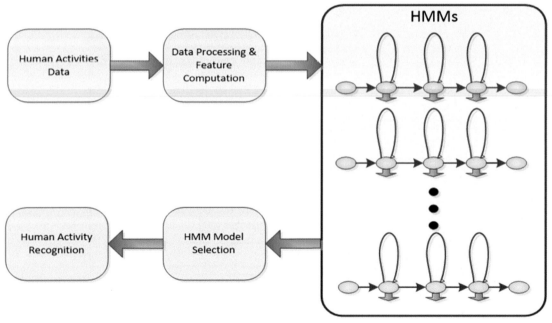

FIG. 5.2 Human Activities Recognition Process by HMM.

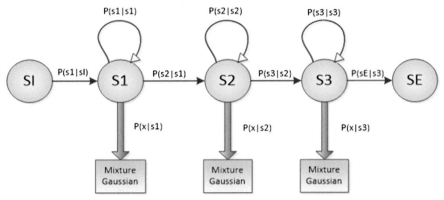

FIG. 5.3 Hidden Markov Model with $M$-component Gaussian Mixture Output Distribution.

convergence. Before applying EM, a few parameters need to be noticed:

$\alpha_t(s_j)$: is the probability of observing the observation sequence $x_1 \cdots x_t$ and being in state $s_j$ at time $t$.

$\beta_t(s_j)$: is the probability of future observations given an HMM is in state $s_j$ at the time $t$. $\gamma_t(s_j)$: The state occupation probability is the probability of occupying state $s_j$ at the time $t$ given the sequence of observations.

$\xi_t(s_i, s_j)$: is the probability of being in $s_i$ at the time $t$ and $s_j$ at $t + 1$ given the observation.

In E step: first, recursively compute the forward probabilities $\alpha_t(s_j)$ and backward probability $\beta_t(s_j)$. Then, compute the state occupation probability $\gamma_t(s_j)$ and $\xi_t(s_i, s_j)$. In M step: based on the estimated parameters, reestimate the HMM and GMM parameters: the mean vector $\mu^j$, covariance matrices $\Sigma^j$, and transition probabilities $a_{ij}$.

In this step, an open-source MATLAB-based HMM toolbox is used. This toolbox is developed by Kevin Murphy that supports inference and learning for HMMs with discrete output, Gaussian output, and mixture Gaussian output. Detailed information about this toolbox can be found in https://www.cs.ubc.ca/~murphyk/Software/HMM/hmm.html. The performance of the HMM recognizer is shown in Fig. 5.4 and Table 5.2. Fig. 5.4 is the ROC of the six HMMs representing six kinds of human activities. Table 5.2 illustrates the statistical performance of the HMM recognizer. As HMM is a type of probability graph model, each time the recognition performance will be different even with the same model structure and parameters. The performance will be influenced by the prior, transition matrix, and observation distribution at each time. Therefore the performance is averaged after running the HMM 10 times. Besides, by using this toolbox, the model structure, especially, the mixture

component quantity and the number of hidden states need to be chosen first. The probable range for hidden states number and mixture components are set between 2 and 12, and an algorithm is designed to find the best combination of these two values at this scope. For each combination, the result is the average value for 10 times of evaluation. Finally, a good combination is determined (cannot say this is the best combination) with six hidden states and nine mixture components.

From the earlier graphs, it can be seen that the classification performance of SVM is better than that of HMM. However, both classifiers achieve acceptable recognition accuracy. For these two kinds of classifiers, human laying action is the easiest action to be classified while walking downstairs and walking are relatively difficult to be recognized because a large overlap occurs with neighbor actions.

## FEATURE ENGINEERING IN DRIVER BEHAVIOR RECOGNITION
### Driver Behavior Overview

Driver behavior normally can be classified into intended and nonintended behaviors. The intended behavior of the driver is the extension of the driver's mental thought, which can be used to infer the driver's inner mental states. Driver behaviors have been widely studied in previous studies. Normal driver behaviors include driver head rotation, eye-gaze dynamics, hand motion and gestures, body movement, and foot dynamics. These driver behavior signals have been successfully used in the estimation of driver fatigue, driver distraction, and driver attention. A better understanding of driving-related driver behaviors will contribute to a more accurate intention inference system.

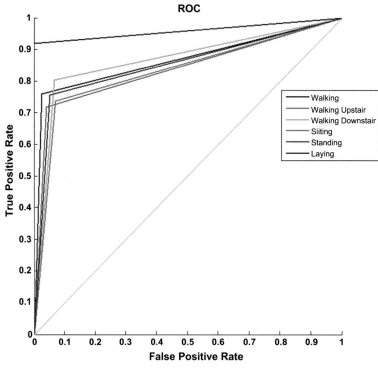

FIG. 5.4 ROC Graph of Hidden Markov Model Classifier.

**TABLE 5.2**
**Comparison of Six Activities Hidden Markov Model Classifiers.**

| Predict<br>True | C1 | C2 | C3 | C4 | C5 | C6 | Total<br>Number | Accuracy |
|---|---|---|---|---|---|---|---|---|
| C1 | 377 | 33 | 86 | 0 | 0 | 0 | 496 | 0.760 |
| C2 | 47 | 385 | 39 | 0 | 0 | 0 | 471 | 0.817 |
| C3 | 18 | 65 | 337 | 0 | 0 | 0 | 420 | 0.802 |
| C4 | 0 | 2 | 0 | 407 | 80 | 2 | 491 | 0.837 |
| C5 | 0 | 0 | 0 | 93 | 439 | 0 | 532 | 0.855 |
| C6 | 0 | 0 | 0 | 43 | 0 | 494 | 537 | 0.919 |
| | | | | | | Total Accuracy | | 0.831 |

In Ref. [56], an integration method of driver's hand, head, and eye for driver activity recognition was proposed. The two cameras mounted in front on the top of the driver were used for head and hand detection, respectively. Rezaei and Klette [23] introduced an intelligent driver assistance system to prevent ego-vehicle from rear-end crashes based on both driver monitoring and front vehicle detection. A driver's facial model was first constructed with an asymmetric active appearance model. Next, the head pose was estimated based on the proposed face appearance model and a 3D head model mapping. In Ref. [24], a driver drowsiness alerting system was proposed according to driver head and eye dynamic features. The driver head pose was estimated based on the Euler angle comparison between a single head region image and a 3D head model with known rotations.

In Ref. [25], driver mirror-checking behaviors during normal driving tasks and secondary tasks were analyzed. The authors reported that mirror-checking behaviors are one of the most important driving situation perception processes and reflect the attention level of human drivers. Besides, mirror-checking behaviors were highly detectable maneuvers with machine-learning methods and can achieve a 95% detection accuracy. In Ref. [26], a seven skeleton points-based driver upper body tracking system using Kinect depth image was applied. The proposed system is efficient in the detection of driver merging, turning behaviors according to detected body pose and arms motion. The system can also be used for the analysis and comparison of driving maneuver style from different drivers. In Ref. [44], a comprehensive on-vehicle perception system for driver surveillance and assistance was proposed. Multimodel sensors were fused to integrate the driver's major physical cues as well as traffic situations. In-vehicle vision systems that monitor the driver's head pose, hand, and foot dynamics were fused with outer vehicle sensors such as Radar, Lidar, and cameras. In this section, the utilization of the driver's head pose estimation and body joints signals into driver intention inference are described in detail. The driver's head pose estimation is based on a rotation regression using machine-learning methods. Driver body joints estimation is detected with a low-price Microsoft Kinect sensor.

## Driver Head Pose Estimation

Human head pose estimation has been widely studied in previous literature. A comprehensive literature review of head pose estimation methods could refer to Ref. [16]. In Ref. [16], head pose estimation methods were classified into different categories. In this project, the method of head pose regression based on machine learning is used. This method requires no precalibration and complex modeling procedures, which is robust and able to process a larger scene image. The general process of head pose estimation based on machine-learning regression is to extract some significant features from the captured image first, and then a rotation regression model is built based on the selected features and the corresponding rotation ground truth values. The head pose is at 3 degrees that contains the pitch, yaw, and

roll angles of the head. Therefore three regression models are usually being modeled to estimate each angle rotation.

Existing regression-based methods can be classified into 2D and 3D categories. 2D means that the extracted features only contain 2D features that are extracted from the image while 3D estimation takes depth information into consideration and estimates the rotation with both 2D and 3D features. In Ref. [57], a feature extraction method based on scale-invariant feature transform (SIFT) and histogram of oriented gradients (HOG) was used to construct the feature vector of the pose estimation. The integrated features along with depth image captured with Kinect were fed into a random forest algorithm and final results achieved 8-degree errors on average. Eye gaze and head pose estimation method for driver eye-off-road detection was proposed in Ref. [58].

One CCD camera and one infrared camera are mounted behind the driver as the monitoring devices. Facial landmarks and head pose were first detected with a parameterized appearance model, and next, the eye-gaze direction is determined according to the face model and eye geometry calculation. A deep CNN with four convolutional layers and one fully connected layer was applied in Ref. [59] for facial pose estimation. Face positions were detected with Viola-Jones (VJ) detector and then scaled into the format of $32 \times 32 \times 3$ before CNN estimation. The proposed algorithm was tested with the CMU Pose, Illumination, and Expression (CMU-PIE) dataset. The final results showed that the CNN algorithm can detect facial pose with an accuracy of 99.4% with a low mean absolute error of 0.135 degrees$0.135°$.

In Ref. [60], a head pose estimator based on RF is proposed. The RF is trained with the CMU-PIE dataset to recognize seven discrete pose angles. To improve the illumination invariance properties, a binary pattern run-length matrix, which was a combination of binary pattern and run-length matrix was introduced for facial feature extraction. In Ref. [61], a novel Hough network that combines Hough forest and convolution neural network together to simultaneously estimate head pose and detect facial landmark was introduced. The proposed CNN-based network is able to do both classification and regression like Hough forest and does not rely on handcrafted features. In Ref. [62], a coarse eye-gaze zone estimation based on head pose is proposed. The eye-gaze zones were separated into eight regions in the vehicle. A multiview camera system was used for head and gaze feature extraction and the ground truth gaze information was annotated by human experts through the video frames. The authors concluded that with an accurate in-vehicle gaze zone estimation, the coarse gaze detection was sufficient for many applications.

2D estimation methods can be less robust and have an identity issue that causes a bad generalization. Therefore depth information is augmented into the feature vectors to enhance the system performance and robustness property. A combination of RGB images and depth image for human head pose detection is proposed in Ref. [63]. RGB and depth images were captured with a Kinect sensor. The head position was first located with a Viola–Jones face detector, which was applied twice to find a rough and precise face location, respectively. Then, the grayscale and depth face features were extracted with multiple feature extractors and concatenated together for pose regression. Finally, an SVR is applied to continuously estimate the head pose frame by frame.

Face detection and head pose estimation methods that combine 2D and 3D algorithms were proposed in Ref. [64]. Head and face were first detected with VJ detector. Then, the points cloud map of the face region was generated with the depth measurement given by Kinect. The head pose was determined based on an iterative closest points algorithm that compares two adjacent cloud maps and returns the transformation matrix. In Ref. [65], A 3D face template was constructed with the RGB-D images online and used for head pose estimation. The points cloud map was created based on the RGB and depth image. Instead of building a face template offline, in this work, the template and head pose tracking were realized by introducing the iterative closest point algorithm. The authors divided the head pose tracking and face template into two loosely independent procedures that only need two threads to implement. The final processing speed can reach 21 ms per frame.

In Ref. [66], a temporal probabilistic graph model for head pose estimation in video frames and unconstrained places was proposed. Different from most pose estimation algorithms that return an explicit pose direction, the proposed algorithm returns a probability density function over the range [−90 degrees, +90 degrees] $-90°, +90°$. The hierarchical graph model contains three levels. The lower two levels capture face features such as patch, edge, region, and landmarks and calculate the pose distribution at a certain frame, while the top-level estimates the head rotation in the video sequence according to the temporal information given by a belief propagation network. The performance of the proposed head pose estimation method was reported to be more accurate than the relevant algorithms when applied to a video sequence in uncontrolled environments.

### Driver Head Pose Estimation Using Head Features

In this part, two different feature extractors are used to extract the head rotation features, namely, Gabor filter and HOG.

Gabor filter has been widely accepted and regarded as an efficient facial feature extraction algorithm in the past decades. From the very beginning, the oriented two-dimensional Gabor filters are an approximate model of the simple cells in the Mammalian vision system. In 1987, Jones and Palmer [67] proved that the real part of Gabor function performs a good fit for the receptive field mechanism, which was found in the simple cells in a cat striate cortex because the frequency and orientation representation are similar to the human visual system. Because of this, Gabor filters are thought to be a reasonable model that mimics the ways of how humans distinguish texture; thus, it has become an efficient algorithm to recognize textures in image processing. In image processing scope, Gabor filter can be viewed as a band-pass filter as it filters a certain band of signals. When applying Gabor filter to images, the filter gives the highest response at areas of edges and the places where textures change. The impulse response is defined by a sinusoidal wave multiplied by a Gaussian function. A complex Gabor filter is defined as the product of a Gaussian kernel function and a sinusoidal function. Given a 2D spatial Gabor filter, it can be expressed as

$$g(x, y) = s(x, y)G(x, y) \qquad (5.6)$$

where $s(x, y)$ is a complex sinusoidal function called the carrier and $G(x, y)$ is the Gaussian kernel function that is called the Gaussian envelope. The filter consists of a real component and an imaginary component that form the filter into a complex number formation. A more detailed Gabor filter can be described as follows [68]:

The complex form of Gabor filters in the spatial domain:

$$g(x, y, \lambda, \theta, \varphi, \sigma, \gamma) = exp\left(-\frac{x'^2 + \gamma^2 y'^2}{2\sigma^2}\right) exp\left(i\left(2\pi\frac{x'}{\lambda} + \varphi\right)\right)$$
$$(5.7)$$

The complex form of the filter can also be expressed separately with the real part and imaginary part.Real part:

$$g(x,\gamma,\lambda,\theta,\varphi,\sigma,\gamma)=\exp\left(-\frac{x'^2+\gamma^2\gamma'^2}{2\sigma^2}\right)\cos\left(2\pi\frac{x'}{\lambda}+\varphi\right)$$

(5.8)

Imaginary part:

$$g(x,\gamma,\lambda,\theta,\varphi,\sigma,\gamma)=\exp\left(-\frac{x'^2+\gamma^2\gamma'^2}{2\sigma^2}\right)\sin\left(2\pi\frac{x'}{\lambda}+\varphi\right)$$

(5.9)

where

$$x'=x\cos\theta+y\sin\theta$$

(5.10)

and

$$y'=-x\sin\theta+y\cos\theta$$

(5.11)

In these equations, parameter $\lambda$ represents the wavelength of the sinusoidal factor, $\theta$ is the orientation of Gabor function, $\varphi$ represents the phase offset, $\sigma$ is the deviation of Gaussian envelope, and $\gamma$ represents the spatial aspect ratio. Fig. 5.5 shows the composition of

a Gabor filter with 0 orientation sinusoidal wave and 1 derivation Gaussian wave.

The ratio of $\sigma$ and the wavelength $\lambda$, $\frac{\sigma}{\lambda}$ determine the bandwidth of the Gabor filter. The bandwidth can be described as

$$b=\log_2\frac{\frac{\delta}{\lambda}\pi+\sqrt{\frac{\ln 2}{2}}}{\frac{\delta}{\lambda}\pi-\sqrt{\frac{\ln 2}{2}}}$$

(5.12)

The study in Ref. [69] evaluates the effect of the different filter banks on the performance of the Gabor filter. They proved that filter parameter selection is one of the major factors that influence filter performance, and different tasks need to accordingly select different parameters that meet the requirement of the processing issues. Among the parameters, the most important parameters contain the number of frequencies or scales, the number of orientations, and the central frequency at the highest frequency.

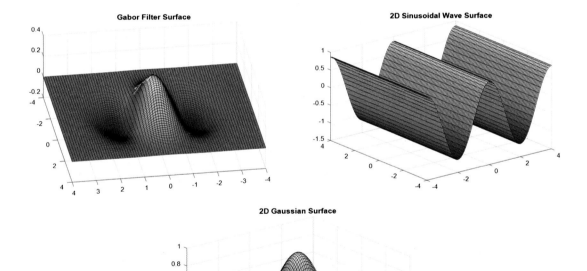

FIG. 5.5  3D surface of the Gabor filter with sinusoidal and Gaussian function.

According to Ref. [70], the wavelength along with the Gaussian parameter in the Gabor filter determines the sensitivity to the frequencies, and the variation of orientation $\theta$ influences the sensitivity to the edge and texture orientation. It has also been proved that the Gaussian derivation depends on the orientation, which stands for the spatial frequency.

A combination of multiple Gabor filters with different frequency and orientation can be used as feature extractors to represent different features in the image. Ensemble Gabor filters play a similar rule as a single convolutional layer as described in CNN. Given a grayscale image with size $M \times N$, denoted as I(M, N), and a Gabor function $G_{i,j}(M, N)$, where $i$ equals to the number of the scale of the filter and $j$ equals to the number of orientations, the image passes each filter bank with the following transform equations:

$$O_{i,j}(m, n) = \sum_{a=1}^{M} \sum_{b=1}^{N} I(a, b)\overline{G}_{i,j}(m - a, n - b) \quad (5.13)$$

where $\overline{G}_{i,j}$ denotes the complex form of Gabor filter. The output $O_{i,j}(m, n)$ is the product of complex convolution between the input image and the Gabor filter. Then, the output can be further decomposed into a real component and imaginary component as follows [71]:

Real Component:

$$R(m, n) = \mathrm{Re}\left[O_{i,j}(m, n)\right] \quad (5.14)$$

Imaginary component:

$$I(m, n) = \mathrm{Im}\left(O_{i,j}(m, n)\right) \quad (5.15)$$

Once the real part and imaginary part of the output have been evaluated, the phase feature and magnitude feature of the output can be described as follows:

Phase Feature:

$$\varnothing(m, n) = \arctan(I(m, n) / R(m, n)) \quad (5.16)$$

While Magnitude Feature:

$$A(m, n) = \sqrt{R^2(m, n) + I^2(m, n)} \quad (5.17)$$

Due to the fact that the phase response is very sensitive to spatial location and varies dramatically even between few pixels, phase features are always ignored to avoid unstable performance and only magnitude features are selected. This is because, contrary to phase features, magnitude features are less unstable and vary slightly with the spatial position and hence become the primary choice of Gabor filter response.

One of the popular parameter combinations of Gabor filter is to choose four to six scales, which are also known as wavelength, and eight orientations from 0 degrees to 180 degrees. This is because the absolute response of the filter with orientation larger than 180 degrees would be the same as those between 0 and 180 degrees. So, in this step, the scales and orientation of the Gabor filter are simply chosen 5 and 8, separately. It should be noted that other combinations may lead to a better result, and there are many methods to find the optimal combinations such as grid searching and generic algorithm; however, this is beyond the scope of this head pose estimation work. The feature visualization of the Gabor filter is shown in Figs. 5.7 and 5.8, based on the raw image shown in Fig. 5.6.

HOG is a popular feature extractor that can extract the shape and edge of a given image. It was first introduced in 2005 as a SIFT similar feature extractor. The

FIG. 5.6 Face Detection in Real-world Vehicles.

Magniture Features of Gabor Filter

FIG. 5.7 Magnitude Feature of the Gabor Filter.

Real Part features of Gabor Filter

FIG. 5.8 Real Part Feature of the Gabor Filter.

main idea behind HOG is that the local shape and appearance can be well described by the image pixel density gradient or edge directions [72]. HOG has been widely used in previous computer vision studies, especially in pedestrian detection and many other object detection problems. HOG is a statistical method that works on local regions instead of a whole image. Due to these reasons, it has a good characteristic that is scale and illumination invariant. The basic component in the HOG detector is some units called cell, the gradient magnitude and orientation of each pixel will be calculated within the cell. Then, a certain number of cells composed of a larger area is called a block. The raw image contains a certain amount of blocks that can be either overlapped with each other or nonoverlapped.

A standard process of HOG feature extraction normally contains gamma normalization, pixel gradient computation, spatial and orientation computation within cells, orientation binning, block contrast normalization, and HOG feature collection. The detail process is described later:

1. Gamma normalization. This process is not always necessary according to Dalals and Trigges [72]. The main idea here is to modify the image contrast and reduce the negative influence of shadow and illumination to the local region of the image as well as decrease noise influence. Suppose we have a grayscale $I(x, y)$, gamma normalization is described as

$$I(x, y) = I(x, y)^{Gamma} \qquad (5.18)$$

where Gamma is the compression factor and normally selected to be 0.5.

2. Compute the gradient of each pixel. The gradients of each pixel are computed in two directions, which are along the horizontal and vertical directions. These two-direction gradient values are obtained with convolving $[-1\ 0\ 1]$ and $[-1\ 0\ 1]^T$ kernels for the raw image and then the gradient is calculated as

$$G_x(x, y) = I(x+1, y) - I(x-1, y) \qquad (5.19)$$

$$G_y(x, y) = I(x, y+1) - I(x, y-1) \qquad (5.20)$$

where $G_x(x, y)$, $G_y(x, y)$ represent the gradient values of each pixel in $x$ and $y$ directions (horizontal and vertical).

3. Spatial and orientation computation. After obtaining the two gradient values of each pixel, the relative gradient magnitude and direction can be obtained directly:

$$M(x, y) = \sqrt{G_x(x, y)^2 + G_y(x, y)^2} \qquad (5.21)$$

$$\theta = \arctan \frac{G_y(x, y)}{G_x(x, y)} \qquad (5.22)$$

where $G(x, y)$ and $\theta$ are the magnitude and orientation of the gradient in each pixel, and the direction of the gradient is constrained to be between 0 and 180 degrees.

4. Orientation binning. As the original orientation range is between $-90$ degrees and 90 degrees, the negative values have to sum 180 to become positive. The reason is that in most studies, the orientation is treated as unsigned values with a range from $(0.-180)$. After obtaining the gradient orientations, they can be classified with a certain number of bins. Then, using magnitude to weight the number and we can get the histogram of each cell. The bin number is noted as $B$. Sometimes an orientation value next to a bin boundary may lead to an adverse effect on the histogram statistics if we strictly contribute the orientation to one bin. To prevent these artifacts, each pixel in a cell will contribute to two adjacent bins and a technique called bilinear interpolation can be used for efficient voting. Given bin number $B$, the width of each bin in degrees can be evaluated as

$$w = \frac{180}{B} \qquad (5.23)$$

For a bin $i$ with boundary $[wi, w(i+1)]$, the center for this bin is

$$C_i = w\left(i + \frac{1}{2}\right) \qquad (5.24)$$

Given a pixel in a cell with gradient magnitude $m$ and orientation $\theta$, the voting for bin $i$ and bin $i+1$ can be expressed separately as follows:

$$v_i = m \frac{C_{i+1} - \theta}{w} \qquad (5.25)$$

$$v_{i+1} = m \frac{\theta - C_i}{w} \qquad (5.26)$$

where $v_i$ and $v_{i+1}$ is the voting for bin $i$ and $i+1$.

5. Block contrast normalization. Construct the block with multiple cells and concatenate the histogram features of the cells that lead to the block features. After block histogram features are connected, it will be normalized to reduce the effect of illumination and shadow to each region. The normalization method contains L2 norm, L1 norm, and their extension evaluation. Normally, the block features can be normalized as follows:

$$V = \frac{V}{\sqrt{V^2 + \varepsilon}} \qquad (5.27)$$

where $V$ is the concatenate block histogram feature vector and $\varepsilon$ is a small positive constant that prevents zero division.

6. The final process of HOG feature extraction is to connect all the block histogram features together and concatenate them into a bigger feature vector. The HOG feature visualization is shown in Fig. 5.9.

## Head Pose Estimation Using Random Forest

Random forest is a set of ensemble decision trees, which can be more accurate and efficient in dealing with large scale data and has the ability to overcome the overfitting problem of the decision tree. A head pose estimation process is shown in Fig. 5.10. In this part, a bagging tree is trained to predict the head orientation based on 2D and 3D depth features. The Biwi face orientation dataset is used to assess the proposed method [73]. Biwi dataset contains 15k images sampled from 20 subjects

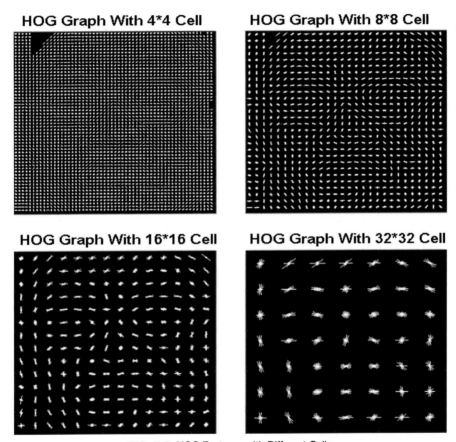

**FIG. 5.9** HOG Features with Different Cells.

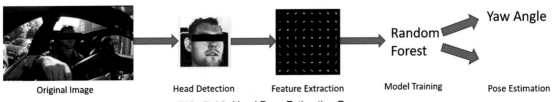

Original Image     Head Detection     Feature Extraction     Model Training     Pose Estimation

**FIG. 5.10** Head Pose Estimation Process.

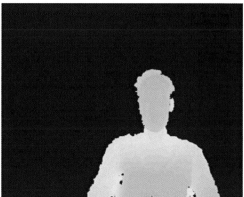

FIG. 5.11 RGB and Depth Images are Given by Kinect.

with continuous head orientation. The subjects 2 and 18 are used for testing, and the rest data are used for training. Fig. 5.11 illustrates the RGB image and the depth image with the detected head position in the RGB image. For each image, the head position is detected and the corresponding facial features are extracted. Then, these features are used as inputs to train the random forest model.

To evaluate the bagging trees method, the number of trees that we used are selected from 10 to 100 with 10 intervals. Therefore each feature vector will be used to train 10 different random forest models with different bagged trees. The regression accuracies are illustrated in Fig. 5.12.

As shown in Fig. 5.12, the regression model with concatenating features achieves the best accuracy in the yaw direction and the Gabor filter achieves the lowest accuracy. The $x$-axis is the number of trees used in the forest tree model, while the $y$-axis is the mean absolute error between the estimation and ground truth. The best estimation performance is shown in Table 5.3.

### Driver Body Detection

Driver body movement is another behavior signal that can reflect the driver's mental state. Driver body movement and body detection will benefit the understanding of the human driver and the prediction of driver actions. Besides, the body signals are the indicators that can be used to determine whether the driver is concentrating on the driving task or not. Sometimes the driver will perform some secondary tasks such as playing mobile phones and the on-board video system. With the body movement features, it is able to detect whether the driver is distracted or not. As driver distraction is closely related to driver intention, body movement and body status will contribute to the prediction of

driver intention. If the driver is distracted, he is less likely to have a driver-oriented intention. Therefore if the driver is detected as distracted and performing the secondary tasks, the predicted driver's intention has to be rechecked and reevaluated.

In this part, the driver's upper body is detected and tracked with the help of Microsoft Kinect 2 sensors. It can detect and track up to four people at the same time. In this part, only the upper body of the driver will be tracked. The Kinect sensor can detect 23 human body joints in total, as shown in Fig. 5.13. As the driver is seated on the driving seat, the upper body joints used in this project consists of the shoulder, elbow, wrist, hand, and thumb joints of both left and right hand.

The body joint positions in RGB images and the corresponding depth values are extracted. The signals captured with the Kinect sensor consist of the RGB images, the driver's head pose in three dimensions, and the driver's upper body joints. All the signals are synchronized with the timestamp. Fig. 5.14 illustrates the head and joints detection result and the depth image with the Kinect sensor. In the near future, these signals will be used to train machine-learning models that are used to distinct driver secondary tasks from the normal driving behavior.

### DRIVER BEHAVIORS RECOGNITION EXPERIMENTAL DESIGN AND DATA ANALYSIS

#### Overall System Architecture

The procedure taken to construct a behavior recognition model is described in this section. The driver monitoring system architecture is shown in Fig. 5.15. The general structure of this study consists of three parts. First, the driver's head and body data

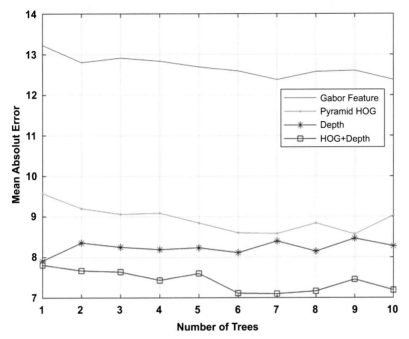

FIG. 5.12 Estimation Result Given by Different Feature Vectors.

**TABLE 5.3**
**Estimation Performance in 3 Degrees**

| | Tree No. | 10 | 20 | 30 | 40 | 50 | 60 | 70 | 80 | 90 | 100 |
|---|---|---|---|---|---|---|---|---|---|---|---|
| Yaw | Gabor | 13.23 | 12.80 | 12.91 | 12.84 | 12.69 | 12.59 | 12.37 | 12.57 | 12.60 | 12.37 |
| | HOG | 9.575 | 9.199 | 9.060 | 9.086 | 8.843 | 8.601 | 8.581 | 8.841 | 8.569 | 9.028 |
| | Depth | 7.919 | 8.355 | 8.250 | 8.189 | 8.234 | 8.112 | 8.395 | 8.146 | 8.463 | 8.275 |
| | H+D | 7.813 | 7.672 | 7..642 | 7.432 | 7.602 | 7.112 | 7.100 | 7.166 | 7.454 | 7.191 |
| Pitch | Gabor | 8.448 | 8.793 | 8.753 | 8.243 | 8.418 | 8.442 | 8.575 | 8.570 | 8.702 | 8.512 |
| | HOG | 7.703 | 7.716 | 7.801 | 7.265 | 7.718 | 7.591 | 7.188 | 7.303 | 7.418 | 7.656 |
| | Depth | 6.277 | 6.107 | 5.684 | 6.059 | 5.941 | 5.808 | 5.741 | 5.819 | 5.716 | 5.620 |
| | H+D | 6.233 | 6.199 | 6.340 | 6.180 | 6.119 | 5.903 | 5.978 | 5.951 | 5.951 | 5.893 |
| Roll | Gabor | 3.822 | 3.622 | 3.135 | 3.281 | 3.022 | 3.187 | 3.084 | 3.104 | 3.150 | 3.099 |
| | HOG | 4.044 | 3.846 | 3.716 | 3.480 | 3.527 | 3.340 | 3.506 | 3.452 | 3.353 | 3.272 |
| | Depth | 3.821 | 3.489 | 3.430 | 3.137 | 3.367 | 3.223 | 3.196 | 3.225 | 3.168 | 3.247 |
| | H+D | 3.728 | 3.575 | 3.476 | 3.402 | 3.452 | 3.444 | 3.299 | 3.477 | 3.408 | 3.332 |

are collected and time stamped. Then, the signals are smoothed, and the noise is filtered. Second, feature importance prediction is proposed using a combination of random forests and maximal information coefficient (MIC), and the feature importance given by the two algorithms shows strong consistency. The "model selection" block in Fig. 5.15 processes the feature evaluation based on the feature importance provided by the "feature importance estimation" block. Meanwhile, the influence of depth, head, and body features to the driver status detection will be studied. Then, real-time driver behavior identification will be

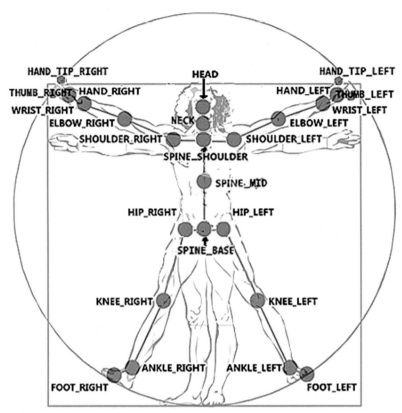

FIG. 5.13 23 Body Joints Detected with Kinect 2. (https://developer.microsoft.com/en-us/windows/kinect/).

FIG. 5.14 Head, Body Detection in Depth Image.

conducted using a feedforward neural network (FFNN) model with leave-one-out (LOO) cross-validation. Finally, the performance with different features is analyzed, and a behavior classification

performance comparison between different algorithms will be proposed.

### Inner-Vehicle Experiment Setup and Data Collection

In this part, the experiment setup and data collection methods are introduced. Driver behavior data are collected using the low-cost range camera Kinect, which was developed by Microsoft. In this study, the second version of Kinect (V2) was adopted. Kinect is a consumer camera that supports color images, depth images, audio, and infrared information. It was first designed for indoor human interaction with computers and has been successfully applied in vehicles for driver monitoring [35,36].

Kinect supports tracking the head and the body skeletons of as many as six individuals. In this work, the head and upper body joint detection functions are integrated for collecting driver head and body signals. The head detection provided by the Kinect requires tracked body information. Therefore to use the Kinect inside a vehicle, it must be mounted above the dashboard to have a full

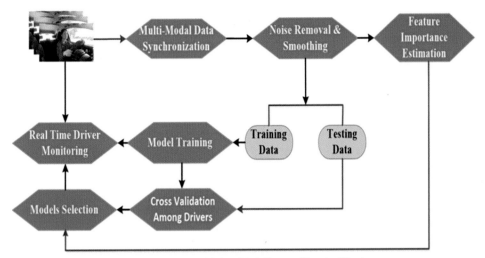

FIG. 5.15 Proposed Driver Task Recognition Architecture.

vision of the driver's body. Considering the mounting requirements in Ref. [37], the Kinect is mounted in the middle of the dashboard, facing the driver, which does not interfere with the driver's field of view and allows for monitoring of the driver's entire upper body. Fig. 5.16 illustrates the detected head center and upper body joints using Kinect and an example depth image.

In this section, the head and body signals and color and depth images are collected and synchronized with a time stamp. The sampling rate is eight frames per second. The data are sampled with an Intel Core i7 2.5 GHz computer and the code is written in C++ based on the Windows Kinect SDK and OpenCV. The size of the color image captured using a Kinect is 1920 × 1080. However, to increase computational efficiency, the stored color image was compressed to 640 × 360. According to Ref. [25], short-term driver mirror-checking actions last from 0.5 s to 1 s. Therefore the sampling frequency is fast enough to capture normal driver actions and behavior. The three-

dimensional head rotation vector contains yaw, pitch, and roll angles. The upper body joints are recorded using $X$ and $Y$ coordinates in the color image and the corresponding depth value in the depth map. The 42 signals collected are shown in Table 5.4.

## DATA PROCESSING

The Kinect data processing methodologies used in this study are described in this section. The two data processing steps are head rotation calibration with an orientation sensor and noise removal and smoothing based on a combination of a median filter and an exponential filter.

## KINECT SENSOR-BASED HEAD ROTATION DATA CALIBRATION

In Ref. [38], Kinect head rotation data were evaluated and compared with a high-precision head rotation

FIG. 5.16 Driver Body Joints Detection with Kinect.

**TABLE 5.4**
Multimodal Features Given by Kinect.

| MULTIMODAL FEATURES FROM KINECT (42 FEATURES) | |
|---|---|
| Head (12 Features) | Head pitch angle (pitch) Head yaw angle (yaw) Head roll angle (roll) Left eye (X, Y, Z) Right eye (X, Y, Z) Nose (X, Y, Z) |
| Body (30 Features) | Left hand (X, Y, Z) Right hand (X, Y, Z) Left wrist (X, Y, Z) Right wrist (X, Y, Z) Left elbow (X, Y, Z) Right elbow (X, Y, Z) Left shoulder (X, Y, Z) Right shoulder (X, Y, Z) Left-hand tip (X, Y, Z) Right-hand tip (X, Y, Z) |

detection device. The author concluded that the average errors in absolute yaw, pitch, and roll angles were $2.0 \pm 1.2$ degrees, $7.3 \pm 3.2$ degrees, and $2.6 \pm 0.7$ degrees, respectively. However, the experiment and data calibration were proposed for indoor environments in standard conditions. However, in this work, the Kinect V2 was implemented inside a vehicle, which is a more challenging environment. During the experiment, the Kinect detection signals inside the vehicle have more noise and are less stable than the signals collected inside the room. Therefore the first step was to calibrate the Kinect head rotation data with a high-precision head rotation sensor. As driver head rotation is a very important signal for determining the driver's attention and distraction status, only the head rotation signals were evaluated in this study and the detected body positions provided by the Kinect were not calibrated.

To calibrate the estimated head rotation results of the Kinect, a head-mounted head tracker was used and 3-degree rotation data from the head tracker were used as the ground truth. The head tracker is based on an Arduino microcontroller board and an intelligent nine-axis absolute orientation sensor (BNO055)

designed by BOSCH. The sampling frequency of the orientation sensor is up to 100 Hz. The rotation sensor and Arduino data-recording sensor are fixed on a head-mounted harness belt strap, as shown in Fig. 5.17. Moreover, seven driver behaviors studied in this research are shown in Fig. 5.18.

The Kinect sensor is mounted in the middle of the front dashboard. The optical axis of the Kinect camera is not perpendicular to the yaw axis of the driver's head, which will influence the detected yaw angles. The rotation angle of the Kinect sensor in world coordinates is reflected by a constant bias of the detected yaw angle, as shown in Fig. 5.19. The blue line is the original yaw angle. The yellow line is the shifted yaw angle, which shifts the original signal by a constant offset (30 degrees). The red line shows the ground truth results of the head tracker. The calibrated Kinect signal and ground truth have similar variations, which means that the head rotation angle detected by Kinect is reliable and can be used for further analysis.

The data recording frequency for the head tracker is 30 Hz, which is approximately three times greater than Kinect; therefore, the head tracker yaw angle shown in Fig. 5.19 is the smoothed version of the original signal. Finally, the mean error and standard deviation between the calibrated Kinect signal and the head tracker for yaw, pitch, and roll angles are $1.93 \pm 11.55$ degrees, $1.47 \pm 5.98$ degrees, and $1.44 \pm 6.98$ degrees, respectively.

**Noise Removal and Data Smoothing**
The temporal spikes due to noise can cause more serious problems. The body and head detection results using the Kinect can be influenced by lighting conditions or the location and distance to the driver and human gesture or body pose can influence joint detection, especially inside the vehicle. Due to the less precise detection results using the Kinect, an integrated signal process scheme combining two different filtering techniques is adopted in this study.

FIG. 5.17 Experimental Setup for Driver Behavior Detection.

Normal Driving          Right Mirror Checking          Rear Mirror Checking

Left Mirror Checking          Texting          Answering Mobile Phone

Using Radio Device

FIG. 5.18   Illustration of Seven Driver Behaviors.

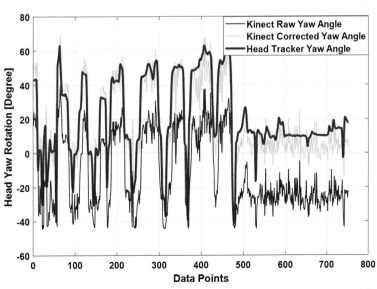

FIG. 5.19   Head Yaw Angle Detection Results Given by Kinect and the Orientation-based Head Tracker.

Specifically, abnormal data removal and exponential smoothing filter are applied to the raw signals to smooth and track the detection results

$$\widehat{x}_n = \begin{cases} x_n, & x_n \neq 0 \\ mean(X_{pre}), & x_n = 0 \end{cases} \tag{5.28}$$

where $\widehat{x}_n$ is the filtered data value, $x_n$ is the raw data, and $X_{pre}$ represents all the nonzero data before step $n$. The exponential smoothing filter is defined as follows (2):

$$\begin{cases} s_0 = x_0 \\ s_t = \alpha \sum_{i=0}^{W} (1 - \alpha)^i x_{W-i} \end{cases} \qquad (5.29)$$

where $s_t$ is the smoothed version of raw signal $x_t$, $W$ is the sliding window size that depends on the number of previous inputs used for smoothing, $\alpha$ is called the dampening factor, which controls the weight of previous inputs, and $0 \leq \alpha \leq 1$.

As shown in Fig. 5.18, the driver's right arm is partially blocked by the steering wheel, which causes the Kinect to detect inaccurate body joints. During the data recording process, some data points will be lost or unreasonable due to the driver pose, lighting conditions, or Kinect detection algorithms. First, these data points are recorded as zeroes to indicate abnormal detection status. Then, the data are fed into the hierarchical filter module to smooth the original signals. To track the signals, an abnormal data removal algorithm is applied. The zero data points are replaced by the mean value of the nonzero data. Then, the exponential smoothing filter is applied to further smooth the noisy signal. Fig. 5.20 shows the smoothing result of the right wrist signal.

## TASKS IDENTIFICATION ALGORITHMS DESIGN

In this section, driver feature evaluation is proposed to study the relationship between driver features and driver behavior estimation. The most relevant features for driver behavior recognition are detected. Then, a feedforward neural network is adopted as the driver behavior classifier to identify the driver actions based on the selected feature vectors.

### Feature Importance Evaluation Using Random Forest and Maximal Information Coefficient

For some machine-learning tasks, feature vector dimensions can be very high (hundreds or thousands, or even larger). Although machine-learning methods are particularly suitable for modeling large datasets, they are always viewed as a black box where it is difficult to analyze the intrinsic structure. Therefore it is important to understand how the input features influence or are associated with the model output. In this study, to understand how the driver signals influence behavior detection, the relationship between body signals and driver behavior is analyzed. Such feature evaluation and selection enable a subjective understanding of the relationship between driver body signals and behavior.

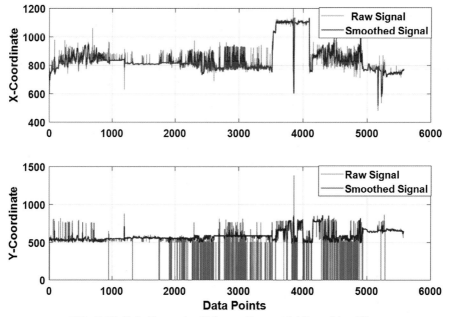

FIG. 5.20 Date Processing Using an Exponential Smoothing Filter.

Feature selection is a major research area of feature engineering. By selecting a subset of feature vectors, machine-learning models can be trained more efficiently, and better results can be obtained. In this section, to understand how driver features influence the corresponding behavior detection and which features are important for the behavior recognition task, two distinct feature selection methods are applied and compared. First, a random forest was used to estimate the driver feature importance with an out-of-bag (OOB) dataset. Second, MIC is used as another indicator of the association between features and the behavior class. The final conclusion of feature importance will be summarized according to the results given by these two distinct algorithms.

### Random forests for feature importance estimation

Random forests, introduced by Breiman in 2001, were built on classification and regression trees [39]. It has proven to be a powerful machine-learning tool for many applications: In Ref. [40], the author evaluated the RF classification performance on 121 public datasets and the RF algorithm achieved the best classification result among 179 algorithms. RF is an ensemble learning machine that integrates multiple decision trees. One decision tree is constructed with one root node and multiple middle leaf nodes. The prediction ability for a single tree is limited and given a large dataset, overfitting is common for a single decision tree. According to the drawbacks of a single decision tree, RF combines multiple decision trees and uses average or voting schemes to calculate the final results.

Random forest is a specific application of ensemble learning. The RF is built with multiple decision trees. To increase the diversity of each tree in the forest, RF is trained using a bootstrap aggregating (Bagging) technique. Specifically, the number of trees $B$ in the RF is selected. Then, according to this number, $B$ separate training datasets are chosen from the original dataset. As Bagging is a random sampling technique with replacement, approximately one-third of the data is not used for training each subtree. The remaining dataset for each tree is the OOB dataset. Normally cross-validation is not necessary for training RF as the OOB can be used to evaluate the model performance by evaluating the OOB errors [39]. Moreover, the OOB dataset can be used to evaluate the feature importance for model accuracy. The Bagging technique causes the initial training dataset that only contains 63.2% samples, and the rest 36.8% of samples will construct the OOB dataset. To train each decision tree, it is necessary to record the training sample for the tree. Consider the

training set as $D = \{(x_1, y_1), (x_2, y_2), \cdots (x_m, y_m)\}$, $D_t$ as the training data for tree $h_t$, and $H^{oob}(x)$ represents the OOB estimation result on sample $x$, hence

$$H^{oob}(x) = \operatorname*{argmax}_{y \in Y} \sum_{t=1}^{T} I(h_t(x) = y) \qquad (5.30)$$

And the generalizing error for the OOB data is

$$\varepsilon^{oob} = \frac{1}{|D|} \sum_{(x,y) \in D} I\left(H^{oob}(x) \neq y\right) \qquad (5.31)$$

The base learner for RF is the single decision tree, and the training method for RF relies on the utilization of the Bagging technique. Moreover, the trees among the forest are expected to be as diverse as possible to that the model has a good generalization ability. Therefore a random branch choosing is applied during the training for each tree. Specifically, for each node of the base decision tree learner, a subset that contains $k$ properties are chosen from the total property set. Then, the optimal property will be selected in this subset for generating a branch. The parameter $k$ will control the randomness of the random forest and normally it can be selected as $k = \log_2 d$ [39].

To increase the diversity of each tree in the forest, RF is trained using a bootstrap aggregating (Bagging) technique. Specifically, the number of trees $B$ in the RF is selected. Then, according to this number, $B$ separate training datasets are chosen from the original dataset. As Bagging is a random sampling technique with replacement, approximately one-third of the data is not used for training each subtree. The remaining dataset for each tree is the OOB dataset. Normally, cross-validation is not necessary for training RF because the OOB can be used to evaluate the model performance by evaluating the OOB errors [39]. Moreover, the OOB dataset can be used to evaluate the feature importance for model accuracy. To obtain the feature importance, for each variable $X_i$, the variable is randomly permuted. The feature importance is calculated as follows:

$$I(X_i) = \frac{1}{B} \sum_{t}^{B} \widetilde{OOBerr}_{t^i} - OOBerr_t \qquad (5.32)$$

where $X_i$ is the permuted $i$th feature in the feature vector $X$, $B$ is the number of trees in the random forest, $\widetilde{OOBerr}_{t^i}$ is the model prediction error of the perturbed OOB sample with the permuted feature $X_i$ for tree t, and $OOBerr_t$ is the untouched OOB data sample with the permuted variable.

The concept of permutation feature importance is that a large importance value indicates the feature is influential in the prediction and permuting the feature

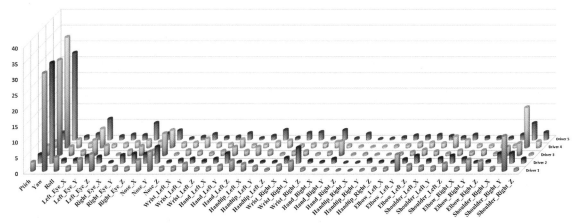

FIG. 5.21 Feature Importance Prediction Using Random Forests Based on the Permutation Method.

value will influence the model prediction. In contrast, a small influential feature will have no or less impact on model prediction. The predicted feature importance for the 42 driver signals using RF is illustrated in Fig. 5.21. From the importance estimation results, the driver yaw angles are extremely important for action classification for all five drivers. To verify the prediction results given by RF, the next section proposes another feature evaluation technique called the maximal information coefficient, which uses a completely different method to estimate feature importance.

### Maximal information coefficient for feature importance estimation

The MIC is designed to efficiently solve the mutual information estimation problem for continuous variables and continuous distributions. The MIC provides an equitable measurement for the linear or nonlinear strength association between two variables. The MIC introduced a maximal mutual information searching technique by varying the grid drawn on a scatterplot of two variables [41]. Mutual information usually can be used to evaluate the mutual dependence between different variables and assess the amount of information the two variables share, or more generally, the correlation between the joint distribution of the two variables and the product of the independent distribution of the two variables [42]. The mutual information for two discrete vectors is defined as

$$MI_D(X, Y) = \sum_{y \in Y} \sum_{x \in X} p(x, y) \log\left(\frac{p(x, y)}{p(x)p(y)}\right) \quad (5.33)$$

where $MI_D$ is the mutual information of two discrete vectors, $p(x, y)$ is the joint probabilistic distribution of $x$ and $y$. $p(x)$ and $p(y)$ are the marginal probability

distribution functions of $x$ and $y$, respectively. For continuous variables, the mutual information format is slightly changed to

$$MI_C(X, Y) = \iint p(x, y) \log\left(\frac{p(x, y)}{p(x)p(y)}\right) dxdy \quad (5.34)$$

where $MI_C$ is the mutual information for two continuous vectors, and $p(x, y)$, $p(x)$, and $p(y)$ represent the corresponding probabilistic density functions.

As shown in Eq. (5.34), calculating the mutual information of continuous variables is difficult. Therefore the maximal information coefficient technique, which concentrates on the optimal binning method, is applied to assess the mutual information of the continuous case. Meanwhile, MIC enables the mutual information score to be normalized into the range [0,1], which makes assessing the dependency and corelationship between two variables more convenient. In this study, in addition to the first three continuous head rotation angles, the remaining features are discrete image coordinates and depth values. Therefore the MIC can be efficiently used for feature association prediction.

### Comparison of the feature importance prediction

To evaluate the prediction results of feature importance using the two algorithms, the 10 most important features for each subject are extracted and compared. Specifically, for each driver, the 10 most important features are selected. Then, five selected feature vectors are fused into 42 bins and the count in each bin represents the number of occurrences for each feature of the five subjects. Therefore the highest value, 5, indicates the feature is one of the 10 most important features for all five drivers. The importance estimation for the 42 features given by two feature importance estimators (RF and MIC) are illustrated in Fig. 5.22. As shown in Fig. 5.22, although the results are not the same, the important estimation results given by the two algorithms show the consistency. For example, the head rotation angles, as well as the arm positions, are estimated as the most important features for the two algorithms while some other features like wrist and hand tips are regarded as less important by the two algorithms.

The statistical results are shown in Fig. 5.22. Blue bars represent the results of RF while red bars represent the standard for the maximal information coefficient method.

As shown in Fig. 5.22, although the prediction results of the two algorithms are not identical, there is some consistency in the results of the two algorithms. For example, the driver yaw and *y*-coordinate of the right shoulder features (Nos. 2 and 41) are both significant. According to Fig. 5.22, the 12 most important features (marked as the 10 most important features by at least two drivers) are listed in Table 5.5.

According to Table 5.5, the importance of predictions given by RF and MIC are similar. The most important features are the head rotation angles (yaw, pitch, and roll), eye and nose position, shoulder position, and hand position. The remaining features such as the wrist, hand tip, and elbow positions are less likely to influence the behavior detection result. A quantitative analysis of the feature impact on behavior recognition

based on a feedforward neural network is proposed in the next section.

## Feedforward Neural Network for Driver Behavior Classification

In this section, an ANN is used for driver behavior pattern recognition. Specifically, a one-way FFNN is adopted. The FFNN passes the input vectors to the output layer by layer without any feedback connections. FFNN is a powerful tool for solving complex nonlinear mapping problems. By learning the neuron parameters

**TABLE 5.5**
Feature Importance EEstimation Result.

|  | Importance Order | Features |
|---|---|---|
| RF | 1–5 | Yaw, Left_eye_Y, Red_eye_Y, Nose_Y, Right_shoulder_Y |
|  | 6–12 | Pitch, roll, Left_eye_Z, Nose_X, Left_hand_Y, Right_wrist_Y, Right_shoulder_X |
| MIC | 1–5 | Yaw, Left_eye_X, Left_eye_Y, Nose_X, Right_shoulder_Y |
|  | 6–12 | Right_eye_Y, Right_eye_Z, Nose_Y, Nose_Z, Right_hand_X, Left_shoulder_X, Right_shoulder_X |

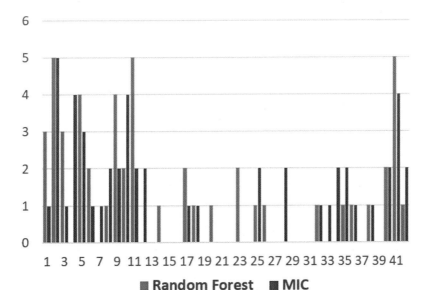

■ **Random Forest** ■ **MIC**

FIG. 5.22 Feature Importance Prediction Results Using Random Forest OOB Permutation.

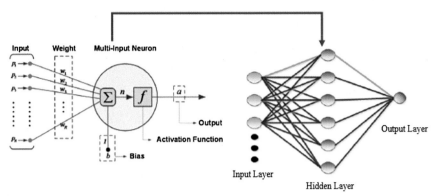

FIG. 5.23  Illustration of Multiinput Neuron and Multilayer FFNN.

and the connection width, the FFNN model can construct a nonlinear mapping between the input and output (Fig. 5.23).

The artificial neural network model is inspired by the biological neural networks that constitute the animal brains. The basic units of the ANN model are called the neurons, each ANN model is constructed with multiple neurons, which connected loosely through the weighted connections. The connection between the artificial neurons is called "edges". The neurons will compute the weighted inputs with some nonlinear activation functions. Normally the activation function can be represented as a *sigmoid* or *tanh* function. Once the computed result exceeds a certain threshold, then this neuron is called to be activated, otherwise, output the zero value. Usually, the ANN is structured with multiple layers and each layer contains multiple neurons. The layers between the input layer and the output layer are called hidden layers of the ANN model. Normally, the more hidden layers the ANN model has, the more powerful the model will be. However, increasing the number of hidden layers will significantly increase the computational burden and the requirement for data volume. Therefore deeper ANN models are more easily to be overfitted. The training process of the ANN model is to judge the weights of the edges and bias in each neuron to minimize the total loss of the output layer. In this part, the feedforward neural network is used, which is a specific ANN model with all the edges that have a feedforward connection.

The left part of Fig. 5.20 shows the structure of a single neuron and the right part indicates the architecture of a multilayer FFNN with one hidden layer. The neuron unit will weigh the $R$ inputs with a bias $b$ to form the activation input n, which has the following form:

$$n = \sum_{j}^{R} w_j p_j + b = Wp + b \qquad (5.35)$$

Then $n$ passes through the activation function $f$ to generate the neuron output $a$.

$$a = f(n) \qquad (5.36)$$

The total FFNN model can be approximately represented as follows:

$$y = f(X, \theta) + \varepsilon \qquad (5.37)$$

where $y$ is the output of FFNN, $f()$ is the learned model mapping function with a model parameter $\theta$, $X$ is the input data vectors, and $\varepsilon$ is the bias between the actual output and the target.

For the FFNN, parameter $\theta$ represents the set of activation function parameters and the width set between neurons. In this study, a multilayer FFNN with one hidden layer is used to train the driver behavior, recognition model. The sigmoid transfer function in the hidden layer is chosen. The sigmoid activation function for a single neuron is represented as

$$f = \frac{1}{1 + e^{-X}} \qquad (5.38)$$

where $f$ is the neuron output and $X$ is the neuron input, which has the following form:

$$X = \sum_{i=1}^{N} \omega_i x_i + b \qquad (5.39)$$

where $\omega_i$ is the weight of the $i$th input, and normally each neuron has a bias parameter $b$. An important reason for using the sigmoid activation function is the computation efficiency during model training. The back propagation requires the derivative of the neuron transfer function to be calculated. The sigmoid function has the convenient derivative form [43]:

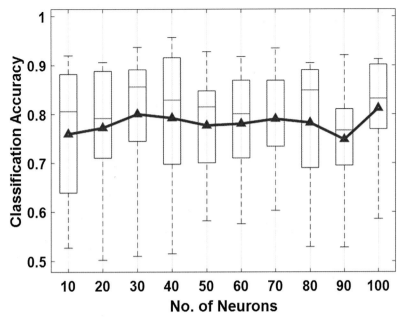

**FIG. 5.24** Boxplot of the FFNN Classification Results with Cross-validation of the Neuron Numbers in the Hidden Layer.

$$f' = f \cdot (1 - f) \qquad (5.40)$$

Although the sigmoid function will cause the loss of a gradient problem in most scenarios, it is not a serious problem in this shallow network case. In this case, the supervised FFNN is trained with driver head and body signals as the input and an output of the corresponding behavior among the seven actions. Different from some existing research that uses time-series models, the FFNN used in this study does not consider the previous step status of the driver. The reason for this is that humans can normally distinguish the current driving behavior using one image and do not require video sequences. Different from the inner mental states of the driver, which is a long-term process and depends on previous states, the outer behaviors can be considered a transient state and are not highly dependent on prior information. Therefore the FFNN is applied to detect the driving tasks frame by frame based on the collected driver's body information.

As time information is not considered in the model construction procedure, the training and testing dataset are reordered randomly. For model training, cross-validation is used. Specifically, the LOO method is adopted. For the five-driver dataset, data of four drivers are used for model training and validation,

the data of the remaining driver are used to test the classification performance. The general classification accuracy is the average of the five classification results. Another hyperparameter for FFNN is the number of neurons in the hidden layer. To evaluate the influence of neuron quantity on classification performance, different neuron numbers and cross-validation are studied. A boxplot of the classification results is shown in Fig. 5.24. The neuron numbers vary from 10 to 100 with an interval of 10. The red line represents the mean accuracy of the five drivers with different neurons. As shown in Fig. 5.24, variation in the number of neurons does not significantly influence performance. The most accurate detection occurs at the 100-neuron cases, with an accuracy of approximately 81.2%. In the next section, more detailed statistical results using FFNN with 60 neurons are proposed, and the results are compared with multiple machine-learning methods (Fig. 5.24).

## EXPERIMENT RESULTS AND ANALYSIS

In this section, the task recognition results are discussed. Specifically, task classification with FFNN is compared with other machine-learning methods. In addition, the impact of the head, body, and depth information on the classification results will be evaluated separately in part two.

## Behavior Recognition Results

In this section, the identification accuracy for driving and nondriving tasks is analyzed. As mentioned in the previous section, the classification model is trained using the LOO-cross validation method. The prediction results for the five drivers are illustrated in Table 5.6. The first four mirror-checking tasks are divided into driving-related tasks, while the remaining three tasks are divided into nondriving and distraction tasks.

The seven driver tasks are ordered as {Normal Driving, Right Mirror Checking, Rear Mirror Checking, Left Mirror Checking, Using Video Device, Texting, and Answering Mobile Phone}. As shown in the far-right column of Table 5.6, the average classification result (Ave) for each driver is defined as the average of the seven tasks. The mean values shown in the bottom row represent the average classification accuracy for each task of the five drivers. Detection results equal to 1.00 shown in Table 5.6 indicate an accuracy of 100%. The FFNN classification model is trained with 60 neurons using the entire feature vector (42 features). The classification results for driver 2 are much lower than the other four drivers, with an average of only 0.630. This is due to the imprecise detection of the driver skeleton during data collection. To have a clear perspective of detection performance, the confusion matrix for driver 2 is shown in Fig. 5.25.

In the confusion matrix, the green diagonal shows the number of correct detection cases for that class. The bottom row shows the classification accuracy with respect to the target value, and the far-right column shows the classification accuracy with respect to the predicted labels. As shown in Fig. 5.25, the normal driving behavior for driver 2 only achieved 38% detection accuracy and 289 cases are classified into the phone-answering task. This is mainly due to the similar postures between normal driving and phone-answering behavior. Once hand detection is inaccurate, it is very difficult to classify these two tasks only according to head pose. A detailed discussion will be proposed later. In addition, the low detection accuracy means the trained model using the other four drivers is not sufficient to precisely recognize all the behaviors for driver 2 due to the diversity of the drivers. However, once driver 2 is included in the training data, the model will obtain better detection results for the other four drivers. The most accurate detection occurs for driver 4, the relative results are shown in Fig. 5.26.

As shown in Fig. 5.26, the classification results for the seven tasks for driver 4 are much better than for driver 2. False detection between different classes decreased significantly. Similar results are achieved for the remaining three drivers. In conclusion, although very accurate results were not achieved for driver 2 compared with the other drivers, the general classification accuracy for the seven tasks was 82.4% (the mean value of the average column), which indicates efficient classification results.

In Table 5.7, the classification results of FFNN are compared with four other machine-learning methods, which are RF, SVM, NB, and K-nearest neighbor (KNN, K equals 5 in this case). The accuracy in Table 5.7 is defined as the average detection result for the five drivers, that is, the average of the Ave column in Table 5.6. Meanwhile, to evaluate the driver distraction detection performance, the seven classification tasks are merged into a binary classification. Here, the negative group is defined as the combination of the first four normal driving tasks, and the true distraction group consists of the remaining three distracted driving tasks.

$$TPR = \frac{TP}{P} \qquad (5.41)$$

**TABLE 5.6**
**Classification Results Using FFNN With Entire Features.**

|  | Driving Tasks | | | | Non-Driving Tasks | | | |
|---|---|---|---|---|---|---|---|---|
|  | T1 | T2 | T3 | T4 | T5 | T6 | T7 | Ave |
| D1 | 0.905 | 0.856 | 0.843 | 0.925 | 0.686 | 0.896 | 1.00 | 0.883 |
| D2 | 0.380 | 0.557 | 0.498 | 0.985 | 0.877 | 0.617 | 0.684 | 0.630 |
| D3 | 0.985 | 0.976 | 0.690 | 0.998 | 0.988 | 1.00 | 0.662 | 0.898 |
| D4 | 0.720 | 0.973 | 0.994 | 0.999 | 1.00 | 1.00 | 0.858 | 0.927 |
| D5 | 0.583 | 0.977 | 0.801 | 1.00 | 0.798 | 0.969 | 0.991 | 0.871 |
| Mean | 0.715 | 0.838 | 0.747 | 0.981 | 0.867 | 0.884 | 0.838 | 0.824 |

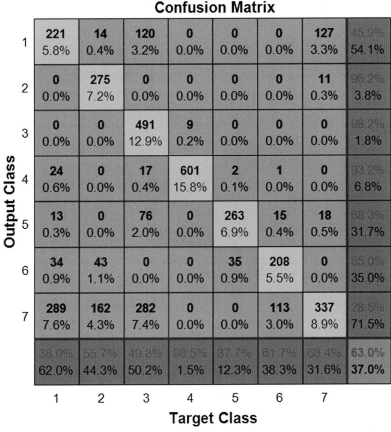

FIG. 5.25  Confusion Matrix of Driving Task Classification Results for Driver 2.

and

$$FPR = \frac{FP}{N} \qquad (5.42)$$

where *TP* is the number of correctly detected distracted cases, *P* is the total number of distracted cases, which is the total quantity of the three distracted cases. *FP* is the number of false detections. In this case, it represents the number of normal driving tasks that are classified in the abnormal driving group. Finally, *N* is the total amount of normal driving cases.

According to Table 5.7, FFNN binary classification outperforms the other four models, indicating that FFNN is a powerful model suitable for driver behavior modeling. Note that there are no optimization algorithms used in the other four models. These models are used with their default setup in MATLAB. The RF is constructed with 100 decision trees, and SVR uses a radial-based kernel. Better results may be obtained with parameter tuning and optimization, however,

this is beyond the scope of this study. The binary classification model can distinguish normal driving behavior and distracted behavior. From the perspective of safety, although it may annoy the driver, it is safe to classify normal driving behavior into distracted behavior and warn the driver. On the other hand, if the model classifies distracted behavior into the normal driving group, it is more dangerous than the previous case and this misclassification should be avoided. In the real world, in terms of nondriving tasks, the time constants are always much longer than normal driving tasks, texting or answering a phone can last for a few minutes. However, mirror-checking actions usually last for 1−2 s. These time properties of the different tasks can be adapted to predict the correct states in the future.

### Feature Evaluation for Behavior Classification Performance

In this section, the impact of the driver's head and body features on driving task classification will be analyzed.

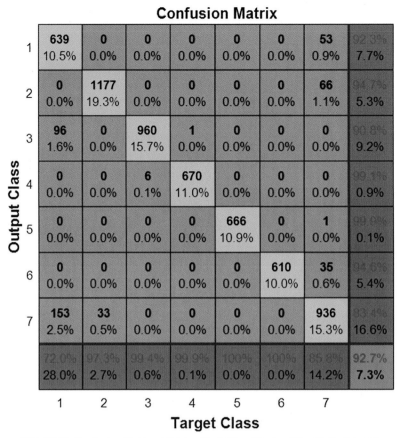

FIG. 5.26  Confusion Matrix of Driving Task Classification Results for Driver 4.

**TABLE 5.7**
**Classification Results Using Different Machine-Learning Methods.**

|  | Accuracy | TPR | FPR | Training Cost (s) | Testing Cost (s) |
|---|---|---|---|---|---|
| FFNN | 0.824 | 0.939 | 0.088 | 4.92 | 0.05 |
| RF | 0.736 | 0.900 | 0.144 | 33.55 | 0.41 |
| SVM | 0.747 | 0.913 | 0.177 | 2.85 | 0.03 |
| NB | 0.767 | 0.922 | 0.171 | 0.188 | 0.02 |
| KNN | 0.623 | 0.771 | 0.090 | 0.049 | 1.08 |

The TPR and FPR in Table 5.4 represent the true-positive rate (sensitivity) and false-positive rate, respectively. TPR and FPR are calculated as

The feature evaluation is divided into three parts. First, the depth information of the detected joints and facial landmarks (eyes and nose) are evaluated. Then, task classification using only head signals or only body signals is proposed. The classification results for these three parts are illustrated in Table 5.5.

First, the 2D-only case in Table 5.8 represents a feature set only consisting of the head rotation and joint coordinates ($X$ and $Y$ coordinates), and depth information is not used. As shown in Table 5.8, the model trained with 2D information achieves similar accuracy results compared with the model trained with the entire

**TABLE 5.8**
**Tasks Classification Based on Different Features.**

| | | Driving Tasks | | | | Non-Driving Tasks | | | |
|---|---|---|---|---|---|---|---|---|---|
| | | T1 | T2 | T3 | T4 | T5 | T6 | T7 | Ave |
| 2D Only (29) | D1 | 93.8 | 48.1 | 60.5 | 88.0 | 100 | 91.7 | 70.4 | 79.4 |
| | D2 | 66.8 | 79.6 | 41.4 | 88.2 | 62.7 | 84.6 | 37.1 | 62.7 |
| | D3 | 58.7 | 64.4 | 100 | 100 | 97.8 | 96.0 | 48.9 | 77.6 |
| | D4 | 69.3 | 73.1 | 96.9 | 99.7 | 99.1 | 96.7 | 91.0 | 87.6 |
| | D5 | 68.5 | 54.3 | 39.3 | 100 | 32.7 | 95.5 | 99.5 | 71.1 |
| | Mean | 71.4 | 63.9 | 67.6 | 95.2 | 78.5 | 92.9 | 69.4 | 75.7 |
| Head Only (3) | D1 | 78.9 | 92.3 | 99.7 | 96.8 | 5.4 | 24.9 | 0.7 | 61.8 |
| | D2 | 0.0 | 85.4 | 15.7 | 66.7 | 93.0 | 90.8 | 26.0 | 44.6 |
| | D3 | 39.3 | 4.2 | 16.6 | 100 | 98.5 | 96.2 | 40.5 | 52.7 |
| | D4 | 94.6 | 99.8 | 77.2 | 100 | 33.2 | 99.3 | 24.8 | 74.8 |
| | D5 | 52.8 | 91.1 | 33.5 | 99.6 | 97.4 | 99.6 | 50.0 | 76.6 |
| | Mean | 53.1 | 74.5 | 48.5 | 92.6 | 65.5 | 82.1 | 28.4 | 62.1 |
| Body Only (30) | D1 | 76.2 | 0.0 | 2.0 | 0.0 | 90.2 | 90.0 | 96.0 | 47.0 |
| | D2 | 12.0 | 0.0 | 26.6 | 0.0 | 40.3 | 1.2 | 32.9 | 16.3 |
| | D3 | 1.2 | 100 | 0.0 | 7.4 | 96.9 | 94.0 | 53.5 | 48.2 |
| | D4 | 55.4 | 45.0 | 0.0 | 97.0 | 97.9 | 95.1 | 62.1 | 58.9 |
| | D5 | 44.0 | 2.1 | 0.0 | 97.3 | 35.3 | 75.0 | 99.3 | 49.2 |
| | Mean | 37.8 | 29.4 | 5.72 | 40.3 | 72.1 | 71.1 | 68.8 | 43.9 |

feature set (Table 5.7). The results indicate that depth information has a very limited impact on the model classification task.

The second block in Table 5.8 illustrates driving task classification using only head pose information. Specifically, the three head rotation angles: yaw, pitch, and roll are used to construct the feature set. The classification accuracy using the head pose is much less than the accuracy in previous cases. For the left mirror-checking and texting tasks, which have significantly different characteristics than other tasks, the detection is accurate. However, for the other tasks, using only head pose information is not sufficient for accurate detection. For example, the driver rear mirror-checking behavior (T3) is like the task of using a video device. Moreover, without considering body information, the phone-answering behavior cannot be detected accurately as the driver is usually looking forward to the road and the head pose is very similar to normal driving.

The confusion matrix for driver 5, which has the most accurate results among the five drivers, is shown in Fig. 5.27. In terms of driver 5, FFNN is not able to accurately distinguish tasks 1, 3, and 7, which are normal driving, rear mirror checking, and answering the phone. Approximately one-third of normal driving cases (224 samples) are classified into the phone-answering task. For rear mirror checking, more samples are falsely detected as the video device using tasks. It is obvious from the confusion matrix that, without using body features and using only head pose features, it is difficult to identify the actual driver behavior. The reason may because when using the head only features, the training data from other drivers can be representative to the fifth driver. Meanwhile, after analyses the testing data from driver 5, it was found that the data are less noisy compared with the other drivers as fewer missing points were founded. Therefore the system performance is expected to be improved with a clearer dataset or a more efficient noise removal filter.

The third block indicates the behavior detection using only body features. There are 30 total features used, containing the $X$, $Y$, and $Z$ coordinates of the hand, wrist, elbow, and shoulder joints. As shown in Table 5.9, ignoring the 3D head pose features and the eyes and nose location information, the detector fails to identify the mirror-checking behaviors. By using

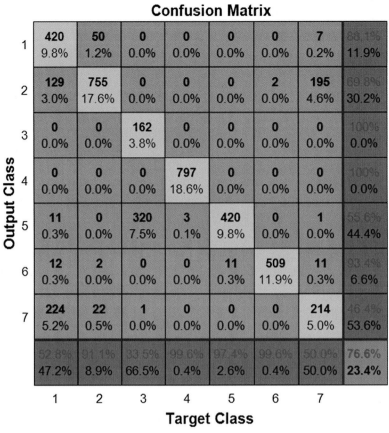

FIG. 5.27 Confusion Matrix of the Classification of the Seven Tasks for Driver 5 Using 3D Head Pose Features Only.

**TABLE 5.9**
**Classification Result Using Ffnn With 18 Selected Features.**

| | Driving Tasks | | | | Non-Driving Tasks | | | |
|---|---|---|---|---|---|---|---|---|
| | T1 | T2 | T3 | T4 | T5 | T6 | T7 | Ave |
| D1 | 90.6 | 91.8 | 98.2 | 98.1 | 100 | 88.3 | 100 | 95.1 |
| D2 | 1.4 | 20.9 | 25.5 | 91.1 | 97.3 | 60.2 | 93.5 | 49.3 |
| D3 | 97.8 | 98.7 | 82.5 | 99.8 | 98.1 | 100 | 28.5 | 85.0 |
| D4 | 65.2 | 78.1 | 96.7 | 99.9 | 98.3 | 97.9 | 80.4 | 86.2 |
| D5 | 51.4 | 100 | 89.9 | 99.9 | 87.9 | 94.9 | 100 | 88.0 |
| Mean | 61.3 | 80.0 | 78.6 | 97.8 | 96.3 | 88.2 | 80.5 | 80.7 |

only body features, the distraction behavior can be detected with a certain degree of accuracy, while the detection accuracy for the four mirror-checking behaviors is quite low. The worst case is the rear mirror-checking behavior, which only achieved 5.72% accuracy in general.

Based on the above evidence, to obtain a better understanding of the tasks that the driver is undergoing, both the head and body features are necessary. From the feature comparison, head pose features are more useful than body features as the 3D head pose information leads to better detection results (62.1% average)

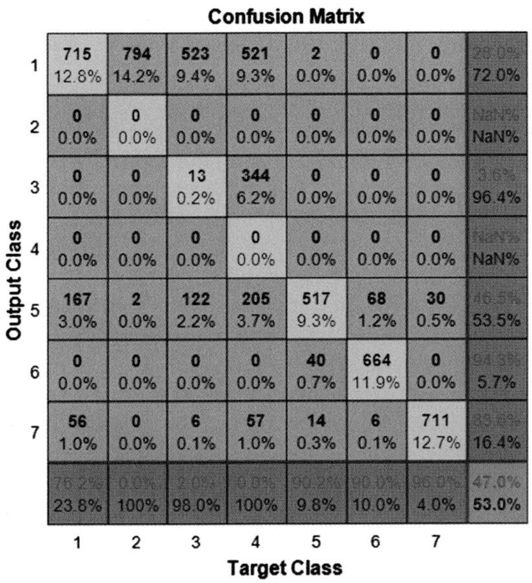

**Confusion Matrix**

| Output Class | 1 | 2 | 3 | 4 | 5 | 6 | 7 | |
|---|---|---|---|---|---|---|---|---|
| 1 | 715 12.8% | 794 14.2% | 523 9.4% | 521 9.3% | 2 0.0% | 0 0.0% | 0 0.0% | 28.0% 72.0% |
| 2 | 0 0.0% | 0 0.0% | 0 0.0% | 0 0.0% | 0 0.0% | 0 0.0% | 0 0.0% | NaN% NaN% |
| 3 | 0 0.0% | 0 0.0% | 13 0.2% | 344 6.2% | 0 0.0% | 0 0.0% | 0 0.0% | 3.6% 96.4% |
| 4 | 0 0.0% | 0 0.0% | 0 0.0% | 0 0.0% | 0 0.0% | 0 0.0% | 0 0.0% | NaN% NaN% |
| 5 | 167 3.0% | 2 0.0% | 122 2.2% | 205 3.7% | 517 9.3% | 68 1.2% | 30 0.5% | 46.5% 53.5% |
| 6 | 0 0.0% | 0 0.0% | 0 0.0% | 0 0.0% | 40 0.7% | 664 11.9% | 0 0.0% | 94.3% 5.7% |
| 7 | 56 1.0% | 0 0.0% | 6 0.1% | 57 1.0% | 14 0.3% | 6 0.1% | 711 12.7% | 83.6% 16.4% |
| | 76.2% 23.8% | 0.0% 100% | 2.0% 98.0% | 0.0% 100% | 90.2% 9.8% | 90.0% 10.0% | 96.0% 4.0% | 47.0% 53.0% |
| | 1 | 2 | 3 | 4 | 5 | 6 | 7 | |

**Target Class**

FIG. 5.28 Confusion Matrix of the Classification of the Seven Tasks for Driver 1 Using Only Body Features.

compared with the 3D body features (43.9%). Fig. 5.28 shows the model classification results for driver 1 when the model is only trained with body features. The three distraction behaviors are accurately detected using body features, while the four mirror checking detections are difficult to identify.

Finally, the important features for driver task classification are selected according to the integrated feature extraction technique in the last section and these features are input to the FFNN model. In total, 18 features are selected as important features. The feature set contains the following features: {yaw, pitch, roll, nose ($X$, $Y$, $Z$), left hand ($X$, $Y$, $Z$), right hand ($X$, $Y$, $Z$), left shoulder ($X$, $Y$, $Z$), and right shoulder ($X$, $Y$, $Z$)}. The classification results are shown in Table 5.6. The overall accuracy of task detection is 80.7%, which is slightly less than the model trained with the entire feature set. However, the selected 18 features still yield an acceptable accurate detection; also, the time cost of the training and testing process is less than when using the entire feature set. Therefore, the driver tasks can be detected using the small feature set.

## DISCUSSION AND FUTURE WORK

Based on the results shown in the previous section, driver task recognition can be achieved with a feedforward neural network. The FFNN could reasonably

detect seven tasks for different drivers and achieved high-precision detection for secondary tasks. The FFNN has advantages for driver task detection over other machine-learning methods. Classification for different tasks resulted in different detection accuracy. The results indicate that for tasks such as texting and left and right mirror checking, which has obvious distinct features, the detection results are accurate. However, for tasks that have similar postures, the model can be confused. In this study, normal driving behavior has similar characteristics to rear mirror-checking behavior and phone-answering tasks; therefore, the detection results for these behaviors are slightly worse than for other behaviors. In addition to the similar characteristics of these behaviors, another reason for less accurate detection results is the driving style. Although accurate detection results can be achieved for some drivers, the FFNN cannot obtain a universal accuracy for all drivers. For example, task detection for driver 2 is less accurate than for other drivers due to driving style and sensor noise. A driver has a unique driving and mirror-checking style. Some drivers prefer to use significant head and body movement during mirror checking while others may try to use less body movement and use eye movement to capture information. Therefore the following aspects are discussed and can be improved to achieve higher task detection accuracy.

First, the driver head and body signals captured with a Kinect are very noisy. Sometimes the detection is less precise, and the detected joint positions are shifted and unreasonable. This phenomenon is particularly worse for the seated driver inside the vehicle. In this study, a simple integrated tracking and smoothing technique is used, which consists of a jitter removal filter and an exponential filter. Although the integrated filter can recover unreasonable detection and smooth the signals, important information can be lost, and the filter can be further improved by using more advanced filters such as the Kalman filter or particle filter for joint position tracking. Therefore the quality of Kinect signals, as well as the model detection results, can be further improved. Moreover, in this study, only color and depth images are collected; however, Kinect also supports audio recording. Therefore in the future, audio information in the cabin can be captured as another important data source to assist in the detection of nondriving-related tasks.

Second, in this study, the feature selection and extraction methods are constructed based on random forests and the maximal information coefficient technique. This integrated method estimates the importance of the driver's body features and the FFNN using these features achieved accurate detection results for some drivers. However, detection accuracy decreased significantly for the second driver for a few reasons. To obtain universal accurate task detection results, more drivers must be studied in the future. Increasing the dataset volume and data diversity is an efficient way to solve the problem. Meanwhile, more driver features can be used. In the study, only the position and depth information for the eyes are used. The driver gaze movement and gaze tracking technique have been successfully adopted in some research on driver fatigue, inattention, and distraction monitoring. Gaze information can be very useful when drivers prefer not to move their bodies when performing mirror-checking tasks.

Finally, on-road data collection can be performed in the future for the study of real-time driver behavior detection within normal driving environments. Currently, for safety considerations, the drivers were asked to perform the experimental tasks without driving the vehicle because secondary tasks such as texting and playing a video device are extremely dangerous when driving and should be avoided. Therefore the most naturalistic data are difficult to collect. However, in the future, with the help of advanced driver assistant systems (ADAS) and the midlevel-automated vehicle technique, drivers are allowed to remove their hands from the steering wheel. Therefore more distraction behaviors can be collected and the study for real-time driver distraction detection in a real vehicle can be performed. The real-time driver monitoring study will significantly improve the driving safety for both conventional vehicles and highly automated vehicles.

## CONCLUSIONS

In this study, driving behaviors for different drivers are studied. The driving behaviors are classified into two categories: normal driving tasks and distracting tasks. A feedforward neural network is trained to distinguish the four mirror-checking behaviors from the three secondary driving tasks. Both depth information and the 2D location of the body joints are collected using Kinect. The noisy data are processed with an integrated filtering system. Then, the importance of each driver feature to behavior recognition is evaluated using random forests and maximal information efficiency. The feature importance prediction with these two feature evaluation techniques shows consistent results. The most important driver features for driver behavior among all the drivers are determined. The FFNN has been proven to have advantages for behavior detection tasks over other popular machine-learning methods.

The model achieved an average of greater than 80% accuracy for the five drivers. With the evaluation of feature importance and their influence on the classification task, the head pose feature, hand position, and shoulder positions for the driver are selected as the most important features. In addition, based on the evaluation of the depth, head, and body features, it is found that the depth information for the body joints and facial markers has a very limited influence on behavior recognition. Meanwhile, the head and body features should be combined with a comprehensive driver behavior understanding because only using the head or body features will lead to large false detection rates.

The conclusion is made that for future driving monitoring and behavior understanding, the head and body signals are equally important and necessary. Future works will focus on the collection of a more real-world dataset and recognize more sophisticate driver behaviors. This study will benefit future ADAS design and improve driving safety by real-time driver status monitoring.

## REFERENCES

[1] F.-Y. Wang, Computational social systems in a new period: a fast transition into the third axial age, IEEE Transactions on Computational Social Systems 4 (3) (2017) 52−53.

[2] N. Arbabzadeh, M. Jafari, A data-driven approach for driving safety risk prediction using driver behavior and roadway information data, IEEE transactions on intelligent transportation systems 19 (2) (2017) 446−460.

[3] Y. Xing, et al., Driver workload estimation using a novel hybrid method of error reduction ratio causality and support vector machine, Measurement 114 (2018) 390−397.

[4] J.A. Michon, A critical view of driver behavior models: what do we know, what should we do, Human behavior and traffic safety (1985) 485−520.

[5] F.-Y. Wang, et al., Parallel driving in CPSS: a unified approach for transport automation and vehicle intelligence, IEEE/CAA Journal of Automatica Sinica 4 (4) (2017) 577−587.

[6] G. Castignani, et al., Driver behavior profiling using smartphones: a low-cost platform for driver monitoring, IEEE Intelligent Transportation Systems Magazine 7 (1) (2015) 91−102.

[7] C. Lv, et al., Simultaneous observation of hybrid states for cyber-physical systems: a case study of electric vehicle powertrain, IEEE transactions on cybernetics 48 (8) (2017) 2357−2367.

[8] N. Foroutan, A. Hamzeh, Discovering the hidden structure of a social network: a semi supervised approach, IEEE Transactions on Computational Social Systems 3 (4) (2016) 151−163.

[9] G. Tong, et al., Effector detection in social networks, IEEE Transactions on Computational Social Systems 4 (1) (2017) 14−25.

[10] C. Lv, H. Wang, D. Cao, High-precision hydraulic pressure control based on linear pressure-drop modulation in valve critical equilibrium state, IEEE Transactions on Industrial Electronics 64 (10) (2017) 7984−7993.

[11] Y. Liang, J.D. Lee, A hybrid Bayesian Network approach to detect driver cognitive distraction, Transportation Research Part C: Emerging Technologies 38 (2014) 146−155.

[12] C. Lv, et al., Analysis of autopilot disengagements occurring during autonomous vehicle testing, IEEE/CAA Journal of Automatica Sinica 5 (1) (2017) 58−68.

[13] M. Sivak, B. Schoettle, Motion Sickness in Self-Driving vehicles. UMTRI-2015-12, University of Michigan Transportation Research Institute, Ann Arbor, 2015.

[14] A. Jain, et al., Car that knows before you do: anticipating manoeuvres via learning temporal driving models, in: Proceedings of the IEEE International Conference on Computer Vision, 2015.

[15] E. Murphy-Chutorian, M.M. Trivedi, Head pose estimation and augmented reality tracking: an integrated system and evaluation for monitoring driver awareness, IEEE Transactions on Intelligent Transportation Systems 11 (2) (2010) 300−311.

[16] E. Murphy-Chutorian, M.M. Trivedi, Head pose estimation in computer vision: a survey, IEEE Transactions on Pattern Analysis and Machine Intelligence 31 (4) (2009) 607−626.

[17] L. Fletcher, Z. Alexander, Driver inattention detection based on eye gaze—road event correlation, The International Journal of Robotics Research 28 (6) (2009) 774−801.

[18] N. Das, E. Ohn-Bar, M. Mohan, Trivedi, On performance evaluation of driver hand detection algorithms: challenges, dataset, and metrics, in: Intelligent Transportation Systems (ITSC), 2015 IEEE 18th International Conference on IEEE, 2015, 2015.

[19] S.Y. Cheng, S. Park, M.M. Trivedi, Multi-spectral and multi-perspective video arrays for driver body tracking and activity analysis, Computer Vision and Image Understanding 106 (2) (2007) 245−257.

[20] H. Du, W. Li, N. Zhang, Vibration control of vehicle seat integrating with chassis suspension and driver body model, Advances in Structural Engineering 16 (1) (2013) 1−9.

[21] C. Tran, A. Doshi, M.M. Trivedi, Modeling and prediction of driver behavior by foot gesture analysis, Computer Vision and Image Understanding 116 (3) (2012) 435−445.

[22] E. Ohn-Bar, et al., Head, eye, and hand patterns for driver activity recognition, in: 22nd International Conference on Pattern Recognition (ICPR), 2014, IEEE, 2014.

[23] M. Rezaei, R. Klette, Look at the driver, look at the road: No distraction! No accident!, in: Proceedings of the IEEE Conference on Computer Vision and Pattern Recognition, 2014.

[24] R.O. Mbouna, S.G. Kong, M.-G. Chun, Visual analysis of eye state and head pose for driver alertness monitoring,

[25] N. Li, C. Busso, Detecting drivers' mirror-checking actions and its application to manoeuvre and secondary task recognition, IEEE Transactions on Intelligent Transportation Systems 17 (4) (2016) 980−992.

[26] A. Kondyli, et al., Computer assisted analysis of drivers' body activity using a range camera, IEEE Intelligent Transportation Systems Magazine 7 (3) (2015) 18−28.

[27] S. Jha, C. Busso, Analyzing the relationship between head pose and gaze to model driver visual attention, in: 19th International Conference on Intelligent Transportation Systems (ITSC), 2016 IEEE, IEEE, 2016.

[28] J. Morton, T.A. Wheeler, M.J. Kochenderfer, Analysis of recurrent neural networks for probabilistic modeling of driver behavior, IEEE Transactions on Intelligent Transportation Systems 18 (5) (2017) 1289−1298.

[29] M. Wollmer, et al., Online driver distraction detection using long short-term memory, IEEE Transactions on Intelligent Transportation Systems 12 (2) (2011) 574−582.

[30] F. Tango, M. Botta, Real-time detection system of driver distraction using machine learning, IEEE Transactions on Intelligent Transportation Systems 14 (2) (2013) 894−905.

[31] Y.-C. Lee, et al., Learning to predict driver behaviorfrom observation. The AAAI 2017 Spring Symposium on Learning from Observation of Humans Technical Report SS-17-06, 2017.

[32] J.L. Harbluk, Y. Ian Noy, M. Eizenman, The Impact of Cognitive Distraction on Driver Visual Behavior and Vehicle Control, No. TP 13889 E, 2002.

[33] Mühlbacher-Karrer, Stephan, et al., A driver state detection system—combining a capacitive hand detection sensor with physiological sensors, IEEE Transactions on Instrumentation and Measurement 66 (4) (2017) 624−636.

[34] B. Ranft, C. Stiller, The role of machine vision for intelligent vehicles, IEEE Transactions on Intelligent Vehicles 1 (1) (2016), 8−1.

[35] S. Gaglio, Lo Re Giuseppe, M. Morana, Human activity recognition process using 3-D posture data, IEEE Transactions on Human-Machine Systems 45 (5) (2015) 586−597.

[36] L.B. Neto, et al., A Kinect-based wearable face recognition system to aid visually impaired users, IEEE Transactions on Human-Machine Systems 47 (1) (2017) 52−64.

[37] M. Azimi. Skelton Joint Smoothing White Paper [Online]. Available: https://msdn.microsoft.com/en-us/library/jj131429.aspx.

[38] J. Darby, et al., An evaluation of 3D head pose estimation using the Microsoft Kinect v2, Gait & Posture 48 (2016) 83−88.

[39] L. Breiman, Random forests, Machine Learning 45 (1) (2001) 5−32.

[40] Do we need hundreds of classifiers to solve real world classification problems, Journal of Machine Learning Research 15 (1) (2014) 3133−3181.

IEEE Transactions on Intelligent Transportation Systems 14 (3) (2013) 1462−1469.

[41] D.N. Reshef, et al., Detecting novel associations in large data sets, Science 334 (6062) (2011) 1518−1524.

[42] T.M. Cover, J.A. Thomas, Elements of Information Theory, Wiley, 1991, ISBN 978-0-471-24195-9.

[43] W. Duch, N. Jankowski, Survey of neural transfer functions, Neural Computing Surveys 2 (1) (1999) 163−212.

[44] E. Ohn-Bar, et al., On surveillance for safety critical events: in-vehicle video networks for predictive driver assistance systems, Computer Vision and Image Understanding 134 (2015) 130−140.

[45] V. Alizadeh, O. Dehzangi, The impact of secondary tasks on drivers during naturalistic driving: analysis of EEG dynamics, in: IEEE International Conference on Intelligent Transportation Systems IEEE, 2016, pp. 2493−2499.

[46] M. Raja, V. Ghaderi, S. Sigg, Detecting Driver's Distracted Behavior from Wi-Fi, in: VTC, Spring 2018, pp. 1−5.

[47] N. Karatas, et al., Sociable driving agents to maintain driver's attention in autonomous driving, in: 2017 26th IEEE International Symposium on Robot and Human Interactive Communication (RO-MAN), IEEE, 2017.

[48] H.S. Kim, et al., Predicting the EEG level of a driver based on driving information, IEEE Transactions on Intelligent Transportation Systems 20 (4) (2018) 1215−1225.

[49] C. Ou, et al., Driver behavior monitoring using tools of deep learning and fuzzy inferencing, in: 2018 IEEE International Conference on Fuzzy Systems (FUZZ-IEEE), IEEE, 2018.

[50] A. Koohestani, et al., Drivers performance evaluation using physiological measurement in a driving simulator, in: 2018 Digital Image Computing: Techniques and Applications (DICTA), IEEE, 2018, pp. 1−6.

[51] L. van der Maaten, E. Postma, J. van den Herik, Dimensionality Reduction: A Comparative Review, 2009, pp. 1−35.

[52] F. Camastra, Data dimensionality estimation methods: a survey, Pattern Recognition 36 (12) (2003) 2945−2954.

[53] E.L.L. Sonnhammer, G. Von Heijne, K. Anders, A hidden Markov model for predicting transmembrane helices in protein sequences, Proceedings of the 2nd International Conference on Intelligent Systems for Molecular Biology. ISMB-94 6 (1998).

[54] J. Yamato, J. Ohya, K. Ishii, Recognizing human action in time-sequential images using hidden markov model, in: Proceedings 1992 IEEE Computer Society Conference on Computer Vision and Pattern Recognition, IEEE, 1992.

[55] A._P. Varga, R.K. Moore, Hidden Markov model decomposition of speech and noise, in: International Conference on Acoustics, Speech, and Signal Processing, IEEE, 1990.

[56] R. Mahdi, R. Klette, Look at the driver, look at the road: No distraction! No accident!, in: Proceedings of the IEEE Conference on Computer Vision and Pattern Recognition, 2014.

[57] B. Wang, et al., Head pose estimation with combined 2D SIFT and 3D HOG features, in: Image and Graphics (ICIG), 2013 Seventh International Conference on IEEE, 2013.

[58] F. Vicente, et al., Driver gaze tracking and eyes off the road detection system, IEEE Transactions on Intelligent Transportation Systems 16 (4) (2015) 2014−2027.

[59] L. Wu, et al., A facial pose estimation algorithm using deep learning, in: Chinese Conference on Biometric Recognition, Springer International Publishing, 2015.

[60] H. Kim, et al., Illumination invariant head pose estimation using random forests classifier and binary pattern run length matrix, Human-Centric Computing and Information Sciences 4 (1) (2014) 9.

[61] G. Riegler, et al., Hough networks for head pose estimation and facial feature localization, Journal of Computer Vision 101 (3) (2013) 437−458.

[62] A. Tawari, M.M. Trivedi, Robust and continuous estimation of driver gaze zone by dynamic analysis of multiple face videos, in: Intelligent Vehicles Symposium Proceedings, 2014 IEEE, IEEE, 2014.

[63] A. Saeed, A. Al-Hamadi, G. Ahmed, Head pose estimation on top of haar-like face detection: a study using the kinect sensor, Sensors 15 (9) (2015) 20945−20966.

[64] G. Peláez, E. Arturo de la, A. Jose, Head pose estimation based on 2D and 3D information, Physics Procedia 22 (2014) 420−427.

[65] S. Li, et al., Real-time head pose tracking with online face template reconstruction, IEEE Transactions on Pattern Analysis and Machine Intelligence 38 (9) (2016) 1922−1928.

[66] M. Demirkus, et al., Probabilistic temporal head pose estimation using a hierarchical graphical model, in: European Conference on Computer Vision, Springer International Publishing, 2014.

[67] J.P. Jones, L.A. Palmer, An evaluation of the two-dimensional Gabor filter model of simple receptive fields in cat striate cortex, Journal of Neurophysiology 58 (6) (1987) 1233−1258.

[68] I. Fogel, D. Sagi, Gabor filters as texture discriminator, Biological Cybernetics 61 (2) (1989) 103−113.

[69] F. Bianconi, A. Fernández, Evaluation of the effects of Gabor filter parameters on texture classification, Pattern Recognition 40 (12) (2007) 3325−3335.

[70] P. Moreno, A. Bernardino, J. Santos-Victor, Gabor parameter Selection for Local Feature detection[C]/Iberian Conference on Pattern Recognition and Image Analysis, Springer Berlin Heidelberg, 2005, pp. 11−19.

[71] BTAS'09 N. Pave, Principal gabor filters for face recognition[C]//Biometrics: theory, applications, and systems, 2009, in: IEEE 3rd International Conference on. IEEE, 2009, pp. 1−6.

[72] N. Dalal, B. Triggs, Histograms of oriented gradients for human detection, in: IEEE Computer Society Conference on Computer Vision and Pattern Recognition CVPR 2005 Vol. 1, IEEE, 2005.

[73] G. Fanelli, J. Gall, L. Van Gool, Real time head pose estimation with random regression forests, in: IEEE Conference on Computer Vision and Pattern Recognition (CVPR), 2011, IEEE, 2011.

# CHAPTER 6

# Application of Deep Learning Methods in Driver Behavior Recognition

## INTRODUCTION

A driver is in the center of the road–vehicle–driver (RVD) loop. Driver decisions and behaviors are the major aspects that can affect driving safety. It is reported that more than 90% of light vehicle accidents are caused by human driver misbehavior in the United States, and the accident rate can be reduced by 10%–20% with a precise driver behavior monitoring system [1–5]. Therefore, the recognition of driver behaviors is becoming one of the most important tasks for intelligent vehicles. For the conventional advanced driver assistance systems (ADAS), the driver is in the center of the RVD loop. The understanding of driver behavior enables the ADAS to generate the optimal vehicle control strategies that are suitable for the current driver states [5–8]. Regarding the intelligent and highly automated vehicles, such as the Level-3 automated vehicles (according to the definition in Society of Automotive Engineers standard J3016), the driver is responsible for taking over the vehicle control under emergencies. At this moment, the real-time driver behavior and activity monitoring system has to decide whether the driver can take over or not. Therefore, in this section, a deep learning-based driver activities recognition system is proposed to monitor and understand the driver behaviors continuously. The recognition models are trained to identify seven common driving-related tasks and also to determine whether the driver is being distracted or not. With this end-to-end approach, intelligent vehicles can better interact with human drivers and properly making decisions and generating human-like driving strategies.

Driver behaviors have been widely studied over the past 2 decades. Previous studies mainly focus on the driver attention [9] and distraction (either physical distraction or cognitive distraction) [10], driver intention [7,11], driver styles [12], driver drowsiness, and fatigue detection[13–15]. The National Highway Traffic Safety Administration (NHTSA) defined driver distraction as a process that the driver shifts their attention away from the driving tasks. Four types of distractions are clarified by the NHTSA, which are visual distraction, auditory distraction, biomechanical distraction, and cognitive distraction [16]. To understand the driver behaviors, most of the studies require capturing the driver status information, such as the head pose [17], eye gaze [8], hand motion [18], foot dynamics [19], and even the physiological signals [20,21].

Specifically, in Ref. [22], the video information for the driver head movement along with the audio signals was collected to identify the secondary driving tasks. In Ref. [23], the driver's head pose, eye gaze direction, and hand movement were combined to identify driver activities. In Ref. [24], the driver's head poses estimation was proposed and applied to the rear-end crash avoidance system. Despite the vision-based feature extraction methods, the physiological signals, such as the electroencephalogram (EEG) and electrooculography (EOG), are also widely used for real-time driver status monitoring. In Ref. [25], EEG signals were collected to predict the driver braking intention. In Ref. [26], the EEG and EOG signals were used to estimate the driver drowsiness and fatigue status. The EEG signals are proved to be closely related to the driver behaviors and can illustrate an earlier response to the human mental states compared with the outer physical behaviors.

However, as aforementioned, most of the existing driver behavior studies require extracting specific features in advance, such as the head pose angle, gaze direction, EEG, and the position of hand and body joints [27]. These features are not always easy to be obtained, and some even require specific hardware devices, which will increase either the temporal or the financial cost. Therefore, in this work, an end-to-end driver activity recognition system is proposed based on the deep convolutional neural network (CNN) models, which is accurate and easy to be implemented. To study the driver distraction behaviors, visual distraction, auditory distraction, and biomechanical distraction are involved. Although the cognitive distraction is not considered, it has been well studied in

Advanced Driver Intention Inference. https://doi.org/10.1016/B978-0-12-819113-2.00006-3

Refs. [28,29], which can be effectively detected with a nonvision-based approach.

Regarding the current development of deep learning techniques, significant progress has been made in the computer vision area due to the development of deeper CNN models, parallel computing hardware, and the large-scale annotated dataset. Deep CNN models have achieved the state-of-art results in object detection, classification, generation, and segmentation tasks. Meanwhile, it has been successfully applied to some driver monitoring tasks [30,31]. In this work, three different CNN models will be evaluated for driver activities recognition and distraction detection tasks. The only sensor required in this section is a low-cost RGB camera. Based on the report in Ref. [32], the seven most common in-vehicle activities for both manual driving and automated driving vehicles, which contain normal driving activities as well as secondary tasks, are selected. The CNN models take the processed images directly without any manual feature extraction procedure. By applying the transfer learning scheme, the pretrained CNN models can be efficiently fine-tuned to satisfy the behavior detection task.

The contribution of this section can be summarized as follows. First, a novel deep learning-based approach is applied to identify driver behaviors. Different from existing studies that require complex algorithms to estimate the driver status information, the proposed algorithm takes merely the color images as the input and directly outputs the driver behavior information. With the deep CNN models, the manually feature extraction process can be replaced by an automatic feature learning process.

Second, transfer learning is applied to fine-tune the pretrained deep CNN models. The models are trained to deal with both the multiple classification tasks and the binary classification task. The algorithm is proved as a practical solution for nonintrusive driver behavior detection. Besides, this section also shows that transfer learning can successfully transfer the domain knowledge that is learned from the large-scale dataset to the small-scale driver behavior recognition task.

Finally, an unsupervised Gaussian mixture model (GMM)-based segmentation method is applied to process the raw images and extract the driver's body region from the background. It is found that by applying a segmentation model before the behavior detection network, the detection accuracy on the driving activities recognition can increase significantly.

## EXPERIMENT AND DATA COLLECTION

This section describes the experimental design of this section. Fig. 6.1 illustrates the general system architecture. The raw images are firstly collected using Kinect. Then, the cropped images are segmented using the GMM algorithm. Finally, the CNN models are applied to the activities recognition task.

In this section, the driver behavior images are collected using Kinect. Kinect is able to collect multimodal signals, such as the color image, depth image, and audio signals. It was originally designed for indoor human interaction and has been successfully applied to driver monitoring systems [33,34]. As described in Ref. [27], the driver's head poses and upper body joints also can be detected using Kinect, while, in this part, only the RGB images are captured.

According to the Kinect mounting requirements [35], it is mounted in the middle of the front window, facing the upper body of the driver so that not to

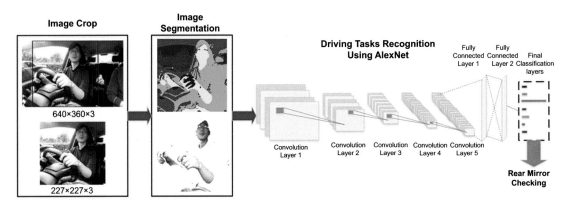

FIG. 6.1 Overall system architecture for driver behavior detection using AlexNet.

interfere driver's field of view while driving. The device setup is shown in Fig. 5.3. The sampling rate for the image collection is eight frames per second. According to the study in Ref. [22], short-term driver behaviors like mirror checking last from 0.5 to 1 s. Therefore, the sampling rate is fast enough to capture these behaviors. The data are recorded with an Intel Core i7 2.5 GHz CPU, and the codes are written in C++ based on the Windows Kinect SDK and OpenCV. To store the images, the raw images are compressed to $640 \times 360 \times 3$ format to increase the computation efficiency.

Ten drivers are involved in the experiment. They were asked to perform seven activities, which consist of four normal driving tasks (normal driving, left mirror checking, right mirror checking, and rear mirror checking) and three secondary tasks (using in-vehicle radio/video device, answering the mobile phone, and texting). It took about 20−30 min for each driver to finish all these tasks, and about 34,000 images were captured in total. In this experiment, five drivers were asked to perform these tasks during driving in a testing field, while the remaining five drivers were asked to mimic the driving tasks and not drive the vehicle. This is because, during normal driving, it is dangerous to perform secondary tasks so that secondary data are limited. However, the steady scenario can be used to collect enough secondary behavior data safely. The number of images for each task and the quantitative comparison between normal driving and secondary tasks is shown in Fig. 6.2. Different from some human activity recognition studies that require temporal information, in this section, no temporal information is considered, and each image is processed individually. The reason is that during driving the driver outer behaviors are always explicit, such as mirrors checking and performing secondary tasks. Therefore, no temporal information is required for inferring driver behaviors because most of the images carry enough information for activity recognition.

## END-TO-END RECOGNITION BASED ON DEEP LEARNING ALGORITHM

This section describes the algorithms used in this section. Specifically, Section IIIA introduces the image preprocessing and segmentation based on the GMM algorithm. 1 describes the two deep CNN frameworks and transfer learning methods.

### Image Preprocessing and Segmentation

As aforementioned, the raw images are stored in the format of $640 \times 360 \times 3$. To speed up the CNN training process and increase the classification accuracy, the raw images are cropped in the beginning. A region of interest (ROI) that mainly contains the driver is selected. Fig. 6.1 indicates the raw image and the selected ROI. After the raw images are cropped, these images are transformed into the size of $227 \times 227 \times 3$ to satisfy the input requirement of AlexNet and $224 \times 224 \times 3$ for the GoogLeNet, respectively.

Before directly putting the raw color images into the CNN model, the raw images are firstly processed using the GMM segmentation. The driver's body will be detected based on this unsupervised machine learning method in the case to have a more precise driver body estimation and behavior recognition. The main reason for using GMM for image segmentation is the unsupervised machine learning process requires no manual labeling, which is more convenient to use than the supervised learning methods. Although the unsupervised learning results can be less accurate than the supervised learning methods that with accurate ground truth labeling, the classification accuracy of the GMM can be controlled by choosing the proper number of clustering components. GMM is an unsupervised machine learning method, which can be used for data clustering and data mining. GMM is a probability density function represented by a weighted sum of sub-Gaussian components [36]. As an unsupervised learning method, one of the advantages of using

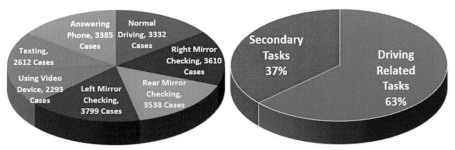

FIG. 6.2 Illustration of the collected dataset.

GMM to segment the images is GMM requires no training labels and very flexible to select a different number of clusters. To train a GMM for segmentation, each image is represented by a feature vector according to the pixel intensity. The feature vector for the GMM is a three-dimensional vector that contains the RGB intensity of each pixel.

$$x_k = [R(u,v), \ G(u,v), \ B(u,v)] \qquad (6.1)$$

where $x$ is the feature vector of a cropped image, $k$ is the index of $x$ which has a maximum value of $227 \times 227$ for AlexNet and $224 \times 224$ for GoogLeNet. $(u,v)$ is the image coordinates.

The GMM can be represented as the following equation:

$$p\left(x_i|\theta\right) = \sum\nolimits_{k=1}^{K} \pi_k N\left(x_i|\mu_k, \sum\nolimits_k\right) \qquad (6.2)$$

where $x_i$ is the three-dimensional feature vector, $\theta$ is the parameter of GMM, $K$ is the total number of components in the model (five in this case), $\pi_k$ are the weight of each component Gaussian distribution function and the sum of $\pi_k$ equals to one. $\mu$ and $\sum$ are the mean and covariance parameter of multivariate Gaussian function. $N\left(x_i|\mu_k, \sum\nolimits_k\right)$ is the univariate Gaussian distribution function in this case with the following form.

$$N\left(x|\mu, \sum\right) = \frac{1}{(2\pi)^{D/2}\left|\sum\right|^{1/2}} \exp\left[-\frac{1}{2}(x-\mu)^T\sum\nolimits^{-1}(x-\mu)\right] \qquad (6.3)$$

where $D$ is the dimension of the data vector. A complicate GMM contains three parameters and can be represented as

$$\theta = \left\{\pi_k, \mu_k, \sum\nolimits_k\right\} \qquad (6.4)$$

The most common method for GMM training is the expectation-maximization (EM) maximum likelihood estimation algorithm. It computes the maximization of the cost function in an iterative manner. The detailed description of the EM algorithm can be found in Ref. [37]. Fig. 6.3 illustrates the image segmentation results using GMM, driver body and skin can be identified using GMM. After GMM segmentation, only the head and hand pixels have remained for CNN model training. As shown in the next section, this is an efficient method to increase activity recognition accuracy.

Then, the GMM algorithm is applied to segment the images and extract the driver's body region from the background. One of the advantages of using GMM to unsupervised segmentation of the images is it requires no manual labeling and can be flexible to modify the model by adjusting the cluster centers [37]. To train a GMM-based segmentation model, each image is represented by a feature vector according to the pixel intensity. The feature vector for the GMM is a three-dimensional vector that contains the RGB intensity of each pixel.

FIG. 6.3 Image segmentation results using GMM.

Fig. 6.3 illustrates the segmented images of the 10 drivers for model training and testing. The driver head and body region can be identified with the GMM segmentation method. As the camera is fixed inside the vehicle cabin, the driver's seat position and the corresponding head position will be fixed within a certain area. The driver body region can be determined based on a set of predefined points that are located around the driver's head position. The points around the head position and the corresponding label will be used to indicate the driver regions. In the future, the manual selection method can be replaced by using an automatic detection method. For example, a precise driver head position can be first detected using the head detection algorithms and then the driver body regions can be determined directly or using a simple semantic segmentation network. As shown in the next section, the segmentation-based method can dramatically increase the model recognition accuracy.

## Model Preparation and Transfer Learning

Currently, the deep convolutional neural network has gained a tremendous improvement in the domain of computer vision. One of the major reasons is the distribution of ImageNet dataset [38]. ImageNet is a large-scale dataset, which contains more than 15 million high-resolution annotated natural images of over 22,000 categories. A large number of annotated images benefit the training of deeper and more accurate CNN models. In this work, two deep convolutional neural network models namely AlexNet and GoogLeNet were chosen as the basic model structures for the recognition of driver behavior. The AlexNet was first proposed by Alex Krizhevsky in 2012 [39]. The model won the ImageNet Large Scale Visual Recognition Challenge (ILSVRC12) with a much more accurate detection accuracy compared with other models at that time. The model was trained for the classification of 1000 categories. There are five convolutional layers and three fully connected neural network layers with nonlinearity and pooling layers between the convolutional layers. In total, AlexNet contains 60 million parameters and 650,000 neurons. The simplified model structure for AlexNet is shown in Fig. 6.1.

GoogLeNet is another deep CNN model, which won the ILSVRC14 [40]. GoogLeNet is significantly deeper than AlexNet and achieved more accurate classification results on the ImageNet dataset. Despite the model depth, the main contribution of GoogLeNet is the utilization of inception architecture. As mentioned in Ref. [40], the most common ways of increasing CNN

model performance are to increase the network size (either the depth or the width of the model). However, it gives rise to the requirement for the larger-scale dataset and a more computational burden. They introduce the Inception layers into the CNN model to increase the sparsity among the layers and reduce the number of parameters. Each Inception layer consists of six basic convolution filters and one max pooling filter. With different scales, the parallel-arranged convolution filters will have more accurate detailing and a wider representation of the information from previous layers. A dimension reduction inception layer is shown in Fig. 6.4. In total, there are two traditional convolutional layers at the lower level of the GoogLeNet and nine inception layers are concatenated at a higher level. With the application of inception layers, the general quantity of parameters in GoogLeNet is 12 times less than that in the AlexNet.

Recent evidence has shown that the network depth is of importance to the feature representation and generalization [41]. It is common to see that simply stacking the convolutional layers to increase the depth of the model cannot give better training and generalization performance [42]. Accordingly, in Ref. [43], Kaming et al. introduced a novel deep CNN model, namely residual networks (ResNet) to enable the construction of deeper convolutional neural networks. By introducing the residual learning scheme, the ResNet achieved the first place on the ILSVRC 2015 classification competition and won the ImageNet detection, ImageNet localization, COCO detection 2015, and COCO segmentation.

As shown in the left part of Fig. 6.5, the underlying mapping function for the basic residual block can be assumed as $H(x)$. The $x$ represents the inputs to the first layer. The residual network supposes that an explicit residual mapping function $F(x)$ exists such that $F(x) = H(x) - x$, and the original mapping can be represented as $F(x) + x$. The core idea behind the residual network block is that although both $H(x)$ and the $F(x) + x$ mapping is able to approximate the desired functions asymptotically, it is much easier to learn the mapping of $F(x) + x$. The added layers through the shortcut connection are the identity mapping. The right graph in Fig. 6.5 represents the full structure of a deep residual network, where residual learning is performed for every few stacked layers. By introducing the identity mapping and copying the other layers from the shallower model, the deep residual network can efficiently solve the model degradation problem when the models getting deeper [43].

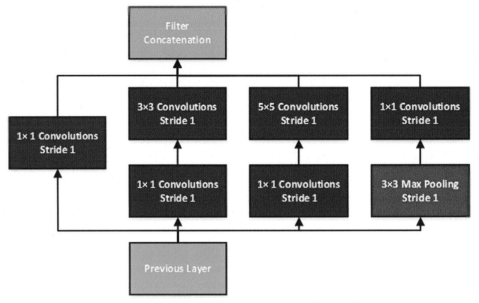

FIG. 6.4 Illustration of the inception layer of GoogLeNet.

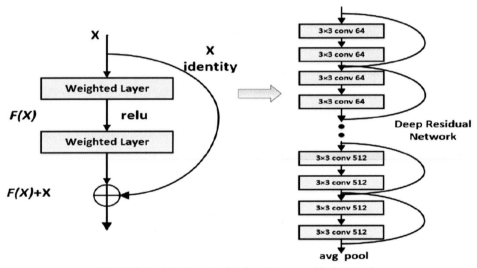

FIG. 6.5 Residual learning block and deep residual network.

To train the deep convolutional neural networks like AlexNet and GoogLeNet from scratch, a large-scale annotated dataset like ImageNet is required. However, in general, large-scale annotated datasets are not always available. Therefore, common ways to use the pretrained deep CNN model are to treat the model as a fixed feature extractor without tuning the model parameters or fine-tune the pretrained model parameters with the limited dataset. In this section, the CNN models are used in the second manner, which is to fine-tune a few layers of the models with the driver behavior dataset. As the two models are trained to classify 1000 categories, in the beginning, the last few layers have to be modified so that the model can satisfy the seven objects' classification tasks. In this work, the last three layers of the AlexNet and GoogLeNet are modified.

Specifically, the original last fully connected layer and the output layer, which generate the probabilities for the 1000 categories, are replaced by a new fully connected layer and softmax layer that output the probabilities for the seven categories. The basic structure and properties of the convolutional layers have remained so that these layers can keep their advantages in the feature extraction. Meanwhile, the knowledge learned from the large-scale ImageNet dataset can be transferred to the driver behavior domain. For transfer learning, a small initial learning rate is selected to slow down the updating rate of the convolutional layers. On the contrary, a larger learning rate factor for the last fully connected layer is chosen to speed up the learning rate in the final layers. With this kind of combination, the new models can be trained to satisfy the new classification tasks.

## EXPERIMENT RESULTS AND ANALYSIS

In this section, the classification results for the driving tasks are proposed. The system performances are evaluated on three major aspects: the impact of GMM-based image segmentation on the recognition of driver behaviors, the classification results compared with the feature extraction methods, and the binary classification results on the distracted behaviors.

### The Impact of GMM Image Segmentation on Driving Tasks Recognition

First, the activity recognition for the five participants is evaluated. The seven driving-related tasks are ordered as {normal driving, right mirror checking, rear mirror checking, left mirror checking, using radio/video device, texting, and answering mobile phone}. Table 6.1 illustrates the seven objects classification results using AlexNet, and Table 6.2 indicates the results given by GoogLeNet. Specifically, the upper parts of Tables 6.1 and 6.2 indicate the classification results with GMM image segmentation, while the lower part illustrates the classification results using the raw RGB images. T1 to T7 represent the seven tasks and D1 to D5 represent the five drivers. The models are trained using MATLAB deep learning toolbox. The model is evaluated using leave-one-out (LOO) cross-validation. To get the activity identification results for a certain driver, the images for this driver are only used as testing images, whereas the images for the remaining four drivers are used as training images. Therefore, for these drivers, their images are completely new to the testing CNN models and the identification performances equal to the model generalization on the new dataset.

As shown in Table 6.1, the general identification accuracy for the segmentation-based AlexNet achieved an average of 81.4% accuracy. The raw-image-based AlexNet was also tested, which achieved only 69.2% recognition accuracy. In Table 6.1, the average performance in the rightmost column is defined as the average detection results for each driver, while the mean accuracy in the bottom row represents the average detection rate for each task. Regarding the detection accuracy for each

**TABLE 6.1**
**Classification Results for Driving Tasks Recognition Using AlexNet.**

| | GMM-BASED ALEXNET | | | | | | | |
|------|-------|-------|-------|-------|-------|-------|-------|-------|
| No. | T1 | T2 | T3 | T4 | T5 | T6 | T7 | Ave |
| D1 | 0.825 | 0.929 | 0.011 | 0.225 | 0.840 | 1.0 | 0.972 | 0.771 |
| D2 | 0.875 | 0.234 | 0.571 | 0.229 | 0.516 | 0.928 | 0.836 | 0.813 |
| D3 | 0.564 | 0.684 | 0.0 | 0.711 | 0.747 | 0.983 | 0.983 | 0.908 |
| D4 | 0.825 | 0.469 | 0.927 | 0.399 | 0.0 | 0.958 | 0.994 | 0.786 |
| D5 | 0.797 | 0.20 | 0.10 | 0.843 | 0.60 | 0.959 | 0.996 | 0.843 |
| D6 | 0.957 | 0.928 | 0.852 | 0.977 | 0.783 | 0.926 | 0.999 | 0.928 |
| D7 | 0.993 | 0.921 | 0.915 | 0.951 | 0.913 | 0.290 | 0.981 | 0.878 |
| D8 | 0.990 | 0.989 | 0.417 | 1.0 | 0.991 | 0.996 | 0.736 | 0.880 |
| D9 | 0.353 | 0.994 | 0.229 | 0.813 | 1.0 | 0.982 | 0.979 | 0.752 |
| D10 | 0.528 | 0.724 | 0.447 | 0.798 | 0.274 | 1.0 | 0.995 | 0.684 |
| Mean | 0.786 | 0.896 | 0.545 | 0.802 | 0.771 | 0.932 | 0.945 | 0.816 |

**TABLE 6.2**
**Classification Results for Driving Tasks Recognition Using GoogLeNet.**

| No. | GMM-BASED GOOGLENET | | | | | | | |
|---|---|---|---|---|---|---|---|---|
|  | T1 | T2 | T3 | T4 | T5 | T6 | T7 | Ave |
| D1 | 0.917 | 0.619 | 0.0 | 0.325 | 0433 | 1.0 | 0.968 | 0.768 |
| D2 | 0.892 | 0.362 | 0.0 | 0.042 | 0.230 | 0.784 | 0.815 | 0.767 |
| D3 | 0.883 | 0.563 | 0.0 | 0.073 | 0.840 | 1.0 | 0.994 | 0.739 |
| D4 | 0.740 | 0.453 | 0.848 | 0.986 | 0.758 | 0.663 | 1.0 | 0.755 |
| D5 | 0.970 | 0.200 | 0.233 | 0.325 | 0.078 | 0.959 | 0.988 | 0.799 |
| D6 | 0.951 | 0.966 | 0.807 | 0.936 | 0.967 | 0.075 | 1.0 | 0.829 |
| D7 | 1.0 | 0.886 | 0.436 | 0.990 | 0.890 | 0.248 | 0.963 | 0.737 |
| D8 | 0.301 | 0.995 | 0.178 | 1.0 | 1.0 | 0.990 | 0.998 | 0.789 |
| D9 | 0.562 | 0.245 | 0.949 | 0.997 | 1.0 | 0.990 | 0.843 | 0.792 |
| D10 | 0.990 | 1.0 | 1.0 | 0.685 | 0.882 | 0.012 | 1.0 | 0.810 |
| Mean | 0.835 | 0.766 | 0.648 | 0.796 | 0.819 | 0.678 | 0.948 | 0.786 |

**TABLE 6.3**
**Classification Results for Driving Tasks Recognition Using ResNet50.**

| No. | GMM-BASED RESNET50 | | | | | | | |
|---|---|---|---|---|---|---|---|---|
|  | T1 | T2 | T3 | T4 | T5 | T6 | T7 | Ave |
| D1 | 0.944 | 0.389 | 0.120 | 0.125 | 0.219 | 1.0 | 0.963 | 0.746 |
| D2 | 0.872 | 0.284 | 0.0 | 0.729 | 0.066 | 0.918 | 0.926 | 0.921 |
| D3 | 0.919 | 0.938 | 0.195 | 0.040 | 0.814 | 0.998 | 0.993 | 0.753 |
| D4 | 0.975 | 1.0 | 0.924 | 0.514 | 1.0 | 0.639 | 0.882 | 0.801 |
| D5 | 0.907 | 0.255 | 0.133 | 0.874 | 0.473 | 0.930 | 0.996 | 0.856 |
| D6 | 0.790 | 0.992 | 0.941 | 0.791 | 0.504 | 0.509 | 0.985 | 0.750 |
| D7 | 0.996 | 0.857 | 0.629 | 0.922 | 0.950 | 0.301 | 0.973 | 0.786 |
| D8 | 0.528 | 0.567 | 0.192 | 0.641 | 0.988 | 0.944 | 0.715 | 0.638 |
| D9 | 0.346 | 0.245 | 0.713 | 0.997 | 0.735 | 0.693 | 0.829 | 0.655 |
| D10 | 0.002 | 0.999 | 0.058 | 0.991 | 0.782 | 0.219 | 1.0 | 0.589 |
| Mean | 0.728 | 0.652 | 0.391 | 0.662 | 0.653 | 0.715 | 0.926 | 0.749 |

task, the answering mobile phone activity gets the most accurate detection results among the 10 drivers for all three models. The worst result happens in the rear mirror checking (T3) case for the three models. One explanation is that the rear mirror checking behavior requires few bodies and head movement, which can be easily misclassified into the normal driving task.

Another evidence that can be drawn from Tables 6.1–6.3 are the CNN model achieved better detection results on the secondary tasks in general. This is mainly because when performing the secondary tasks, the driver has to move his/her body and hands instead of only rotating her/his head, which is more distinct and easier to be detected.

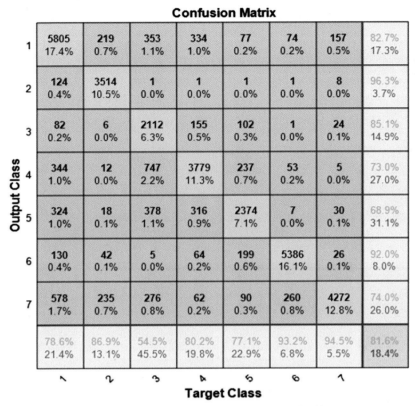

FIG. 6.6 Confusion matrix for driver 1 using AlexNet.

Table 6.2 indicates the activity classification results given by the GoogLeNet. The general detection results are similar to the results in Table 1 except that the overall detection accuracy for the 10 drivers is slightly lower. The GoogLeNet does not achieve better classification results than the AlexNet as it does on the ImageNet dataset. However, the classification results for the GoogLeNet trained with raw images are better than those in the AlexNet case. The general classification results for the GoogLeNet with the raw image is 74.7% accuracy, which is 5% higher than that for the AlexNet. Table 6.3 illustrates the activity classification results given by the ResNet. Similar to the GoogLeNet, the ResNet does not show its advantage on the activity classification task. Instead, the precision is the lowest among these three models. The general classification accuracy is 74.9% for the GMM-ResNet and 61.4% for the raw image-based ResNet. Discussions on the results will be proposed in the next section.

Fig. 6.6 illustrates the confusion matrix for the 10 drivers using the AlexNet model with GMM

segmentation. The green diagonal shows the correct detection cases for the class. The bottom row shows the classification accuracy with respect to the target class, while the rightmost column shows the classification accuracy with respect to the predicted labels. As shown in Fig. 6.6, all of the driving tasks except the third task (rear mirror checking) achieved reasonable detection rates. There are 353 cases of rear-mirror checking that are misclassified into normal driving and 747 cases are misclassified into the left mirror checking.

## Visualization of Deep CNN Models

To have a better understanding of the CNN models and how the model response to the segmented images, the convolutional layer activation and its features are analyzed in the following. Fig. 6.7 shows the raw image and the segmented image for the in-vehicle texting behavior of one participant. Fig. 6.8 shows different activation maps from the CNN models based on the images in Fig. 6.7. As shown in the left part of the dashed

FIG. 6.7 Illustration of texting behavior of driver 3.

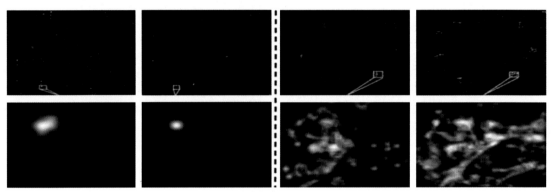

FIG. 6.8 Activation map of the two CNN models. The left part is given by the AlexNet model and the right part is given by the GoogLeNet. From left to right: the Relu5 layer activation of the GMM-AlexNet model, the Relu5 layer of the Raw-AlexNet model, the conv2-Relu layer of the GMM-GoogLeNet, and the same layer activation map for the Raw-GoogLeNet model. The lower part images are the corresponding strongest activation channels in the activation maps for the GMM-AlexNet and GMM-GoogLeNet models.

line in Fig. 6.8, the top activation maps are given by the relu5 layer of the AlexNet models. Specifically, the left one is based on the AlexNet with GMM segmentation (G-AlexNet model), whereas the right one is based on the AlexNet with raw images (R-AlexNet model). The bottom images are the corresponding strongest activation channel. As shown in Fig. 6.7, the relu5 layer for the G-AlexNet model contains many more features than that in the R-AlexNet. The GMM segmentation extracts the driver from the background so that the CNN model can maintain more relevant features for the driver. The strongest activation for the G-AlexNet model keeps the driver head rotation features and other channels keep the arm position information. Fig. 6.8 shows the visualized features for the AlexNet models. The G-AlexNet model learns more contrast filters, while

the filters in R-AlexNet are oversmooth, especially can be found in the last fully connected layer.

As GoogLeNet is much deeper than the AlexNet and the deeper inception layers in the GoogLeNet are complex, in this part, only the activation from the second convolutional layer is analyzed. The right part of Fig. 6.7 indicates the activation of the conv2-relu layer in the GoogLeNet. As shown in the figure, the GoogLeNet trained with segmented images (G-GoogLeNet) maintains more driver-related features instead of the background features. The GoogLeNet model trained with raw images (R-GoogLeNet) cannot give a clear driver position. As driver-related features are not well maintained in the beginning layers, the deeper inception layers also cannot learn a representative feature for the driver. Based on this, the activation maps explain why the GMM

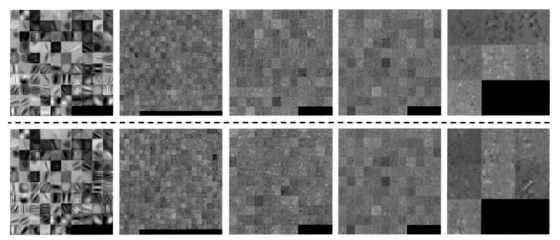

FIG. 6.9 Feature visualization for AlexNet.

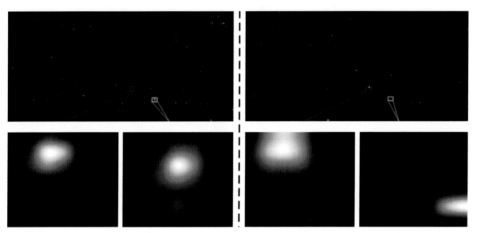

FIG. 6.10 Activation map of the ResNet. The left part indicates the activation of the final Relu layer for the G-ResNet, while the right part is for the R-ResNet. The red boxes in the upper images are the strongest activation for the Raw-ResNet, while the green boxes represent the strongest activation for the G-ResNet. The lower four images are the corresponding activation channels.

segmentation-based CNN models lead to better classification results than the raw image-based models.

The left parts of Fig. 6.8 are given by the AlexNet model and the right part is given by the GoogLeNet. From left to right, the first activation map is the relu5 layer activation of GMM-AlexNet model, the second map is the relu5 layer of Raw-AlexNet model, the third one is the conv2-relu layer of GMM-GoogLeNet, and the last one is the same layer activation map for the Raw-GoogLeNet model. The lower part images are the corresponding strongest activation channels.

The upper part of Fig. 6.9 indicates the features for the model trained with segmented images while the lower part is given by the model trained with raw images. From left to right, the feature maps are given by the conv1 layer, conv3 layer, conv5 layer (1−128 filters), conv5 layer (129−256), and final fully connected layer, respectively.

Similar results can be found in the ResNet case that is shown in Fig. 6.10. The upper images show the activation maps for the G-ResNet and R-ResNet, respectively. The lower part indicates the corresponding strongest activation map. Specifically, the green box in the top row images is the strongest activation channel of the final ReLu layer of the G-ResNet, and its corresponding channel in the R-ResNet map. The red boxes in the top images are the strongest activation channel of the R-ResNet, and its corresponding

| TABLE 6.4 Training and Testing Time Cost for Each Model. | | | | |
|---|---|---|---|---|
| | **G-AlexNet** | **R-AlexNet** | **G-GoogLeNet** | **R-GoogLeNet** |
| Training time (s) | ~1100 | ~1200 | ~2400 | ~2400 |
| Testing time per image (ms) | ~13 | ~12.5 | ~45 | ~45 |

channel in the G-ResNet. As can be seen in the lower part of Fig. 6.10, the strongest activation channel of the G-ResNet is able to capture the head rotation and position features, while the R-ResNet fails to learn a precise representation for this behavior. Based on the model visualization results, it can be seen that with prior image segmentation, the CNN model can learn more representative driver status features.

Finally, the time cost for the training and testing of different models are indicated in Table 6.4. As shown in Table 6.4, the general training for GoogLeNet is two times longer than the AlexNet training time. It takes about 12 ms to process one image for the AlexNet while the testing time for the GoogLeNet is 45 ms. The model training and testing are based on Core i7 2.5 GHz CPU and NVIDIA Quadro K1100M 2 GB GPU.

### Results Comparison Between Transfer Learning and Feature Extraction

As discussed in the last section, a pretrained CNN model normally can be used in two different manners. The first way is to fine-tune the last few layers so that the model can satisfy the new task. The second common usage is treating the pretrained CNN model as a feature extractor [44]. As introduced in Ref. [45], the histogram of oriented gradient (HOG) and support vector machine (SVM) method has been one of the most successful methods for some image-based feature extraction and classification tasks. Therefore, HOG and SVM is adopted as one of the conventional methods. Meanwhile, as some of the studies have used the pretrained CNN as the feature extractor, in this part, the feature set from the conv5 layer of AlexNet is also adopted. The conventional feature extraction and classification scheme will be compared with the proposed end-to-end driver behavior classification algorithm in the case to show the advantage of the proposed method. The deep CNN models are trained on the large-scale dataset, which makes the convolutional layers have a strong representation ability of the objects. The lower level of the convolutional layers is more concentrated on the local features, such as edges, colors, and corners. In contrast, the deeper convolutional layers will focus on the higher-level features. The combination of feature extraction and conventional machine learning algorithms like SVM is more convenient to use [44]. Therefore, in this part, the performance of the vision-based feature extraction method for driving activities recognition will be evaluated and compared with the transfer learning methods.

Two different features are extracted based on the segmented images. The first feature set is generated using the HOGs [45]. A pyramid HOG feature extractor is used, which concatenates two different scale HOG extractors. The block size for the HOG feature extractors is $2 \times 2$ and the cell sizes are $8 \times 8$ and $16 \times 16$, respectively. The visualization for the HOG features is illustrated in Fig. 6.9. The second feature set is generated by the fifth convolutional layer of the AlexNet. The principal component analysis (PCA) algorithm is used to reduce the feature set dimension for the two feature sets. In this work, the dimension for the feature vector is reduced to 50.

Meanwhile, two different classification models are evaluated, which are the SVM and feedforward neural network (FFNN). The classification results are shown in Table 6.5. In Table 6.5, S+C means SVM and Conv5 feature. S+H is SVM and HOG features. Similarly, F+C means FFNN and Conv5 feature. F+H represents FFNN and HOG features. As shown in Table 6.5, the feature extraction methods are unable to accurately identify the driving tasks as the transfer learning method does. The average results for the four combinations are much lower compared with the transfer learning method. Hence, transfer learning is proved to be more suitable for this task. Graph (A) of Fig. 6.11 indicates the original segmented image; graph (B) represents the HOG feature map with $8 \times 8$ cell, and the graph (C) is the feature map with $16 \times 16$ cell.

To further evaluate our method, additional experiments are made to compare the proposed method with conventional hand-craft feature extraction and shallow CNN methods. Specifically, the approaches used for comparison include the following:

**FC7 + ANN:** The method proposed in Ref. [44], which extracts the posture features with a pretrained

| TABLE 6.5 | | | | | | |
|---|---|---|---|---|---|---|
| **Classification Results Using Feature Extraction.** | | | | | | |
| | **D1** | **D2** | **D3** | **D4** | **D5** | **Ave** |
| S+C | 0.497 | 0.344 | 0.331 | 0.236 | 0.249 | 0.331 |
| S+H | 0.390 | 0.215 | 0.184 | 0.240 | 0.196 | 0.245 |
| F+C | 0.304 | 0.298 | 0.537 | 0.250 | 0.289 | 0.335 |
| F+H | 0.428 | 0.480 | 0.454 | 0.173 | 0.136 | 0.334 |

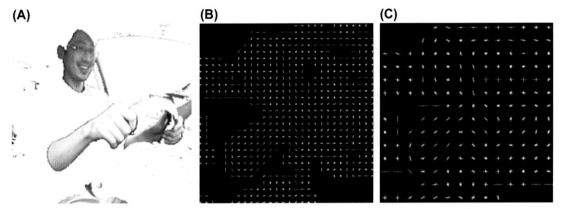

FIG. 6.11  HOG feature visualization map.

AlexNet CNN model. In this part, the activation of the "fc7" layer of AlexNet is extracted, and an FFNN ANN model with 300 neurons is constructed based on the feature set. The dimension for each of the "fc7" feature vector is 4096.

**PHOG+SVM:** The pyramid histogram of oriented gradients (PHOGs) followed by a SVM method. A pyramid HOG feature extractor, which concatenates two different scale HOG extractors, is used. The block size for the HOG feature extractors is $2 \times 2$, and the cell sizes are $8 \times 8$ and $16 \times 16$, respectively. The dimension for each of the PHOG feature vector is 32,328.

**OP+ANN:** The method proposed in Refs. [46,47], which recognizes the motion with optical flow. Specifically, the optical flow of the video sequence is extracted with the Lucas–Kanade method, and a 51,529-dimensional feature vector is concatenated for each image. Then, another FFNN model with the same structure of the one proposed in Ref. [27] is used.

**OPsCNN:** Based on the magnitude of the optical flow, a shallow multiclass CNN is proposed. Three convolution layers are used following with three fully connected layers. The input images are rescaled into the size of $120 \times 120$. The filter size is selected as $5 \times$ 5 for the first two convolutional layers and $3 \times 3$ for the third convolutional layer. Batch normalization, nonoverlap pooling, and ReLu nonlinearity layers are applied between the convolutional layers. The number of neurons for the three FC layers is 512, 128, and 7, which is similar to the architecture used in Ref. [48].

**GMMsCNN:** The shallow CNN with the GMM-segmented color images are also tested. Finally, as the dimensions of the feature vectors given by different algorithms are too high, a PCA algorithm is used to reduce the feature dimension and reduce the training cost for the first three models. The dimension for each feature vector is reduced to 500. The model comparisons are illustrated in Table 6.6. As shown in, Table 6.6 the PHOG and optical flow features are unable to accurately represent the driving tasks and far less precise than the transfer learning method. The bold emphasis represents the overall performance of the proposed method, which is significantly more accurate than the other baseline methods. The recognition results of the optical flow-based and shallow CNN-based methods are slightly better than the feature extraction methods. The high-level features from the FC7 layer of AlexNet with FFNN gives better results than the remaining

**TABLE 6.6**
**Activity Recognition Compared With Other Approaches.**

|          | T1    | T2    | T3    | T4    | T5    | T6    | T7    | Mean  |
|----------|-------|-------|-------|-------|-------|-------|-------|-------|
| FC7 + ANN | 0.478 | 0.343 | 0.113 | 0.249 | 0.311 | 0.803 | 0.631 | 0.497 |
| PHOG+SVM | 0.573 | 0.059 | 0.024 | 0.394 | 0.108 | 0.437 | 0.473 | 0.354 |
| OP+ANN   | 0.404 | 0.044 | 0.093 | 0.121 | 0.209 | 0.561 | 0.506 | 0.347 |
| OPsCNN   | 0.537 | 0.369 | 0.085 | 0.273 | 0.205 | 0.572 | 0.669 | 0.443 |
| GMMsCNN  | 0.423 | 0.242 | 0.096 | 0.109 | 0.212 | 0.598 | 0.531 | 0.400 |
| Proposed | 0.786 | 0.869 | 0.545 | 0.802 | 0.771 | 0.932 | 0.945 | **0.816** |

four methods. However, the average results for the 10 drivers are still significantly lower compared with the proposed method.

In Table 6.7, the proposed method is also compared with relevant studies in the literature. It should be noticed that difficulties exist in making a precise cross-platform comparison between the existing studies as different algorithms, platforms, and experimental methods were used. Based on Table 6.7, some researchers have tried to analyze the driver distracted behaviors with either real vehicle and simulated data. For example, in Ref. [22], Li et al. proposed a machine learning framework for the detection of driver mirror checking behaviors and secondary tasks. The general framework follows a standard machine learning application procedure, which consists of feature extraction, model training, and testing. The detection rates for secondary tasks like radio operating and phone talking are around 75%−80%. We believe this should attribute to the absence of driver body features. In Ref. [27], the driver's behaviors are detected with a Kinect device, which enables the analysis of both the driver's head and upper body features. However, that work also heavily rely on the complex feature extraction and analysis, which is time consuming and requires extra hardware for calibration. Similar work can be found in Ref. [48], where the authors evaluated the performance of different types of CNN models on 10 different driving activities. Although high detection accuracies are achieved in the study, the data are collected on the driving simulator and did not stand for the real-world in-vehicle performance. Another reason that can significantly influence the model accuracy is the evaluation method. In, Table 6.7 the *LOO* method is more strict than the *cross-validation* method as it indicates the model generalization capability on the unseen dataset. If we use the cross-validation method and simply separate the data into training and testing group, the GMM-

AlexNet in this section can achieve 98.9% accuracy for multitasks detection. However, we still suggest using the *LOO* method as it can reflect the performance variance on different subjects.

Based on the comparison with existing studies, the proposed method in this section show three advantages. First, a naturalistic in-vehicle dataset is collected for the fine-tuning and validation of the deep CNN models. The fine-tune method is very efficient in real-world applications as it is hard to collect large scale annotated driver distraction data. Although some studies use side-view images as the images show clear driver body features [48−50], the side-view method is less efficient and robust compared with the front-view method in the real vehicle as the side view can be occluded by the passengers. Second, the LOO model evaluation is used so that the results illustrate an independent performance on the different drivers. Third, the segmentation-based CNN models achieved state-of-art detection accuracy on distracted behaviors such as the phone answering (93.2%) and texting (94.5%) with naturalistic data.

## Driver Distraction Detection Using Binary Classifier

In this section, the three CNN models are modified and trained to detect whether the driver is distracted or not. In this case, the first four tasks are grouped together, while the last three tasks constitute another group. The CNN models are fine-tuned to solve the binary classification problem. The distraction detection results for the AlexNet, GoogLeNet, and ResNet are shown in Table 6.8. As shown in Table 6.8, the segmentation image-based AlexNet leads to the most accurate results. The general classification accuracy for the G-AlexNet based model is 91.4%. The general classification accuracy for the GoogLeNet and ResNet methods is 87.5% and 83.0%, which are slightly lower than the results

**TABLE 6.7**
Classification Results Using Feature Extraction.

| ID | Input | Tasks | Model | Validation | Platform | Subjects | Computation cost |
|----|-------|-------|-------|-----------|----------|----------|------------------|
| [27] | Kinect RGB and depth, body, Head, eye | 7 tasks: 4 mirror checking, radio, phone call, texting | Random forest and FFNN | LOO, recognition: 82.4% | Real vehicle | 5 drivers | 8 fps data collection |
| [49] | Kinect RGB and depth image (eye, arm, head, and facial features) | 5 tasks: phone call, drinking, message, looking object, normal driving | Sequential model with AdaBoost and HMM | LOO, recognition: 85.0%, detection: 89.8% | Simulator | 8 drivers | 1 fps screenshot of the monitors |
| [50] | Triple-view fusion | 7 tasks: gear, driving, phone call, phone pick, control, looking left/right | CNN+RNN sequential feature extractor with SVM classifier | Cross Validation, 90% in average | Simulator | 3 drivers | 15 fps data collection |
| [22] | CAN and cameras with 268D features | 4 mirror tasks and radio, GPS operating and following, phone operating and call, picture, conversation | SVM, KNN, RUSBoost | LOO, recognition rates for different tasks are among 65%–85% | Real vehicle | 20 drivers | 5s window size with 10 samples per window |
| [51] | Side-view images with face and hand detection | 10 tasks: drinking, radio, normal driving, makeup, reach behind, conversation, phone call, texting | Transfer learning with AlexNet and inception V3 | CV, 75% training data and 25% testing, 95.98% in average | Real vehicle | 31 drivers | AlexNet 182 fps and inception V3 72 fps with GTX TITAN |
| [52] | Side-view images | 10 tasks: drinking, radio, normal driving, makeup, reach behind, conversation, phone call, texting | Transfer learning with VGG16, AlexNet, GoogLeNet, and ResNet | LOO, recognition accuracy in the range of 86% and 92% | Simulator | 10 drivers | Frequency in the range of 8 and 14 Hz with Jetson TX1 |
| [48] | Side-view images | 4 tasks: normal driving, operating shift gear, phone call, eating/smoking | Sparse filter and CNN model | CV, 80% training and 20% testing, 99.47% in average | Real vehicle | 20 drivers | – |
| Ours | Front-view images | 7 tasks: 4 mirror checking, radio, phone call, texting | GMM segmentation and transfer learning | LOO, recognition: 81.6%, Detection: 91.4% | Real vehicle | 10 drivers | 14 fps with Nvidia MX150 GPU |

given by the AlexNet. The bold emphasis represents the overall performance of the three deep learning models. It should be noticed that there are no smoothing algorithms applied to the distraction warning module. In real-world situations, the driver assistance system will only warn the driver if the distraction happens continuously in a short period. Therefore, if applying a short period smoothing or voting techniques, the distraction detection system can be more suitable for the real-world application.

Lastly, the AlexNet and GoogLeNet are modified and trained to detect whether the driver is distracted or not. At this moment, the first four tasks are grouped together, and the last three tasks constitute another group. Different from the previous step where the CNN model is designed for multiclass classification, in this step, the last fully connected layer is tuned to solve the binary classification problem. The distraction detection results for AlexNet and GoogLeNet are shown in Tables 6.9 and 6.10, respectively.

**TABLE 6.8**
**Binary Classification Results Using the Three Models.**

|  |  | D1 | D2 | D3 | D4 | D5 | D6 | D7 | D8 | D9 | D10 | Mean |
|---|---|---|---|---|---|---|---|---|---|---|---|---|
| Binary | Normal | 0.970 | 0.994 | 0.849 | 0.873 | 0.991 | 0.897 | 0.897 | 0.997 | 0.911 | 1.00 | 0.936 |
| AlexNet | Distract | 0.910 | 0.674 | 0.858 | 0.917 | 0.763 | 0.856 | 0.857 | 0.941 | 0.948 | 0.988 | 0.881 |
|  | Ave | 0.948 | 0.867 | 0.852 | 0.903 | 0.856 | 0.882 | 0.882 | 0.976 | 0.927 | 0.996 | **0.914** |
| Binary | Normal | 0.993 | 0.940 | 0.141 | 0.850 | 0.988 | 0.992 | 0.992 | 0.833 | 0.908 | 0.999 | 0.897 |
| GoogLeNet | Distract | 0.784 | 0.577 | 1.00 | 0.870 | 0.857 | 0.692 | 0.847 | 0.884 | 0.898 | 0.934 | 0.841 |
|  | Ave | 0.916 | 0.796 | 0.426 | 0.863 | 0.911 | 0.883 | 0.946 | 0.853 | 0.904 | 0.978 | **0.875** |
| Binary | Normal | 0.908 | 0.950 | 0.656 | 0.609 | 0.989 | 0.820 | 0.982 | 0.871 | 0.06 | 0.985 | 0.798 |
| ResNet50 | Distract | 0.905 | 0.768 | 0.525 | 0.976 | 0.794 | 0.941 | 0.946 | 0.809 | 1.00 | 0.994 | 0.891 |
|  | Ave | 0.907 | 0.878 | 0.612 | 0.852 | 0.874 | 0.864 | 0.971 | 0.847 | 0.955 | 0.988 | **0.830** |

**TABLE 6.9**
**Binary Classification Results Using AlexNet.**

|  |  | D1 | D2 | D3 | D4 | D5 | Mean |
|---|---|---|---|---|---|---|---|
| GMM_AlexNet | Normal | 0.991 | 0.944 | 0.999 | 0.944 | 1.00 | 0.976 |
|  | Distract | 0.776 | 0.997 | 0.715 | 0.959 | 0.992 | 0.888 |
|  | Ave | 0.913 | 0.961 | 0.890 | 0.951 | 0.997 | 0.942 |
| AlexNet | Normal | 0.538 | 0.213 | 0.133 | 0.674 | 1.00 | 0.512 |
|  | Distract | 0.732 | 1.00 | 0.999 | 0.899 | 0.644 | 0.855 |
|  | Ave | 0.608 | 0.462 | 0.467 | 0.771 | 0.886 | 0.639 |

**TABLE 6.10**
**Binary Classification Results Using GoogLeNet.**

|  |  | D1 | D2 | D3 | D4 | D5 | Mean |
|---|---|---|---|---|---|---|---|
| GMM_GoogLeNet | Normal | 0.993 | 0.979 | 0.989 | 0.409 | 0.999 | 0.873 |
|  | Distract | 0.814 | 0.987 | 0.977 | 0.935 | 0.926 | 0.923 |
|  | Ave | 0.928 | 0.982 | 0.985 | 0.634 | 0.976 | 0.898 |
| GoogLeNet | Normal | 0.995 | 0.913 | 0.375 | 0.09 | 1.00 | 0.645 |
|  | Distract | 0.633 | 0.984 | 0.998 | 0.962 | 0.991 | 0.913 |
|  | Ave | 0.863 | 0.935 | 0.615 | 0.464 | 0.997 | 0.775 |

In Table 6.9, the upper part indicates the detection results based on the GMM image segmentation and CNN model. On the contrary, the lower part indicates the detection results using raw images. As shown in Table 6.9, the GMM-CNN-based method leads to a much more accurate result than that using the raw image. The general classification accuracy for the GMM-CNN-based method is 94.2%, which is significantly higher than the raw image-based method (only achieved 63.9% accuracy).

Similar results can be found in Table 6.10, where the GMM-GoogLeNet-based method also achieved a better result than using a raw image-based method. The general classification result for GMM-GoogLeNet-based method achieved 89.8% accuracy, which is slightly lower than the result given by the GMM-AlexNet. However, the distraction detection results (92.3%) for the GoogLeNet are better than those in the AlexNet case (88.8%). However, it should be mentioned that one significant low detection accuracy of the normal driving task for the fourth driver is found in Table 6.10, where the proposed method only

achieved 40.9% accuracy. If ignoring this significant low sample, the detection rate would increase to 99% accuracy. It should be pointed out that the fourth driver is the only female driver in this experiment. The gender issue may be a potential explanation for this low detection rate. If ignoring the fourth driver, the GoogLeNet achieved a better classification result on this binary case. Another difference between GoogLe-Net and AlexNet is that GoogLeNet achieved better detection results (77.5%) for the raw image-based case, which is about 13% higher than that in the Alex-Net case.

The confusion matrix for the GMM-CNN based methods is shown in Figs. 6.12 and 6.13, respectively. It should be noticed that, in these cases, no smoothing schemes are applied to the distraction warning module. In most situations, the driver assistance system will only warn the driver if distraction cases happen continuously in a short period. Therefore if applying a short period smoothing or voting technique, the distraction detection system can be more suitable for real-world applications.

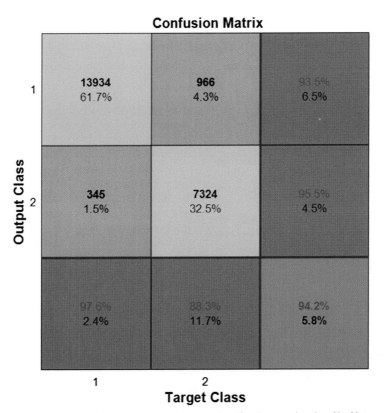

FIG. 6.12 Confusion matrix for the binary classification result using AlexNet.

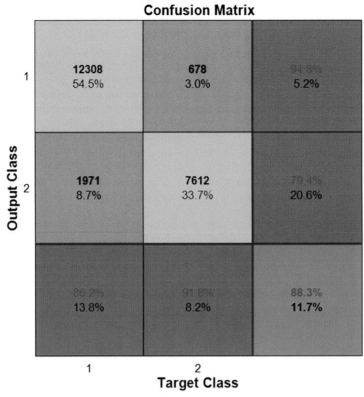

FIG. 6.13 Confusion matrix for the binary classification result using GoogLeNet.

## DISCUSSION
### Transfer Learning Performance

With the analysis of different deep CNN models, it can be found that a deeper CNN model like ResNet50 does not contribute to a higher detection accuracy as it did in the ILSVRC competition. The reasons can be multifold, and we try to explain this phenomenon merely based on the evidence in this section. First, as the GoogLeNet and ResNet are deeper than the AlexNet, the model may need more data to be optimized. Second, the transfer learning approach is different from training from scratch such as using the ImageNet dataset. The fine-tune transfer learning method mainly focuses on the tuning of a few layers while keeping the main characters of the convolutional layers. However, as the model getting deeper, the domain knowledge learned from the much larger dataset may not very suitable for the smaller dataset. This conclusion is only made according to the results in this section and further evaluation is expected. The primary object in this section is not to evaluate the classification performance of different CNN models. However, this section aims to provide an efficient end-to-end approach to understand driver behaviors. Therefore, more experiments and analyses are expected in the future to obtain a more precise explanation.

Although it is essential to understand which features are more critical to the driver behavior recognition, the traditional machine learning framework is less efficient and usually has a specific requirement on the system hardware such as head pose measurement and skeleton tracking. Therefore, in this work, an end-to-end deep learning approach is designed to solve the driver behavior recognition task. The only hardware required for the proposed system is an RGB camera. Meanwhile, as shown in this section, the CNN models can automatically capture the head and body features. The deep learning method achieved competitive detection results compared with the methods that rely on the head and body detection [27,49]. However, the end-to-end process shows its advantages in real-world detection as no complex head pose and body joint estimation algorithms are needed. Besides, this section also evaluated the binary classification results and found that the deep learning approach can provide an accurate estimation of the distraction status. This approach can be easily integrated into most of the current ADAS products due to its efficient and low-cost properties.

### Real-Time Application

In this experiment, the system is implemented to a Windows operating system using MATLAB platform and a single low-cost GPU device. The testing cost of the AlexNet for each image is about 13 ms, also, it cost 50 ms for the GMM to segment each image. The total computational cost for each image is around 60–70 ms, and the general processing ability of the system is about 14 fps. Therefore, the proposed system can satisfy the real-world computational requirement. Meanwhile, regarding the in-vehicle embedded systems in the real world, the Linux platform along with C++/Python programming usually can be more efficient than the MATLAB environment. In addition, as more powerful embedded GPU devices have been published, the in-vehicle graphics processor can provide more powerful parallel computation than the current platform. Hence, the algorithm has no significant limitation in the real-world application.

Next, a sliding window is applied to the detection to smooth the result. The current driver state is selected according to the majority state within the sliding window. The smoothing results of the secondary task detection and distraction detection are shown in Figs. 6.14 and 6.15. The upper images of Figs. 6.14 and 6.15 indicate the comparison between the ground truth label and the predicted values concerning the seven driving activities detection and distraction detection, while the bottom images represent the comparison between the ground truth label and the smoothed version of the predicted values. The sliding window can be used to smooth the result and eliminate some false detection cases. However, it should be noticed that as the sliding window uses the voting scheme, detection delay can happen for the secondary tasks and the horizon of the sliding window will control how much delay the detection system has. A larger horizon of the sliding window will lead to a smoother result; however, it will also cause a larger detection delay. In Figs. 6.14 and 6.15, the window is selected as seven samples, which can cause a 500 ms delay. Considering each task, especially the secondary tasks can last several seconds, this 0.5 s late detection is normally acceptable.

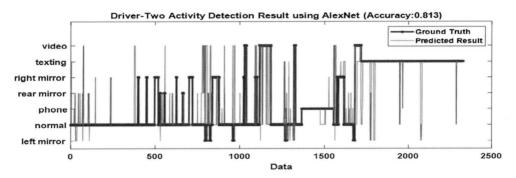

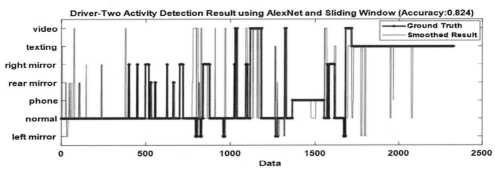

FIG. 6.14 Driver 2 real-time activity detection using AlexNet and sliding window.

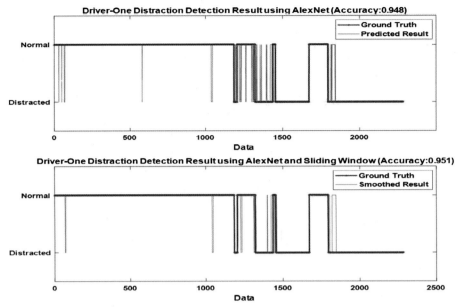

FIG. 6.15 Driver 1 real-time activity detection using AlexNet and sliding window.

## CONCLUSIONS

In this work, a driving-related activity recognition system based on the CNN model is proposed. The deep CNN models are trained with the transfer learning method. Specifically, the pretrained AlexNet and GoogLeNet are fine-tuned to satisfy the seven driving tasks recognition task. To increase the identification accuracy, the raw RGB images are first processed with a GMM-based segmentation algorithm. The GMM method can efficiently identify the drivers from the background environment and remove irrelevant objects. The classification results indicate that the GMM segmentation-based CNN models give a much precise detection result. Another comparison is made between the transfer learning and feature extraction method. Results show that the HOG and CNN features are not able to accurately identify the driver behaviors. Finally, if using the CNN models as a binary classifier, the driver distraction detection rate can achieve 94% accuracy. Although the multiple classification accuracy in this work is no better than the result in Ref. [27], the end-to-end transfer learning scheme only requires RGB images as input. It does not need to complicate the feature extraction and selection process, which makes it much easier to be implemented in real time. Future works will concentrate on the further improvement of the behavior detection rate. More drivers and data are expected to increase system robustness and detection accuracy. The system can be also tested and used for driver/passenger behavior monitoring system on the partially automated vehicles.

## REFERENCES

[1] F.-Y. Wang, et al., Parallel driving in CPSS: a unified approach for transport automation and vehicle intelligence, IEEE/CAA Journal of Automatica Sinica 4 (4) (2017) 577−587.
[2] J. Wang, J. Wang, R. Wang, C. Hu, A framework of vehicle trajectory replanning in lane exchanging with considerations of driver characteristics, IEEE Transactions on Vehicular Technology 66 (5) (2017) 3583−3596.
[3] A. Koesdwiady, et al., Recent trends in driver safety monitoring systems: state of the art and challenges, IEEE Transactions on Vehicular Technology 66 (6) (2017) 4550−4563.
[4] X. Wang, R. Jiang, L. Li, Y. Lin, X. Zheng, F.-Y. Wang, Capturing car-following behaviors by deep learning, IEEE Transactions on Intelligent Transportation Systems (2017).
[5] X. Zeng, J. Wang, A stochastic driver pedal behaviour model incorporating road information, IEEE Transactions on Human-Machine Systems 47 (5) (2017) 614−624.
[6] C. Lv, et al., Analysis of autopilot disengagements occurring during autonomous vehicle testing, IEEE/CAA Journal of Automatica Sinica 5 (1) (2018) 58−68.
[7] V.A. Butakov, I. Petros, Personalized driver/vehicle lane change models for ADAS, IEEE Transactions on Vehicular Technology 64 (10) (2015) 4422−4431.

[8] C. Gou, et al., A joint cascaded framework for simultaneous eye detection and eye state estimation, Pattern Recognition 67 (2017) 23–31.

[9] N. Pugeault, R. Bowden, How much of driving is preattentive? IEEE Transactions on Vehicular Technology 64 (12) (2015) 5424–5438.

[10] A. Tawari, et al., Looking-in and looking-out vision for urban intelligent assistance: estimation of driver attentive state and dynamic surround for safe merging and braking, in: Intelligent Vehicles Symposium Proceedings, 2014 IEEE, IEEE, 2014.

[11] J.C. McCall, et al., Lane change intent analysis using robust operators and sparse bayesian learning, IEEE Transactions on Intelligent Transportation Systems 8 (3) (2007) 431–440.

[12] C.M. Martinez, et al., Driving style recognition for intelligent vehicle control and advanced driver assistance: a survey, IEEE Transactions on Intelligent Transportation Systems (2017).

[13] J. Hu, et al., Abnormal driving detection based on normalized driving behavior, IEEE Transactions on Vehicular Technology 66 (8) (2017) 6645–6652.

[14] R. Chai, et al., Driver fatigue classification with independent component by entropy rate bound minimization analysis in an EEG-based system, IEEE Journal of Biomedical and Health Informatics 21 (3) (2017) 715–724.

[15] B. Mandal, et al., Towards detection of bus driver fatigue based on robust visual analysis of eye state, IEEE Transactions on Intelligent Transportation Systems 18 (3) (2017) 545–557.

[16] T.A. Ranney, et al., NHTSA driver distraction research: past, present, and future, Driver Distraction Internet Forum 2000 (2000).

[17] E. Murphy-Chutorian, M.M. Trivedi, Head pose estimation and augmented reality tracking: an integrated system and evaluation for monitoring driver awareness, IEEE Transactions on Intelligent Transportation Systems 11 (2) (2010) 300–311.

[18] N. Das, E. Ohn-Bar, M. Mohan, Trivedi, On performance evaluation of driver hand detection algorithms: challenges, dataset, and metrics, in: IEEE 18th International Conference on Intelligent Transportation Systems (ITSC), 2015, IEEE, 2015.

[19] C. Tran, A. Doshi, M.M. Trivedi, Modeling and prediction of driver behaviour by foot gesture analysis, Computer Vision and Image Understanding 116 (3) (2012) 435–445.

[20] L. Bi, et al., Queuing network modeling of driver EEG signals-based steering control, IEEE Transactions on Neural Systems and Rehabilitation Engineering 25 (8) (2017) 1117–1124.

[21] T. Teng, L. Bi, Y. Liu, EEG-based detection of driver emergency braking intention for brain-controlled vehicles, IEEE Transactions on Intelligent Transportation Systems (2017).

[22] N. Li, C. Busso, Detecting drivers' mirror-checking actions and its application to manoeuvre and secondary task

recognition, IEEE Transactions on Intelligent Transportation Systems 17 (4) (2016) 980–992.

[23] E. Ohn-Bar, et al., Head, eye, and hand patterns for driver activity recognition, in: 2014 22nd International Conference on Pattern Recognition (ICPR), IEEE, 2014.

[24] M. Rezaei, R. Klette, Look at the driver, look at the road: No distraction! No accident!, in: Proceedings of the IEEE Conference on Computer Vision and Pattern Recognition, 2014.

[25] I.-H. Kim, et al., Detection of braking intention in diverse situations during simulated driving based on EEG feature combination, Journal of Neural Engineering 12 (1) (2014), 016001.

[26] C. Zhang, H. Wang, R. Fu, Automated detection of driver fatigue based on entropy and complexity measures, IEEE Transactions on Intelligent Transportation Systems 15 (1) (2014) 168–177.

[27] X. Yang, C. Lv, D. Cap, et al., Identification and analysis of driver postures for in-vehicle driving activities and secondary tasks recognition, IEEE Transactions on Computational Social Systems 99 (2018) 1–14.

[28] Y. Liao, et al., Detection of driver cognitive distraction: a comparison study of stop-controlled intersection and speed-limited highway, IEEE Transactions on Intelligent Transportation Systems 17 (6) (2016) 1628–1637.

[29] T. Liu, et al., Driver distraction detection using semi-supervised machine learning, IEEE Transactions on Intelligent Transportation Systems 17 (4) (2016) 1108–1120.

[30] T.H.N. Le, et al., DeepSafeDrive: a grammar-aware driver parsing approach to driver behavioral situational awareness (DB-SAW), Pattern Recognition 66 (2017) 229–238.

[31] L. Li, et al., Intelligence testing for autonomous vehicles: a new approach, IEEE Transactions on Intelligent Vehicles 1 (2) (2016) 158–166.

[32] M. Sivak, B. Schoettle, Motion Sickness in Self-Driving Vehicles, UMTRI-2015-12, University of Michigan Transportation Research Institute, Ann Arbor, 2015.

[33] S. Gaglio, Giuseppe Lo Re, M. Morana, Human activity recognition process using 3-D posture data, IEEE Transactions on Human-Machine Systems 45 (5) (2015) 586–597.

[34] L.B. Neto, et al., A Kinect-based wearable face recognition system to aid visually impaired users, IEEE Transactions on Human-Machine Systems 47 (1) (2017) 52–64.

[35] M. Azimi. Skelton Joint Smoothing White Paper [Online]. Available from: https://msdn.microsoft.com/en-us/library/jj131429.aspx.

[36] C.E. Rasmussen, The infinite Gaussian mixture model, Advances in Neural Information Processing Systems (2000).

[37] L. Xu, M.I. Jordan, On convergence properties of the EM algorithm for Gaussian mixtures, Neural Computation 8 (1) (1996) 129–151.

[38] J. Deng, et al., Imagenet: a large-scale hierarchical image database, CVPR 2009. IEEE Conference on, in: Computer Vision and Pattern Recognition, 2009., IEEE, 2009.

[39] A. Krizhevsky, I. Sutskever, G.E. Hinton, Imagenet classification with deep convolutional neural networks, Advances in Neural Information Processing Systems (2012).

[40] C. Szegedy, et al., Going deeper with convolutions, CVPR, 2015.

[41] K. Simonyan, A. Zisserman, Very Deep Convolutional Networks for Large-Scale Image Recognition, arXiv preprint arXiv:1409.1556, 2014.

[42] R.K. Srivastava, K. Greff, Schmidhuber, Highway Networks, arXiv preprint arXiv:1505.00387, 2015.

[43] K. He, et al., Deep residual learning for image recognition, in: Proceedings of the IEEE Conference on Computer Vision and Pattern Recognition, 2016, pp. 770−778.

[44] N. Deo, A. Rangesh, M. Trivedi, In-vehicle hand gesture recognition using hidden Markov models, in: 2016 IEEE 19th International Conference on Intelligent Transportation Systems (ITSC), IEEE, 2016.

[45] N. Dalal, B. Triggs, Histograms of oriented gradients for human detection, in: IEEE Computer Society Conference on Computer Vision and Pattern Recognition CVPR 2005 vol. 1, IEEE, 2005.

[46] H. Jhuang, et al., Towards understanding action recognition, in: Proceedings of the IEEE International Conference on Computer Vision, 2013, pp. 3192−3199.

[47] L. Sevilla-Lara, et al., On the Integration of Optical Flow and Action Recognition, arXiv preprint. arXiv: 1712.08416, 2017, pp. 281−297.

[48] C. Yan, B. Zhang, F. Coenen, Driving posture recognition by convolutional neural networks, IET Computer Vision 10 (2) (2016) 103−114.

[49] C. Craye, F. Karray, Driver Distraction Detection and Recognition Using RGB-D Sensor, arXiv preprint arXiv: 1502.00250, 2015.

[50] R. Kavi, et al., Multiview fusion for activity recognition using deep neural networks, Journal of Electronic Imaging 25 (4) (2016), 043010.

[51] Y. Abouelnaga, H.M. Eraqi, M.N. Moustafa, Real-time Distracted Driver Posture Classification, arXiv preprint arXiv:1706.09498, 2017.

[52] D. Tran, et al., Real-time detection of distracted driving based on deep learning, IET Intelligent Transport Systems 12 (10) (2018) 1210−1219.

# CHAPTER 7

# Longitudinal Driver Intention Inference

## BRAKING INTENTION RECOGNITION BASED ON UNSUPERVISED MACHINE LEARNING METHODS

Recently, the development of the braking assistance system has largely benefited the safety of both drivers and pedestrians. A robust prediction and detection of driver braking intention will enable the driving assistance system respond to traffic situations correctly and improve the driving experience of intelligent vehicles. In this section, two types of unsupervised clustering methods are used to build a driver braking intention predictor. Unsupervised machine learning algorithms have been widely used in clustering and pattern mining in previous research. The proposed unsupervised learning algorithms can accurately recognize the braking maneuver based on vehicle data captured with Controller Area Network (CAN) bus. The braking maneuver, along with normal driving, will be clustered and the results from different algorithms such as K-means and Gaussian mixture model (GMM) will be compared. Additionally, the importance evaluation of features from raw dataset with respect to driving maneuver clustering will be proposed. The experimental data are collected from a pure electric vehicle (EV) in the real world. The final results show that the proposed method can detect the driver's braking intention at the very beginning moment with high accuracy. In addition, the most important features for driving maneuver clustering are selected.

As one of the most important control areas of the vehicle, the longitudinal vehicle control system has been widely studied in the past decades. A variety of commercial products such as the adaptive cruise system [1] and forward-collision assistance system [2,3,15] have been extensively developed. However, previous products either used vehicle state data only or lacked interaction with drivers. Without taking the driver into consideration, these systems are less likely to communicate with drivers efficiently or can even annoy the driver. To develop a driver braking assistance system with higher-level interaction ability, it is important to take the driver status into consideration.

There are many studies focusing on the design of intelligent driver assistance and braking systems [4−10]. In Ref. [11], an adaptive longitudinal driving assistance system, including adaptive cruise control and forward-collision avoidance system, considering driver behavior and characteristics was proposed, and a recursive least squares self-learning algorithm for driver characteristic modeling was introduced. In Ref. [12], a lane departure avoidance system is applied to prevent the driver from performing a lane change without attention based on the combination of the lateral active steering controller and longitudinal differential braking controller. Ref. [13] studies the driver braking intention based on the naturalistic driving data and real vehicle test bed. The authors predict driver braking intention based on a three-dimensional computational model, which uses pedal displacement and its change rate as inputs, and the output is braking intensity. The proposed model can precisely identify emergency and nonemergency braking. In Ref. [14], a method of building a pedestrian automatic emergency braking system is introduced based on two model cars with multiple kinds of sensors. Three different levels of deceleration are proposed to model the braking curve of the tested vehicle. In Ref. [15], two machine learning methods, which are support vector machine and hidden Markov model, are introduced to classify driver braking behavior at intersections. The authors classified driver style as compliant and violating based on whether the drivers will brake safely according to the signs at the intersection or not. The proposed system achieved 80% accuracy for driver behavior classification at the intersection. The authors in Ref. [16] proposed a braking predictive system that not only identifies the necessity of braking action but also is able to determine whether the driver has the braking intention or not. Driver braking intention is represented with probabilities generated by the Bayesian framework. Data are captured from real-world vehicles with multimodal data formats such as CAN bus data, foot and head dynamics, etc. The braking prediction model is trained to be able to predict driver braking intention up to one and a half seconds earlier. Research in Ref. [17] focused

Advanced Driver Intention Inference. https://doi.org/10.1016/B978-0-12-819113-2.00007-5

on the analysis of car-following and the braking characteristics of expert drivers. The braking pattern of expert drivers will be performed according to the proposed index and applied to the braking assistance system.

In addition to the aforementioned research, which focuses on predicting driver braking intention with either driver behaviors or vehicle information, some researchers also try to use electroencephalographic (EEG) signals to identify driver intention directly. Kim and Haufe [18] aim to prove the ability of neural correlated electrophysiology to improve the prediction of emergency braking situations in real-world driving environments. Vehicle parameters as well as EEG and electromyographic (EMG) signals were used to train the classification system. A regularized linear discriminant classifier was trained to identify the emergency braking intention. Their conclusion suggested that the electrophysiologic method can be efficiently used in the braking assistant system. Haufe [19] uses EEG and (EMG) as an input signal to make a prediction of the driver's emergency braking intent. The features from the feature extraction module were fed into a regularized linear discriminant analysis classifier. The simulated systems with EEG showed 130 ms earlier detection than those that rely only on brake pedal signals.

Human driver intention recognition is a difficult task because it is a highly random signal and cannot be measured directly at the moment. It can only be inferred with outer human behaviors or the indirectly measured brain signals. Therefore in previous research, most of the relative research focuses on using machine learning methods, especially supervised learning to train an intention classifier. However, one of the big challenges of using supervised learning methods is hard to define the true intention label. In contrast, unsupervised learning is suitable for those data that we do not know exactly their labels. Hence, it can be used to cluster the data to the group it most likely belongs to and able to find the intrinsic pattern in the data. In this section, instead of identifying human intention with supervised learning methods, we use unsupervised learning methods to recognize the braking intention. In addition, we compare two different unsupervised learning methods, which are K-means and GMM, to evaluate their different performances on the braking intention recognition task. Finally, we also evaluate the contribution of different features to intention recognition and find out the key features.

This section is organized as follows: part 2 presents the basic background of unsupervised learning methods used in this section, part 3 briefly discusses the

experimental environment and design strategy, part 4 illustrates the braking intention identification result of the proposed methods and the features used to train the model have been compared, and finally, part 5 includes a discussion of the research.

## Unsupervised Learning Background

In this section, the basic background of two unsupervised learning methods known as K-means and GMM that used to identify human driver braking intention is given. The reason for using unsupervised learning methods is because human intention is difficult to be labeled unless the experiment is carefully designed. In the following sections, we try to analyze driver braking intention with vehicle CAN bus only and to use unsupervised learning methods to find the cluster information. As these two methods have been widely studied, in this section, only the key ideas are mentioned.

### K-means

K-means algorithm [21] is a popular unsupervised machine learning method that has been widely used in previous research for data clustering. The basic idea behind the K-means algorithm is to minimize the distance between data points and the proposed cluster centers. One of the most important aspects of the K-means algorithm is distance measurement between the data point and there are various methods that can be used. Among them, the two most common ways are known as Euclidean distance and Manhattan distance.

Euclidean distance:

$$D_{i,j} = \sqrt{\sum_{q=1}^{n} |v_{i,q} - v_{j,q}|^2} \qquad (7.1)$$

Manhattan distance:

$$D_{ij} = \sum_{q=1}^{n} |v_{i,q} - v_{j,q}| \qquad (7.2)$$

where $i$ and $j$ are two $n$-dimensional data points denoted as $v$, where $v_i = (v_{i1}, v_{i2} \cdots, v_{i,n})$ and $v_j = (v_{i1}, v_{j2}, \cdots, v_{jn})$ and $D_{i,j}$ represents the distance between two data points.

The object of the K-means algorithm is to minimize the distance between data and their cluster center in each group. Table 7.1 shows the process of the K-means algorithm.

In step 6, the stopping criteria can be either a predetermined loop limitation or achieving an optimal status. In that case, the stopping criterion is a minimum value achieved between the distance of data point and the cluster center within the same group; therefore, the objective of K-means is

**TABLE 7.1**
**K-Means Algorithm Process.**

**K-Means Algorithm**

Inputs: Data, cluster number K
Process:
1. Randomly select the K positions as the center of K groups.
2. Assign cluster centers to the K position.
3. For each data compute the distance between the data and the center.
4. Assign $x_i$ to the cluster with minimum distance.
5. For each cluster center move the cluster center to the mean position in that cluster.
6. If stop criteria are met, then stop; otherwise jump to step 2 and repeat.

$$\min \sum_{i=1}^{K} \sum_{x \in c_i} dist(c_i, x)^2 \tag{7.3}$$

where K is the total number of cluster centers selected in beginning, $x$ are the data points that belong to a cluster $C_i$, and $c_i$ is the cluster center for each group.

*Gaussian mixture model*

Unlike K-means, which minimizes the distance directly, data distribution in GMM [22,23] is viewed as the output of a mixture of the probability density function. Specifically, the probability distribution of the points can be generated by a mixture of Gaussian functions. GMM can be viewed as a soft version of K-means because it uses probability to measure the similarity between points rather than the direct distance measurement. By training the model with these points, we can obtain the distribution of data and handle the uncertainty among them.

The multivariate GMM can be represented as

$$N\left(x \middle| \mu, \sum\right) = \frac{1}{(2\pi)^{d/2} \sum^{1/2}} \exp\left( -\frac{1}{2}(x-\mu)^T \sum^{-1}(x-\mu) \right) \tag{7.4}$$

where $\mu$ and $\sum$ are the mean and covariance of the multivariate Gaussian function, respectively.

The probability distribution of the data given by GMM can be shown as

$$p(x) = \sum_{k=1}^{K} \pi_k N\left(x \middle| \mu_k, \sum_k\right) \tag{7.5}$$

where $K$ is the number of mixture components and $\pi_k$ is the weight of each Gaussian density function, which should meet the normalization and positive requirements:

$$\sum_{k=1}^{K} \pi_k = 1 \tag{7.6}$$

and

$$0 \le \pi_k \le 1 \tag{7.7}$$

The GMM maximum likelihood can be estimated with expectation-maximization (EM) algorithms; detailed information on EM can be found in Ref. [20].

**Experiment Design**

In this section, the key parameters of the testbed EV and driving cycle are illustrated. Then the 12 most relevant parameters that are relevant to braking behaviors are selected from CAN bus signals. Finally, according to the selected parameters, the unsupervised learning methods mentioned earlier are used to figure out the hidden braking rules behind the data.

*Case study vehicle*

The case study vehicle of this work is a pure electric passenger car, which has the most typical powertrain configuration. The car is driven in a front axle by a permanent magnet synchronous motor, which can work in two states as a driving motor or a generator. The battery is electrically connected with the motor, and it can be discharged or charged for motoring or absorbing, respectively, the regenerative power during driving processes. Some key parameters of the case study vehicle are listed in Table 7.2.

*Driving cycle*

To further the study of the impacts of unsupervised learning methods on driver braking intentions, vehicle tests were carried out on chassis dynamometer under typical driving cycles. In this experiment, we use the New European Driving Cycle (NEDC) as a testing protocol. NEDC is a combination of the European Union Urban Driving Cycle (ECE) and the Extra-Urban Driving Cycle (EUDC). It has been widely used for EV energy consumption and regenerative braking

**TABLE 7.2**
Key Parameters of the Case Study Electric Vehicle.

| | Parameter | Value | Unit |
|---|---|---|---|
| Vehicle | Total mass ($m$) | 1360 | kg |
| | Wheel base ($L$) | 2.50 | m |
| | Frontal area ($A$) | 2.40 | m$^2$ |
| | Air resistance coefficient ($C_D$) | 0.32 | — |
| Electric motor | Rated power | 20 | kW |
| | Peak power | 45 | kW |
| | Maximum torque | 144 | Nm |
| Battery pack | Voltage | 326 | V |
| | Capacity | 66 | Ah |
| Transmission and final drive | Final drive ratio | 3.79 | — |
| | Transmission ratio | 2.10 | — |

this section, we choose 12 parameters from the CAN bus; the detailed information is shown in Table 7.3.

### Unsupervised clustering training process

One of the most important parameters for unsupervised clustering methods is cluster numbers, which need to be chosen before learning. However, as we do not know the exact number of cluster centers, we choose a set of numbers to evaluate and compare the performance. In the following section, a number of cluster centers ranging from two to five are applied to K-means and GMM separately. In addition to the algorithm evaluation, we also examine which signals are the most correlated factor for unsupervised clustering.

### Experiment Results

In this section, the clustering performance for both K-means and GMM is illustrated, with cluster centers

**TABLE 7.3**
Selected CAN Bus Signals for Model Training.

| No. | Signal Name | No. | Signal Name |
|---|---|---|---|
| 1 | Command motor torque (N·m) | 7 | Motor speed (rad/s) |
| 2 | Left front brake pressure (MPa) | 8 | Acceleration pedal position |
| 3 | Right front brake pressure (MPa) | 9 | Left front wheel speed (rad/min) |
| 4 | Left rear brake pressure (MPa) | 10 | Right front wheel speed (rad/min) |
| 5 | Right rear brake pressure (MPa) | 11 | Battery state of charge (%) |
| 6 | Velocity (km/h) | 12 | Battery current (A) |

performance testing. As shown in Fig. 7.1, a complete NEDC contains four repeated ECEs following an EUDC section to exhibit a highway driving speed pattern, with the highest speed of 120 km/h.

### Parameter selection

Selecting the most relevant parameters and features is the primary task for unsupervised clustering. Driver braking behavior belongs to the scope of longitudinal behavior; therefore, to recognize the driver's braking maneuver, longitudinal vehicle parameters are selected and chosen from the CAN bus. The brake pressures of four wheels are captured separately, and battery state of charge (SOC) and current are used to test whether varying the battery status influences the detection result [24]. In

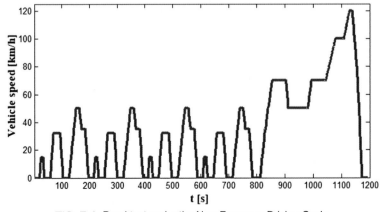

FIG. 7.1 Road test under the New European Driving Cycle.

ranging from two to five. Then the correlation coefficients are calculated to help determine the most important features.

## K-means result

From the clustering result of K-means, as shown in Fig. 7.2, the braking intention can be detected successfully with different cluster numbers. In Fig. 7.2, the x-axis is time with the unit in seconds and the y-axis is the scaled values of the input signals; all the signals have been rescaled between 0 and 1. The blue line represents labels given by K-means, the red line is the left front brake pressure, and the green line represents the velocity of the vehicle. As the value for all the factors has been scaled between 0 and 1, the label given by K-means can also be divided by the relative cluster number for better visualization.

In Fig. 7.2, it can be observed that the braking process (when red line is beyond zero) can be grouped correctly with different cluster numbers. For example, in the first image, once braking happens, which means a brake pressure is detected, the cluster label changed. As the number of cluster centers increases, the cluster becomes more complex and there are more scenarios that can be detected. As shown in the bottom image, which is generated by five cluster centers, the braking intention can be detected. Besides, the label also changes as vehicle velocity varies. K-means with four cluster centers divided the driving cycle into four different scenarios, which are acceleration, constant velocity driving, deceleration, and stopping the process. As we only show the first 200 steps, labels given by K-means with four or five cluster centers may be incomplete. A detailed result is shown in the Appendix. From this result, we can see that once the braking occurs, the label changes accordingly. In most cases, the braking maneuver will last for a short period; once the label changes to the braking group, it can be viewed as the driver is going to decelerate the vehicle in the coming short period and hence the braking assistance system can be initiated to assist the driver.

To verify the impact of different CAN bus signals to the clustering results, we evaluated the correlation coefficient between the input signals with the clustering results given by the two-cluster-center scenario. The correlation coefficient is calculated with the Pearson method to measure the linear dependence between two variable vectors. In this case, the correlation coefficient is calculated separately between the selected 12 CAN bus signals and the cluster label. The results located within the range of $-1$ and 1, where 1 is a total positive correlation, $-1$ is a total negative correlation, and 0 means no correlation between each other. The correlation coefficient is calculated as

$$\rho_{X,Y} = \frac{cov(X,Y)}{\sigma_X \sigma_Y} \tag{7.8}$$

where $\sigma_X, \sigma_Y$ are the standard deviation (STD) of variables $X$ and $Y$, respectively. $cov(X,Y)$ is the covariance between $X$ and $Y$, which can be represented as

$$cov(X,Y) = E[(X - \mu_X)(Y - \mu_Y)] \tag{7.9}$$

where $E[\,]$ and $\mu$ are the expectation calculator and mean value, respectively.

Table 7.4 illustrates the relationship between different features and the clustering result.

From Table 7.4, we can see that some of the signals have a positive or negative strong influence on the clustering result, whereas others have a weak influence. Specifically, the wheel brake pressure signals give a significant contribution to the cluster performance as well as velocity information. As there is a small difference between some similar signals, we can select fewer features to retrain the cluster and increase the computation efficiency of the system. Another interesting result is the acceleration pedal position (signal 8, correlation coefficient value $-0.4003$) is not highly correlated with the clustering result; therefore, we can also ignore this feature, which we think can be important in the beginning.

## Gaussian mixture model result

Similar to the process shown in K-means, in this section, the data are clustered by GMM algorithms, with different cluster numbers ranging from two to five and the correlation coefficients are also evaluated. The performance of GMM methods is illustrated in Fig. 7.3. The cluster label is represented with the blue line and the left front brake pressure and vehicle velocity are shown in red line and green line, respectively. Fig. 7.3 is a short period of the driving cycle, which is in the range between 0 and 200 s like Fig. 7.2.

From Fig. 7.3, we can see that similar to K-means, the GMM algorithm can detect the braking intention accurately. When the braking maneuver occurs, GMM can choose the correct label in the very beginning.

Similarly, the correlation coefficient of GMM is proposed as it is used in K-means, and the Pearson values given by different features are illustrated in Table 7.5.

From Table 7.5, it can be found that the most significant features for GMM clustering are the same as those in K-means. Specifically, the most important factors are brake pressures and the different kinds of velocity information.

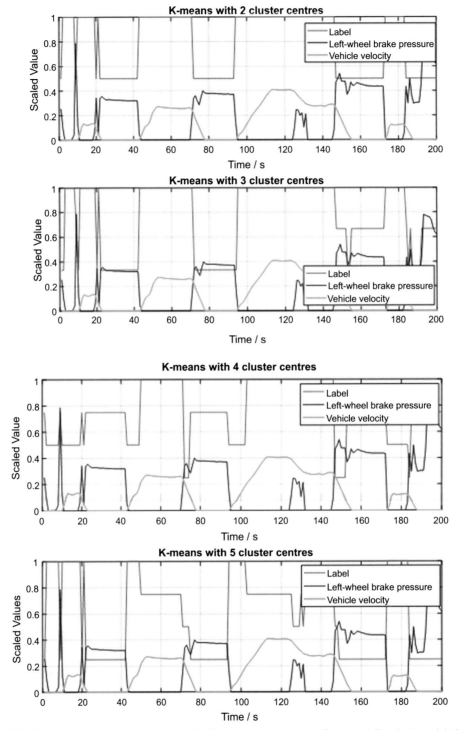

FIG. 7.2 K-means clustering performance with different cluster numbers. From top left to bottom right is K-means algorithms with two to five cluster centers.

**TABLE 7.4**
**Correlation Coefficients of the Selected Signals and K-Means Cluster Result.**

| Signal No. | Correlation Coefficient | Signal No. | Correlation Coefficient |
|---|---|---|---|
| 1 | −0.0229 | 7 | −0.7871 |
| 2 | 0.6470 | 8 | −0.4003 |
| 3 | 0.6470 | 9 | −0.7871 |
| 4 | 0.6470 | 10 | −0.7871 |
| 5 | 0.6470 | 11 | 0.3328 |
| 6 | −0.7870 | 12 | −0.4947 |

### Discussion

From Tables 7.4 and 7.5, it can be seen that the key features for braking intention clustering are the wheel pressure values, vehicle velocity, and motor speed. As some of these signals after rescaled between 0 and 1 have no big difference with each other, for example, vehicle velocity, motor speed, and wheel speed equally contribute to the final result. Therefore we can select one signal among them instead of using all of them. By considering the computation efficiency, we select a subset of the abovementioned features and only left brake pressure and vehicle velocity are chosen as the cluster algorithm inputs. Figs. 7.4 and 7.5 give the performance of K-means and GMM, respectively, with two input signals and two cluster centers.

Evaluation methods for braking intention can be divided into two scopes. The first one is the prediction horizon that measures how much earlier the prediction is made before the maneuver starts. Another method is to test the algorithm accuracy, which checks how many times it detects the braking process successfully. In this work, as we only use CAN bus information, we cannot predict the intention before the driver performs any maneuvers. Therefore we test the detection accuracy of the two algorithms and determine the specific braking moment during the driving cycle. A good performance for the algorithms is to recognize the braking process as accurate as possible.

As can be seen from Figs. 7.4 and 7.5, both K-means and GMM with the two most influential features (brake pressure and vehicle velocity) give a good performance in braking intention prediction. In addition, GMM gives a better and more sensitive recognition than K-means. For example, K-means cannot identify the small increase in velocity between 150 and 250 s and give

braking maintaining signal. However, in terms of GMM, this small velocity increase can be detected and the system is more precise and sensitive to the input variables. In total, the braking process occurs 15 times between 1 and 600 s. GMM recognizes all the 15 braking processes, whereas K-means detects only 8 times. Therefore based on our data, the GMM performance overweighs K-means' performance. Detailed recognition results of the two methods on the whole driving cycles can be found in Figs. 7.6 and 7.7.

### Conclusions

In this section, the driver braking intention recognition algorithm has been proposed with both K-means and GMM. The reason for using unsupervised learning is the difficulty of determining the true moment when driver intention occurs. The supervised knowledge (true moment or true label of intention) cannot be obtained with CAN bus signals only. The unsupervised learning methods automatically group the driving status into a few groups and identify the braking process accordingly. This will further contribute to other classification works such as emergency/nonemergency braking classification. Unsupervised learning methods give a brief illustration of the rules hidden behind the driving cycle. The proposed test is based on a pure EV under the EUDC. The 12 potential signals are selected as the inputs of the clustering algorithm in the beginning, and then with the calculation of the correlation coefficient, the two most relevant features, left front brake pressure and vehicle velocity, are determined. Performances of K-means and GMM are compared. The conclusion is that both the algorithms can efficiently recognize driver braking intention at the very beginning moment. When the cluster label changes to the braking group, it can be viewed as the driver is going to execute a period of braking maneuver so that the braking assistance system can be turned on. In addition, the proposed result indicates that GMM is more sensitive and generates a more precise result than the K-means algorithm.

### LEVENBERG-MARQUARDT BACKPROPAGATION FOR STATE ESTIMATION OF A SAFETY-CRITICAL CYBER-PHYSICAL SYSTEM

As an important safety-critical cyber-physical system (CPS), the braking system is essential to the safe operation of the EV. Accurate estimation of the brake pressure is of great importance for automotive CPS design and control. In this section, a novel probabilistic estimation method of brake pressure is developed for electric

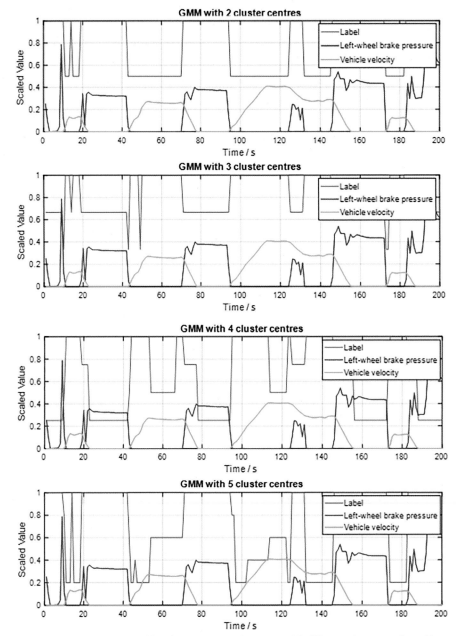

FIG. 7.3 Gaussian mixture model (GMM) clustering performance with different cluster numbers. From top to bottom is the GMM algorithms with two to five cluster centers.

vehicles based on multilayer artificial neural network (ANNs) with Levenberg-Marquardt backpropagation (LMBP) training algorithm. First, the high-level architecture of the proposed multilayer ANN for brake pressure estimation is illustrated. Then the standard backpropagation (BP) algorithm used for training of the feedforward neural network (FFNN) is introduced. Based on the basic concept of BP, a more efficient training algorithm of the LMBP method is proposed. Next, real vehicle testing is carried out on a chassis

**TABLE 7.5**
Correlation Coefficients of the Selected Signals and Gaussian Mixture Model Cluster Result.

| Signal No. | Correlation Coefficient | Signal No. | Correlation Coefficient |
|---|---|---|---|
| 1 | 0.0355 | 7 | −0.5959 |
| 2 | 0.8355 | 8 | −0.4128 |
| 3 | 0.8352 | 9 | −0.5959 |
| 4 | 0.8341 | 10 | −0.5959 |
| 5 | 0.8337 | 11 | 0.2461 |
| 6 | −0.5961 | 12 | −0.4584 |

evaluated and compared with other available learning methods. Experiment results validate the feasibility and accuracy of the proposed ANN-based method for braking pressure estimation under real deceleration scenarios.

CPSs, which are distributed, networked systems that fuse computational processes with the physical world exhibiting a multidisciplinary nature, have become a research focus [25−29]. As a typical application of CPS in green transportation, EVs have been widely studied with different topics by researchers and engineers from academia, industry, and governmental organizations [30−35]. In an EV, the cyber world of control and communication, the physical plant of the electric powertrain, the human driver, and the driving environment are tightly coupled and dynamically interact, determining the overall system's performance jointly [8]. These complex subsystems with multidisciplinary interactions, strong uncertainties, and hard nonlinearities make the estimation, control, and optimization of EVs very difficult [36]. Thus there are still a number of fundamental issues, and critical challenges varying

dynamometer under standard driving cycles. Experimental data of the vehicle and the powertrain systems are collected, and feature vectors for FFNN training collection are selected. Finally, the developed multilayer ANN is trained using the measured vehicle data, and the performance of the brake pressure estimation is

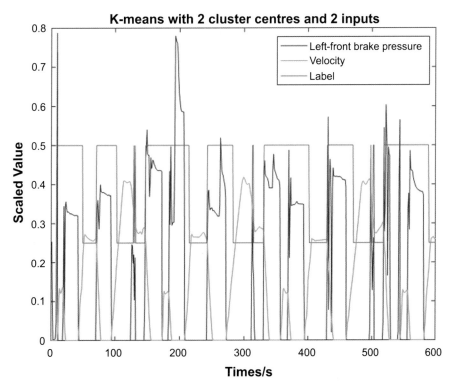

FIG. 7.4 K-means clustering performance with two cluster centers and the selected two input signals in the first 600 s.

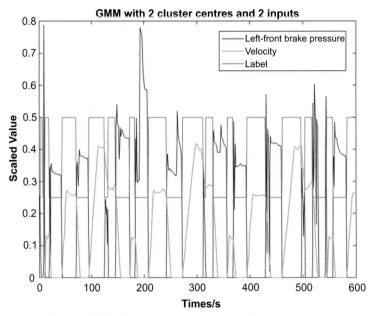

FIG. 7.5 Gaussian mixture model (GMM) clustering performance with two cluster centers and the selected two input signals in the first 600 s.

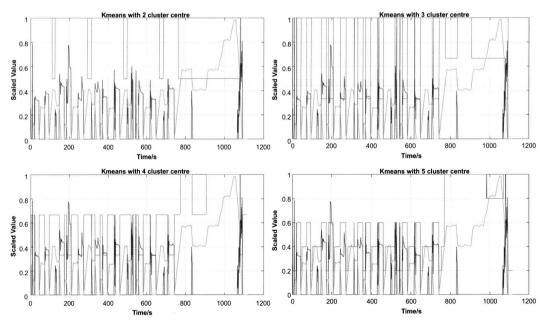

FIG. 7.6 K-means clustering methods with cluster center numbers ranging from two to four of the total driving cycles.

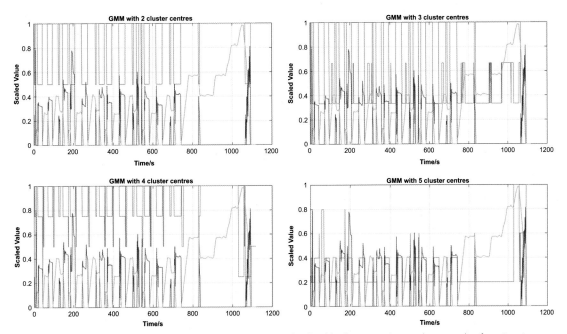

FIG. 7.7 Gaussian mixture model (GMM) clustering methods with cluster center numbers ranging from two to four of the total driving cycles.

in their importance from convenience to safety of EV remain open [36–40].

Among all those concerns in EV CPS, a key one is safety. Safety-critical systems are the ones whose failure or malfunction may result in serious injury or severe damage to people, equipment, or the environment [41]. As one of the most important safety-critical systems in EV, the correct functioning of the braking system is essential to the safe operation of the vehicle [42]. There are a variety of safety standards, control algorithms, and developed devices that help guarantee braking safety for current Evs. However, with increasing degrees of electrification, control authority, and autonomy of automotive CPS, safety-critical functions of the braking system are also required to evolve to keep pace [43].

In the braking system of a passenger car, the braking torque is generated by the hydraulic pressure applied in the brake cylinder. Thus the accurate measurement of the brake pressure through a pressure sensor is of great importance for various braking control functions and chassis stability logics. However, failures of the brake pressure measurement, which may be caused by software discrepancies or hardware problems, could result in the vehicle's critical safety issues. Thus high-precision estimation of brake pressure has become a hot research area in automotive CPS design and control.

Moreover, in order to handle the trade-offs between performance and cost, sensorless observation is required. This makes the study of brake pressure estimation highly motivated.

Based on advanced theories and algorithms from the aspect of control engineering, observation methods of braking pressure for vehicles have been investigated by researchers worldwide. In Ref. [44], a recursive least squares algorithm for the estimation of brake cylinder pressure was proposed based on the pressure response characteristics of the antilock braking system (ABS). In Ref. [45], an extended-Kalman-filter-based estimation algorithm was developed considering the hydraulic model and tyre dynamics. In Ref. [46], an algorithm for online observation of brake pressure was designed through a developed inverse model, and the algorithm was verified in the vehicle's electronic stability program. In Ref. [47], the models of brake pressure increase, decrease, and hold are proposed by using the experimental data. I models can be used for fast online observation of hydraulic brake pressure. In Ref. [48], a brake pressure estimation algorithm was proposed for ABS considering the hydraulic fluid characteristics. In Ref. [49], the estimation algorithm was performed by calculating the volume of fluid flowing through the valve. The amount of fluid is a function of the pressure differential across the valve and the actuation time of the valve. Nevertheless, the existing research on brake pressure estimation was mainly investigated from the perspective of control engineering, while an approach with the probabilistic method, such as machine learning, has rarely been seen.

In this section, an ANN-based estimation method is studied for accurately observing the brake pressure of an electric passenger car. The main contribution of this section lies in the following aspects: (1) an ANN-based machine learning framework is proposed to quantitatively estimate the brake pressure of an EV, (2) the proposed approach is implemented with experimental data obtained via vehicle testing and compared with other methods, and (3)the proposed approaches have a great potential to achieve a sensorless design of the braking control system, removing the brake pressure sensor existing in the current products and largely reducing the cost of the system. Moreover, it also provides additional redundancy for the safety-critical braking functions.

The rest of this section is organized as follows: part 2 describes the high-level architecture of the proposed multilayer ANN for brake pressure estimation, part 3 briefly introduces the standard BP algorithm and illustrates the notations and basic concepts demanded in the Levenberg-Marquardt algorithm, and part 4 presents details of the application of the LMBP method to train the FFNN.

## Multilayer Artificial Neural Network Architecture

In order to achieve the objective of brake pressure estimation, multilayer ANNs are first constructed with the input of vehicle and powertrain states. Details of the high-level system architecture and structure of the component are described in this section.

### System architecture

The system architecture with the proposed methodology is shown in Fig. 7.8. The multilayer ANN receives state variables of the vehicle and the electric powertrain system as inputs and then yields the estimation of the brake pressure through the activation function. The LMBP algorithm is then operated with the performance function, which is a function of the ANN-based estimation and the ground truth of brake pressure. The weight and bias variables are adjusted according to the Levenberg-Marquardt method, and the BP algorithm is used to calculate the Jacobian matrix of the performance function with respect to the weight and bias variables. With updated weights and biases, the ANN further estimates the brake pressure at the next time step. On the basis of the abovementioned iterative processes, the ANN-based brake pressure estimation model is well trained. The detailed method and algorithms are introduced in the following subsection.

### Multilayer feedforward neural network

In this part, a multilayer FFNN is chosen to estimate brake pressure. An FFNN is composed of one input layer, one or more hidden layers, and one output layer. As a neural network (NN) with one hidden layer has the capability to handle most of the complex functions, in this work the FFNN with one hidden layer is constructed. Fig. 7.9 shows the structure of a multilayer FFNN with one hidden layer.

The basic element of an FFNN is the neuron, which is a logical-mathematic model that seeks to simulate the behavior and functions of a biological neuron [50]. Fig. 7.10 shows the schematic structure of a neuron. Typically, a neuron has more than one input. The elements in the input vector $\mathbf{p} = [p_1, p_2, \ldots, p_R]$ are weighted by elements $w_1, w_2, \ldots, w_j$ of the weight matrix $\mathbf{W}$, respectively.

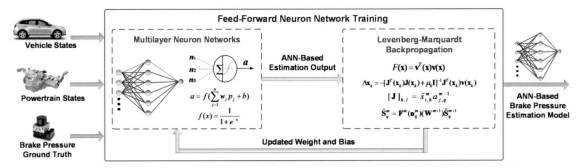

FIG. 7.8 High-level architecture of the proposed brake pressure estimation algorithm based on multilayer artificial neural networks (ANNs).

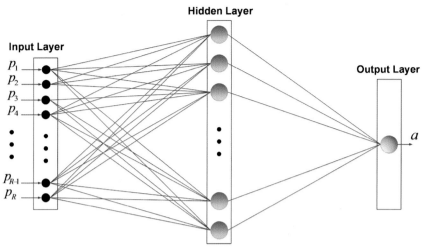

FIG. 7.9 Structure of the multilayer feedforward neural network.

The neuron has a bias $b$ that is summed with the weighted inputs to form the net input $n$, which can be expressed as

$$n = \sum_{j=1}^{R} w_j p_j + b = Wp + b \qquad (7.10)$$

Then the net input $n$ passes through an active function $f$, which generates the neuron output $a$.

$$a = f(n) \qquad (7.11)$$

In this study, the log-sigmoid activation function is adopted. It can be given by the following expression:

$$f(x) = \frac{1}{1 + e^{-x}} \qquad (7.12)$$

Thus the multi-input FFNN in Fig. 7.10 implements the following equation

$$a^2 = f^2 \left( \sum_{i=1}^{S} w_{1,i}^2 f^1 \left( \sum_{j=1}^{R} w_{i,j}^1 p_j + b_i^1 \right) + b^2 \right) \qquad (7.13)$$

where $a^2$ denotes the output of the overall networks. $R$ is the number of inputs, $S$ is the number of neurons in the hidden layer, and $p_j$ indicates the $j$th input. $f^1$ and $f^2$ are the activation functions of the hidden layer and output layer, respectively. $b_i^1$ represents the bias of the $i$th neuron in the hidden layer, and $b^2$ is the bias of the neuron in the output layer. $w_{i,j}^1$ represents the weight connecting the $j$th input and the $i$th neuron of the hidden layer, and $w_{1,i}^2$ represents the weight connecting the

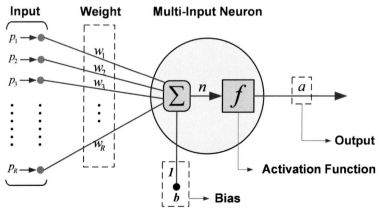

FIG. 7.10 Structure of the multi-input neuron.

$i$th source of the hidden layer to the output layer neuron.

## Standard Backpropagation Algorithm

In order to train the established FFNN, the BP algorithm can be utilized [51]. Considering a multilayer FFNN, such as the one with three layers shown in Fig. 7.9, its operation can be described using the following equation:

$$\mathbf{a}^{m+1} = \mathbf{f}^{m+1}\left(\mathbf{W}^{m+1}\mathbf{a}^m + \mathbf{b}^{m+1}\right) \tag{7.14}$$

where $\mathbf{a}^m$ and $\mathbf{a}^{m+1}$ are the outputs of the $m$th and $(m+1)$-th layers of the networks, respectively. $\mathbf{B}^{m+1}$ is the bias vector of $(m+1)$-th layers of the networks. $m = 0, 1, ..., M - 1$, where $M$ is the number of layers of the NN. The neurons of the first layer obtain inputs

$$\mathbf{a}^0 = \mathbf{p} \tag{7.15}$$

Eq. (7.15) provides the initial condition for Eq. (7.14). The outputs of the neurons in the last layer can be seen as the overall 'etwork's outputs:

$$\mathbf{a} = \mathbf{a}^M \tag{7.16}$$

The task is to train the network with associations between a specified set of input-output pairs $\{(\mathbf{p}_1, \mathbf{t}_1), (\mathbf{p}_2, \mathbf{t}_2), ..., (\mathbf{p}_Q, \mathbf{t}_Q)\}$, where $\mathbf{p}_q$ is an input to the network and $\mathbf{t}_q$ is the corresponding target output. As each input is applied to the network, the network output is compared to the target.

The BP algorithm uses mean square error as the performance index, which is to be minimized by adjusting the network parameters, as shown in Eq. (7.17).

$$F(\mathbf{x}) = E\left[\mathbf{e}^T\mathbf{e}\right] = E\left[(\mathbf{t} - \mathbf{a})^T(\mathbf{t} - \mathbf{a})\right] \tag{7.17}$$

where $\mathbf{x}$ is the vector-matrix of network weights and biases. Using the approximate steepest descent rule, the performance index $F(\mathbf{x})$ can be approximated by

$$\widehat{F}(\mathbf{x}) = (\mathbf{t}(k) - \mathbf{a}(k))^T(\mathbf{t}(k) - \mathbf{a}(k)) = \mathbf{e}^T(k)\mathbf{e}(k) \tag{7.18}$$

where the expectation of the squared error in Eq. (7.17) has been replaced by the squared error at the iteration step $k$.

The steepest descent algorithm for the approximate mean square error is

$$w_{i,j}^m(k+1) = w_{i,j}^m(k) - \alpha\frac{\partial\widehat{F}}{\partial w_{i,j}^m} \tag{7.19}$$

$$b_i^m(k+1) = b_i^m(k) - \alpha\frac{\partial\widehat{F}}{\partial b_i^m} \tag{7.20}$$

where $\alpha$ is the learning rate.

Based on the chain rule, the derivatives in Eqs 7.19 and 7.20 can be calculated as

$$\frac{\partial\widehat{F}}{\partial w_{i,j}^m} = \frac{\partial\widehat{F}}{\partial n_i^m} \cdot \frac{\partial n_i^m}{\partial w_{i,j}^m}, \quad \frac{\partial\widehat{F}}{\partial b_i^m} = \frac{\partial\widehat{F}}{\partial n_i^m} \cdot \frac{\partial n_i^m}{\partial b_i^m} \tag{7.21}$$

We now define $s_i^m$ the sensitivity of $\widehat{F}$ changes in the $i$th element of the net input at layer $m$.

$$s_i^m \equiv \frac{\partial\widehat{F}}{\partial n_i^m} \tag{7.22}$$

Using the defined sensitivity, then the derivatives in Eqs. (7)−(21) can be simplified as

$$\frac{\partial\widehat{F}}{\partial w_{i,j}^m} = s_i^m a_j^{m-1} \tag{7.23}$$

$$\frac{\partial\widehat{F}}{\partial b_i^m} = s_i^m \tag{7.24}$$

Then the approximate steepest descent algorithm can be rewritten in matrix form as

$$\mathbf{W}^m(k+1) = \mathbf{W}^m(k) - \alpha\mathbf{s}^m\left(\mathbf{a}^{m-1}\right)^T \tag{7.25}$$

$$\mathbf{b}^m(k+1) = \mathbf{b}^m(k) - \alpha\mathbf{s}^m \tag{7.26}$$

where

$$\mathbf{s}^m \equiv \frac{\partial\widehat{F}}{\partial\mathbf{n}^m} = \left[\frac{\partial\widehat{F}}{\partial n_1^m}, \frac{\partial\widehat{F}}{\partial n_2^m}, ..., \frac{\partial\widehat{F}}{\partial n_{S^m}^m}\right]^T \tag{7.27}$$

To derive the recurrence relation for the sensitivities, the following Jacobian matrix is utilized.

$$\frac{\partial\mathbf{n}^{m+1}}{\partial\mathbf{n}^m} \equiv \begin{bmatrix} \dfrac{\partial n_1^{m+1}}{\partial n_1^m} & \dfrac{\partial n_1^{m+1}}{\partial n_2^m} & \cdots & \dfrac{\partial n_1^{m+1}}{\partial n_{S^m}^m} \\ \dfrac{\partial n_2^{m+1}}{\partial n_1^m} & \dfrac{\partial n_2^{m+1}}{\partial n_2^m} & \cdots & \dfrac{\partial n_2^{m+1}}{\partial n_{S^m}^m} \\ \vdots & \vdots & & \vdots \\ \dfrac{\partial n_{S^{m+1}}^{m+1}}{\partial n_1^m} & \dfrac{\partial n_{S^{m+1}}^{m+1}}{\partial n_2^m} & \cdots & \dfrac{\partial n_{S^{m+1}}^{m+1}}{\partial n_{S^m}^m} \end{bmatrix} \tag{7.28}$$

Considlr the $i, j$ element in the matrix:

$$\frac{\partial n_1^{m+1}}{\partial n_1^m} = w_{i,j}^{m+1}\frac{\partial a_j^m}{\partial n_j^m} = w_{i,j}^{m+1}\dot{f}^m\left(n_j^m\right) \tag{7.29}$$

Thus the Jacobian matrix can be rewritten as

$$\frac{\partial\mathbf{n}^{m+1}}{\partial\mathbf{n}^m} = \mathbf{W}^{m+1}\dot{\mathbf{F}}^m(\mathbf{n}^m) \tag{7.30}$$

where

$$\dot{\mathbf{F}}^m(\mathbf{n}^m) = \begin{bmatrix} \dot{f}^m(n_1^m) & 0 & \cdots & 0 \\ 0 & \dot{f}^m(n_2^m) & & 0 \\ \vdots & \vdots & & \vdots \\ 0 & 0 & \cdots & \dot{f}^m(n_{S^m}^m) \end{bmatrix} \quad (7.31)$$

Then the recurrence relation for the sensitivity can be obtained by using the chain rule:

$$\mathbf{s}^m = \frac{\partial \widehat{F}}{\partial \mathbf{n}^m} = \left( \frac{\partial \mathbf{n}^{m+1}}{\partial \mathbf{n}^m} \right)^T \frac{\partial \widehat{F}}{\partial \mathbf{n}^{m+1}} \quad (7.32)$$

$$= \dot{\mathbf{F}}^m(\mathbf{n}^m)(\mathbf{W}^{m+1})^T \mathbf{s}^{m+1}$$

This recurrence relation is initialized at the final layer as

$$s_i^M = \frac{\partial \widehat{F}}{\partial n_i^M} = \frac{\partial((\mathbf{t} - \mathbf{a})^T(\mathbf{t} - \mathbf{a}))}{\partial n_i^M} = \frac{\partial \sum_{j=1}^{S^M}(t_j - a_j)^2}{\partial n_i^M} \quad (7.33)$$

$$= -2(t_i - a_i)\frac{\partial a_i}{\partial n_i^M} = -2(t_i - a_i)\dot{f}^m(n_i^m)$$

Thus the recurrence relation of the sensitivity matrix can be expressed as

$$\mathbf{s}^M = -2\dot{\mathbf{F}}^M(\mathbf{n}^M)(\mathbf{t} - \mathbf{a}) \quad (7.34)$$

The overall BP learning algorithm is now finalized and can be summarized in the following steps: (1 propagate the input forward through the network, (2) propagate the sensitivities backward through the network from the last layer to the first layer, and (3) update the weights and biases using the approximate steepest descent rule.

## Levenberg-Marquardt Backpropagation

While BP is the steepest descent algorithm, the Levenberg-Marquardt algorithm is derived from Newton"s method that was designed for minimizing functions that are sums of squares of nonlinear functions [52,53]'

Newton"s method for optimizing a performance index $F(\mathbf{x})$ is

$$\mathbf{x}_{k+1} = \mathbf{x}_k - \mathbf{A}_k^{-1}\mathbf{g}_k \quad (7.35)$$

$$\mathbf{A}_k \equiv \nabla^2 F(\mathbf{x})|_{\mathbf{X}=\mathbf{X}_k} \quad (7.36)$$

$$\mathbf{g}_k \equiv \nabla F(\mathbf{x})|_{\mathbf{X}=\mathbf{X}_k} \quad (7.37)$$

where $\nabla^2 F(\mathbf{x})$ is the Hessian matrix and $\nabla F(\mathbf{x})$ is the gradient.

Assume that $F(\mathbf{x})$ is a sum of squares function:

$$F(\mathbf{x}) = \sum_{i=1}^N v_i^2(\mathbf{x}) = \mathbf{v}^T(\mathbf{x})\mathbf{v}(\mathbf{x}) \quad (7.38)$$

then the gradient and Hessian matrix are

$$\nabla F(\mathbf{x}) = 2\mathbf{J}^T(\mathbf{x})\mathbf{v}(\mathbf{x}) \quad (7.39)$$

$$\nabla^2 F(\mathbf{x}) = 2\mathbf{J}^T(\mathbf{x})\mathbf{J}(\mathbf{x}) + 2\mathbf{S}(\mathbf{x}) \quad (7.40)$$

where $\mathbf{J}(\mathbf{x})$ is the Jacobian matrix

$$\mathbf{J}(\mathbf{x}) = \begin{bmatrix} \frac{\partial v_1(\mathbf{x})}{\partial x_1} & \frac{\partial v_1(\mathbf{x})}{\partial x_2} & \cdots & \frac{\partial v_1(\mathbf{x})}{\partial x_n} \\ \frac{\partial v_2(\mathbf{x})}{\partial x_1} & \frac{\partial v_2(\mathbf{x})}{\partial x_2} & \cdots & \frac{\partial v_2(\mathbf{x})}{\partial x_n} \\ \vdots & \vdots & & \vdots \\ \frac{\partial v_N(\mathbf{x})}{\partial x_1} & \frac{\partial v_N(\mathbf{x})}{\partial x_2} & \cdots & \frac{\partial v_N(\mathbf{x})}{\partial x_n} \end{bmatrix} \quad (7.41)$$

and

$$\mathbf{S}(\mathbf{x}) = \sum_{i=1}^N v_i(\mathbf{x})\nabla^2 v_i(\mathbf{x}) \quad (7.42)$$

If $\mathbf{S}(\mathbf{x})$ is assumed to be small then the Hessian matrix can be approximated as

$$\nabla^2 F(\mathbf{x}) \cong 2\mathbf{J}^T(\mathbf{x})\mathbf{J}(\mathbf{x}) \quad (7.43)$$

Substituting Eqs 7.39 and 7.43 into Eq. (7.35), we achieve the Gauss-Newton method as

$$\Delta \mathbf{x}_k = -\left[\mathbf{J}^T(\mathbf{x}_k)\mathbf{J}(\mathbf{x}_k)\right]^{-1}\mathbf{J}^T(\mathbf{x}_k)\mathbf{v}(\mathbf{x}_k) \quad (7.44)$$

One problem with the Gauss-Newton method is that the matrix may not be invertible. This can be overcome by using the following modification to the approximate Hessian matrix:

$$\mathbf{G} = \mathbf{H} + \mu\mathbf{I} \quad (7.45)$$

This leads to the Levenberg-Marquardt algorithm [54]:

$$\Delta \mathbf{x}_k = -\left[\mathbf{J}^T(\mathbf{x}_k)\mathbf{J}(\mathbf{x}_k) + \mu_k\mathbf{I}\right]^{-1}\mathbf{J}^T(\mathbf{x}_k)\mathbf{v}(\mathbf{x}_k) \quad (7.46)$$

Using this gradient direction, the approximate performance index is recomputed. If a smaller value is the yield, then the procedure is continued with the $\mu_k$ divided by some factor $\vartheta > 1$. If the value of the performance index is not reduced, then $\mu_k$ is multiplied by $\vartheta$ for the next iteration step.

The key step in this algorithm is the computation of the Jacobian matrix. The elements of the error vector and the parameter vector in the Jacobian matrix Eq. (7.43) can be expressed as

$$\mathbf{v}^T = [v_1 v_2 \dots v_N] = \left[ e_{1,1} e_{2,1} \dots e_{S^M,1} \, e_{1,2} \dots e_{S^M,Q} \right] \quad (7.47)$$

$$\mathbf{x}^T = [x_1 x_2 \dots x_N] = \left[ w_{1,1}^1 w_{1,2}^1 \dots w_{S^1,R}^1 b_1^1 \dots b_{S^1}^1 \, w_{1,1}^2 \dots b_{S^M}^M \right] \quad (7.48)$$

where the subscript $N$ satisfies

$$N = Q \times S^M \quad (7.49)$$

and the subscript $n$ in the Jacobian matrix satisfies

$$n = S^1 (R+1) + S^2 \left( S^1 + 1 \right) + \dots + S^M \left( S^{M-1} + 1 \right) \quad (7.50)$$

Making these substitutions into Eq. (7.43), the Jacobian matrix for multilayer network training can be expressed as

$$\mathbf{J}(\mathbf{x}) = \begin{bmatrix}
\frac{\partial e_{1,1}}{\partial w_{1,1}^1} & \frac{\partial e_{1,1}}{\partial w_{1,2}^1} & \dots & \frac{\partial e_{1,1}}{\partial w_{S^1,R}^1} & \frac{\partial e_{1,1}}{\partial b_1^1} & \dots \\[2ex]
\frac{\partial e_{2,1}}{\partial w_{1,1}^1} & \frac{\partial e_{2,1}}{\partial w_{1,2}^1} & \dots & \frac{\partial e_{2,1}}{\partial w_{S^1,R}^1} & \frac{\partial e_{2,1}}{\partial b_1^1} & \dots \\[2ex]
\vdots & \vdots & & \vdots & \vdots & \\[2ex]
\frac{\partial e_{S^M,1}}{\partial w_{1,1}^1} & \frac{\partial e_{S^M,1}}{\partial w_{1,2}^1} & \dots & \frac{\partial e_{S^M,1}}{\partial w_{S^1,R}^1} & \frac{\partial e_{S^M,1}}{\partial b_1^1} & \dots \\[2ex]
\frac{\partial e_{1,2}}{\partial w_{1,1}^1} & \frac{\partial e_{1,2}}{\partial w_{1,2}^1} & \dots & \frac{\partial e_{1,2}}{\partial w_{S^1,R}^1} & \frac{\partial e_{1,2}}{\partial b_1^1} & \dots \\[2ex]
\vdots & \vdots & & \vdots & \vdots &
\end{bmatrix} \quad (7.51)$$

In standard BP algorithm, the terms in the Jacobian matrix are calculated as

$$\frac{\partial \hat{F}(\mathbf{x})}{\partial x_l} = \frac{\partial \mathbf{e}_q^T \mathbf{e}_q}{\partial x_l} \quad (7.52)$$

For the elements of the Jacobian matrix, the terms can be calculated by

$$[\mathbf{J}]_{h,l} = \frac{\partial v_h}{\partial x_l} = \frac{\partial e_{k,q}}{\partial w_{i,j}} \quad (7.53)$$

Thus in this modified Levenberg-Marquardt algorithm, we compute the derivatives of the errors, instead of the derivatives of the squared errors as adopted in standard BP.

Using the concept of sensitivities in the standard BP process, here we define a new Marquardt sensitivity as

$$\widetilde{s}_{i,h}^m \equiv \frac{\partial v_h}{\partial n_{i,q}^m} = \frac{\partial e_{k,q}}{\partial n_{i,q}^m} \quad (7.54)$$

where $h = (q-1)S^M + k$.

Using the Marquardt sensitivity with BP recurrence relationship, the elements of the Jacobian can be further calculated by

$$[\mathbf{J}]_{h,l} = \frac{\partial e_{k,q}}{\partial w_{i,j}^m} = \frac{\partial e_{k,q}}{\partial n_{i,q}^m} \frac{\partial n_{i,q}^m}{\partial w_{i,j}^m} = \widetilde{s}_{i,h}^m a_{j,q}^{m-1} \quad (7.55)$$

If $x_l$ is a bias,

$$[\mathbf{J}]_{h,l} = \frac{\partial e_{k,q}}{\partial b_i^m} = \frac{\partial e_{k,q}}{\partial n_{i,q}^m} \frac{\partial n_{i,q}^m}{\partial b_i^m} = \widetilde{s}_{i,h}^m \quad (7.56)$$

The Marquardt sensitivities can be computed using the same recurrence relations as the ones used in the standard BP method, with one modification at the final layer. The Marquardt sensitivities at the last layer can be given by

$$\widetilde{s}_{i,h}^M = \frac{\partial e_{k,q}}{\partial n_{i,q}^M} = \frac{\partial \left( t_{k,q} - a_{k,q}^M \right)}{\partial n_{i,q}^M} = -\frac{\partial a_{k,q}^M}{\partial n_{i,q}^M}$$

$$= \begin{cases} -\dot{f}^M \left( n_{1,q}^M \right) & \text{for } i = k \\ 0 & \text{for } \quad i \neq k \end{cases} \quad (7.57)$$

After applying the $\mathbf{p}_q$ to the network and computing the corresponding output $\mathbf{a}_q^M$, the LMBP algorithm can be initialized by

$$\widetilde{\mathbf{S}}_q^M = -\dot{\mathbf{F}}^M \left( \mathbf{n}_q^M \right) \quad (7.58)$$

Each column of the matrix should be backpropagated through the network so as to generate one row of the Jacobian matrix. The columns can also be backpropagated together using

$$\widetilde{\mathbf{S}}_q^m = \dot{\mathbf{F}}^m \left( \mathbf{n}_q^m \right) \left( \mathbf{W}^{m+1} \right) \widetilde{\mathbf{S}}_q^{m+1} \quad (7.59)$$

The entire Marquardt sensitivity matrices for the overall layers are then obtained by the following augmentation

$$\widetilde{\mathbf{S}}^m = \left[ \widetilde{\mathbf{S}}_1^m \middle| \widetilde{\mathbf{S}}_2^m \middle| \dots \middle| \widetilde{\mathbf{S}}_Q^m \right] \quad (7.60)$$

# HYBRID-LEARNING-BASED CLASSIFICATION AND QUANTITATIVE INFERENCE OF DRIVER BRAKING INTENSITY

Automated vehicles and intelligent transportation systems have been gaining increasing attention from both academia and industrial sectors [55,56]. Intelligent vehicles have increased their capabilities in high and even fully automated driving, and it is believed that highly automated vehicles are likely to be on public roads within a few years. However, open challenges still

remain because of the strong uncertainties of driver behaviors and cognition [57]. Thus before transitioning to
fully autonomous driving, human driver behavior still
requires to be better understood. This is necessary not
only to enhance the safety, performance, and energy efficiency of the vehicles but also to adjust to the driver's
needs, potentiate driver's acceptability, and ultimately
to meet driver's intentions within a safety boundary.

Driver behaviors and cognition, including operation
actions, driving styles, intention, attention, distraction,
and operation preferences, have been widely investigated by researchers worldwide from different perspectives [58–60]. It has been concluded that driver
behaviors have great impacts on the emission, fuel
economy, ride comfort, and safety for ground vehicles
[8,61–65]. Among various driver operations, braking
maneuver is one of the most significant maneuver
[8,64,65]. As a safety-critical system, the braking system
and its control are of great importance [66]. Therefore a
better understanding of the driver's braking intention,
precise recognition of the braking demand, estimating
the braking intensity, and identifying the braking style
will benefit the active chassis control and energy management by improving vehicle's safety, comfort, and efficiency. For driver's braking maneuvers, existing
research is mainly focused on the following aspects:
braking intention inference, braking style identification,
and braking intensity recognition.

As braking styles and braking actions are of the capability to reflect driver's mental status with the current
driving context, related studies were also conducted.
In Ref. [67] driver's braking styles were grouped into
three classes, namely, the light or no braking, normal
braking, and emergency braking. EEG signals, along
with the vehicle status information from the CAN bus,
were used to infer and classify driver's braking intention. In Ref. [68], to predict driver's braking intention
before the pedal operation, a remaining time to brake
pedal operation estimation method based on the combined unscented Kalman filter and particle filter was
proposed. In Ref. [60], an input-output hidden Markov
model was applied to predict driver's pedal action by
incorporating both road information and individual
driving styles. The maximum efficient prediction horizon can reach up to 60 s.

In existing studies, the driver's braking intensity is
usually estimated as discrete states, which is similar to
the identification of the braking styles. However, in
some situations, a continuous quantification of the
braking pressure can be more useful and efficient for
vehicle control. In Ref. [69], a method of constructing a
longitudinal driver model based on a recursive least

squares self-learning scheme was developed. The driver
model takes vehicle motion states, including time-
headway and time to collision, as inputs, and outputs
the estimated values of throttle pedal position and brake
pressure. In Ref. [45], the estimation method of wheel
cylinder pressure and its relationship with the ABS were
investigated. An extended Kalman filter that combines
two normal pressure estimation models (the hydraulic
model and tyre model) was proposed to predict the
braking pressures. In addition, the impacts of continuous
estimation of braking pressure on the fault-diagnosis and
fault-tolerant control of EVs were also reported [70,71].
Nevertheless, the existing research on driver braking
behavior analysis mainly adopts conventional theories
and approaches from the control engineering with complex mathematic models, and a machine-learning-based
study has rarely been reported.

In order to further advance intelligent control of vehicles and novel design of ADAS, in this section, a
hybrid machine learning scheme is proposed to classify
driver braking intensity levels and quantitatively estimate the brake pressure. An unsupervised GMM
method is applied to automatically label braking intensity, and a supervised random forest (RF) model is
trained to classify the braking action using the output
label of GMM and other vehicle states. Then a regression —ANN—based brake pressure estimation algorithm, which is able to be aware of the current brake
scenario and assist in active chassis control, is proposed.
Real vehicle data are collected, and the hybrid-learning-
based methodology is experimentally verified.

The contributions of this section are as follows. First,
a combination of unsupervised and supervised machine
learning scheme is proposed to automatically label and
infer the driver's braking intensity. Second, beyond the
existing studies in discrete recognition of driver's deceleration actions or intentions, this work proposes a
quantitative approach by continuously estimating the
braking pressure applied by the driver. The proposed
approaches could lead to a sensorless design of the
braking control system, with great potential to remove
the brake pressure sensor existing in the current products, thus largely reducing the system cost. Moreover,
the brake pressure estimation technique provides additional redundancy for the braking system, enhancing
the safety of the system.

The rest of this section is organized as follows: part 2
describes the high-level architecture, methodology, and
detailed algorithms; part 3 describes the testing vehicle
and scenarios, as well as experimental methods; part 4
presents the testing results of the braking intensity classification and brake pressure prediction with proposed

algorithms; part 5 discusses the performance investigation with a reduced order feature vector; and part 6 presents the conclusions.

## Hybrid-Learning-Based Architecture and Algorithms

The recognition of the driver's braking intensity is of great importance for advanced control and energy management of EVs. In this section, the braking intensity is classified into three levels based on novel hybrid unsupervised and supervised learning methods. First, instead of selecting the threshold for each braking intensity level manually, an unsupervised GMM is used to cluster the braking events automatically with brake pressure. Then a supervised RF model is trained to classify the correct braking intensity levels with the state signals of vehicle and powertrain. To obtain a more efficient classifier, critical features are analyzed and selected. Moreover, beyond the acquisition of discrete braking intensity level, a novel continuous observation method is proposed based on ANNs to quantitatively analyze and recognize the brake intensity using the prior determined features of vehicle states. Experimental data are collected in an EV under real-world driving scenarios. Finally, the classification and regression results of the proposed methods are evaluated and discussed. The results demonstrate the feasibility and accuracy of the proposed hybrid learning methods for braking intensity classification and quantitative recognition with various deceleration scenarios.

In order to realize the objectives of classification and quantitative recognition of braking intensity, a high-level methodology and related algorithms are synthesized. The algorithms are mainly composed of three components, namely, the unsupervised labeling of braking intensity level using GMM, supervised classification of braking intensity level using RF, and continuous quantitative recognition of braking intensity based on ANN. The details of the system architecture and algorithms are described as follows.

### High-Level architecture of the Proposed algorithms

The high-level system architecture with the proposed methodology is shown in Fig. 7.11. The GMM receives the brake pressure of the master cylinder as the input and then yields the labels of the braking intensity levels through learning. This label vector will be used as the desired output of the supervised RF learning algorithm. The RF model takes vehicle state information from the CAN bus as the model input and aims to recognize the real braking intensity level without using brake

pressure signals, thus providing a sensorless solution. Furthermore, an FFNN model is used to quantitatively observe the brake intensity, which is reflected by the brake pressure exerted by the driver's operation. The FFNN takes the brake pressure as the model training label with similar input signals as used in the RF model.

In order to achieve the objective of brake pressure estimation, multilayer ANNs are first constructed with the input of vehicle and powertrain states. Details of the high-level system architecture and structure of the component are described in this section.

### Classification of braking intention level using gaussian mixture model

According to the above-proposed methodology, the driver's braking intention is first clustered using a GMM. GMM is a probability density function that is represented by the sum of weighted sub-Gaussian components [72]. In this study, the GMM is adopted to obtain the probability distribution of the brake pressure data with an unsupervised learning approach. The brake pressure was measured in the master cylinder by a hydraulic pressure sensor. One advantage of the above-proposed approach is that the GMM is an unsupervised learning method requiring no training labels and it is very flexible to select the different numbers of clusters. In this work, a total of three clusters are generated by GMM, representing prebraking, moderate braking, and intensive braking.

The series of the brake pressure data can be described as $X = \{x_1 \cdots x_T\}$, where $T$ is the final time index. Then the GMM can be represented as

$$p\left(x_i|\theta\right) = \sum_{k=1}^{K} \pi_k N\left(x_i|\mu_k, \sum\nolimits_k\right) \qquad (7.61)$$

where $x_i$ is the one-dimensional value of brake pressure, $\theta$ is the parameter of GMM, $K$ is the total number of the component in the model (three in this case), and $\pi_k$ is the weight of each component's Gaussian distribution function and the sum of $\pi_k$ equals to 1.

$N\left(x_i|\mu_k, \sum\nolimits_k\right)$ is the univariate Gaussian distribution function, which can be given by

$$N\left(x|\mu, \sum\right) = \frac{1}{(2\pi)^{D/2} \left|\sum\right|^{1/2}} \exp\left[-\frac{1}{2}(x-\mu)^T \sum\nolimits^{-1}(x-\mu)\right]$$

$$(7.62)$$

where $D$ is the dimension of the data vector and it is taken as 1 in this case. A complete structure of GMM, which contains three parameters, can be represented as

$$\theta = \left\{\pi_k, \mu_k, \sum\nolimits_k\right\} \qquad (7.63)$$

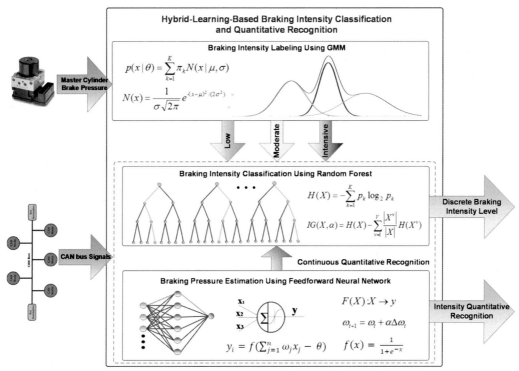

FIG. 7.11 Illustration of the Proposed Hybrid-learning-based Architecture. *CAN*, Controller Area Network; *GMM*, Gaussian mixture model.

Given the training data of brake pressure, the GMM can be trained with both maximum likelihood estimation and maximum a posteriori estimation. The most common method for GMM training is the EM maximum likelihood estimation algorithm [73]. Suppose the training dataset of brake pressure is in the format $X = \{x_1 \cdots x_T\}$, then the overall likelihood of the GMM can be calculated by

$$p(X|\theta) = \prod_{t=1}^{T} p(x_t|\theta) \tag{7.64}$$

Substituting log-likelihood function to Eq. (7.4), the function can be transformed into a much easier one:

$$l(\theta) = \sum_{t=1}^{T} \log p(x_t|\theta) \tag{7.65}$$

Owing to the nonlinear characteristics of the log function, the utilization of Eq. (7.65) makes it hard to figure out the maximization directly. Therefore EM is selected to compute the maximization with an iterative process, which contains two steps, i.e., the E-step and the M-step.

1) E-step: Estimate the *posterior* probability $Pr_{ik}$ of each component $k$ with data point $i$.

$$Pr_{ik} = \frac{\pi_k p\left(x_t \middle| \theta_k^{(t-1)}\right)}{\sum_k \pi_k\, p\left(x_t \middle| \theta_k^{(t-1)}\right)} \tag{7.66}$$

2) M-step: Update the model parameters according to the estimated *posterior* probability in E-step.

$$\pi_k = \frac{1}{T} \sum_{i=1}^{T} Pr_{ik} = \frac{Pr_k}{T} \tag{7.67}$$

$$\mu_k = \frac{Pr_k \cdot x_t}{Pr_k} = \frac{\sum_{i=1}^{T} Pr_{ik} \cdot x_t}{\sum_{i=1}^{T} Pr_{ik}} \tag{7.68}$$

$$\sum_k = \frac{Pr_k (x_t - \mu_k)(x_t - \mu_k)^T}{Pr_k} \tag{7.69}$$

### Braking intention classification using random forest

Following the previous section, to learn the output label from GMM and yield accurate results correspondingly, a supervised machine learning method is required. According to the study in Ref. [74], the RF algorithm achieved the best results on 121 public datasets among

179 classification algorithms. Thus in this study, with a great capability to accurately classify objects, the RF algorithm is used to classify and further infer the braking intention level using vehicle and powertrain signals obtained from the CAN bus. Its performance of the braking intention classification will be analyzed and compared with other existing ones in the following sections.

An RF is an ensemble learning method that is constructed with a combination of multiple weighted decision trees [75]. Decision tree, also known as Classification and Regression Tree (CART), is a popular machine learning method that has been used in many pattern recognition and prediction tasks [76]. One decision tree is constructed with one root node, multiple middle nodes, and leaf nodes. The decision results are represented by the leaf nodes in CART. Specifically, for the classification trees, the outputs are the discrete labels of the classification objects, while the outputs of regression trees are the estimated continuous values. The decision nodes and decision tree structures are determined by minimizing the information entropy $H(X)$ and maximizing the information gain $IG(X, \alpha)$, as shown in Fig. 7.11. However, a single decision tree is a weak learner and has limited ability to deal with those problems with a large amount of data o— large— dimensional feature vectors. The RF, however, parallel combines multiple trees to reduce the overfitting risk and improves the generalization ability with the bootstrap aggregating scheme (often known as Bagging trees) [77].

RF enhances the model prediction performance by introducing the random property during the model training process. Specifically, like other Bagging algorithms, RF randomly selects a subset of data as training data from the original data set to train the subdecision trees. Besides, it introduces a random subset feature selection technique to avoid getting highly correlated predictors and models. Therefore the subdecision trees will be trained with a different dataset as well as different feature vectors, thus efficiently increasing the properties of the whole RF model. The final output of RF is the ensemble of the subdecision trees. There are mainly three ensemble methods, namely, averaging, voting, and learning. A common ensemble classification algorithm is the one with weighted voting, as shown in the following.

$$H(x) = lbl_{argmax} \sum\nolimits_{i=1}^{T} \omega_i h_i^j(x) \tag{7.70}$$

where $H$ is the final output label, $lbl$ is the label set $\{lbl_1, \cdots lbl_j, \cdots lbl_N\}$, $\omega_i$ is learned the weight of each tree output, and $h_i(x)$ is the output of tree $i$. The final

output label of braking intensity will be the voting results of each subdecision tree.

### Brake Pressure Estimation Based on Artificial Neural Network

According to the abovementioned analysis, discrete level estimation of the brake intensity relies on the information of brake pressure. In some modern brake systems, there are pressure sensors directly providing the measurement of brake pressure. However, these sensors add considerable costs to the whole system. If the estimation technique can be achieved by realizing a sensorless braking system, then the system cost could be largely reduced. Meanwhile, the brake pressure estimation technique provides additional redundancy for the safety-critical function of the braking system. Thus in this part, ANNs are used to observe the continuous state of brake pressure based on vehicle and powertrain states. A multilayer FFNN with a hidden layer is trained to observe the brake pressure value. The architecture of the FFNN features that the information flow only has one direction and is transferred from the input layer to the output layer without cycles in the model. As the FFNN is of great ability to theoretically represent any complex polynomial function with the different hidden neurons, it is selected as the model to quantitatively predict brake intention in this work.

As shown in Fig. 7.11, in the training process, the FFNN uses both CAN bus signals and the value of brake pressure as the supervised output response. While during the testing procedure, the FFNN only adopts the CAN bus signals to identify the brake pressure. The NN is construed by basic calculation units called neurons, which is inspired by the human neural system. The neurons located in different layers can have different thresholds as well as distinguished activation functions. Neurons in each layer are interconnected with weightings. After receiving the input signals, the neurons in the hidden layer firstly sums all the weighted signals and compares them with the neuron threshold. If the summation is larger than the threshold, the neuron will be activated and the processed value will be output. The mathematic model of the neuron can be given by

$$y_i = f\left(\sum\nolimits_{j=1}^{n} \omega_j x_j - \theta\right) \tag{7.71}$$

where $y_i$ is the output of neuron $i$ in the hidden layer, $f()$ is the activation function, $x = \{x_1 \cdots x_n\}$ is the input from the input layer, $\omega_j$ is the corresponding weight, and $\theta$ is the neuron threshold. A differentiable sigmoid function is adopted as the activation function, which is in the form

$$f(x) = \frac{1}{1 + e^{-x}} \qquad (7.72)$$

Once the model structure is determined and parameters are initialized, the FFNN is to be trained with the method of backward propagation of error (BP). BP is an iterative NN training scheme, which consists of two basic steps including propagation and weight update. The input signals from CAN bus first propagate forward through the network to generate an initial estimated value, then the yield value is compared with the ground truth based on the loss function. In this study, the mean square error is utilized as the loss function.

$$E = \frac{1}{N} \sum_{i=1}^{n} (t_i - y_i)^2 \qquad (7.73)$$

where $E$ is the mean square error of the actual value and the desired one, $N$ is the total number of data samples, and $t_i$ and $y_i$ are the values of ground truth and actual output, respectively.

After the propagation step, the weights of the network can be updated by calculating the gradient of the loss function. The optimal weights are expected to result in the global minimum of the loss function. Suppose that $\omega_{j,i}$ is the weight between the input neuron $j$ and the hidden neuron $i$, and $y_i$ is the output of the neuron $i$, then the gradient can be calculated using the chain rule presented as follows:

$$\Delta\omega_{j,i} = \frac{\partial E}{\partial \omega_{j,i}} = \frac{\partial E}{\partial y_i} \frac{\partial y_i}{\partial f} \frac{\partial f}{\partial \omega_{j,i}} \qquad (7.74)$$

where $\Delta\omega_{j,i}$ is the weight gradient and $\partial f$ is the gradient of the sigmoid activation function shown in Eq. (7.73).

Finally, the weight can be updated at each step as

$$\omega_{t+1} = \omega_t - \alpha_t \Delta\omega_t \qquad (7.75)$$

where $\omega_{t+1}$ is the updated weight, $\omega_t$ is current weight, and $\alpha_t$ is the learning rate of the FFNN. Eq. 7.75 is the simplest gradient descent updating method, and more sophisticated model training algorithms for ANN can be found in Ref. [78].

## Experimental Testing and Data Preprocessing

In order to validate the hybrid-learning-based methodology and algorithms proposed earlier, vehicle testing with real-world driving scenarios is required to be carried out. In this section, the experimental scenario setup and the test vehicle will be introduced. The collected experimental data will be analyzed through signal processing methods. Also the selected feature vectors and model training process will be illustrated.

*Experiment design*
As shown in Fig. 7.12, the testing is carried out on a chassis dynamometer with an EV operating under typical driving cycles. There are a lot of standards utilized for chassis dynamometer driving cycles [79]. In this work, the NEDC, containing the ECE and the EUDC, is adopted. As shown in Fig. 7.13, the first section of the NEDC, which is composed of four successive ECEs, exhibits a low-speed urban operating condition. The second part, i.e., the EUDC section, represents a highway scenario, with the highest speed at 120 km/h.

During testing, accessory devices such as the heater and air-conditioner need to be switched off. The battery should be fully charged with the SOC being at 100% before the test. The test drive requires repeated NEDCs with a maximum deviation of 2 km/h in the speed profile. Once the vehicle is unable to follow up the target speed due to low SOC or other reasons, the experiment will be terminated.

*Experimental vehicle with brake blending system*
The vehicle utilized in the road tests is an electric passenger car, and the structure of the vehicle with a regenerative and hydraulic blended braking system is shown in Fig. 7.14A. The front wheels of this test EV are driven by a permanent-magnet synchronous motor that can work in two states as a driving motor or a generator. The battery, which is connected to the motor through the d.c. bus, can be discharged or charged for motoring or absorbing, respectively, the regenerative power during the braking process. During deceleration, the blended brake torque complies with the serial regenerative strategy, i.e., the regenerative brake is applied at first, and the hydraulic brake compensates the

FIG. 7.12 The Testing Vehicle with a Chassis Dynamometer During Experiments.

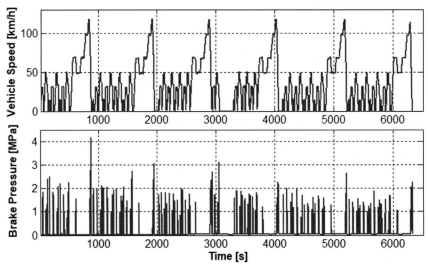

FIG. 7.13 Speed Profile of the Driving Cycle with Corresponding Braking Pressure.

remaining part of the overall braking demand. The co-ordination of the blended brakes is able to guarantee the braking comfort and regeneration efficiency of the vehicle.

A schematic diagram of the hydraulic braking system is shown in Fig. 7.14B. $P_{FW}$, $P_{FW}$, and $P_0$ denote the pressures of the master cylinder, wheel cylinder, and the low-pressure accumulator, respectively. The inlet and outlet valves are respectively PWM controlled. The structure of the wheel cylinder can be simplified to a piston and spring. $k_{FW}$ is the equivalent stiffness of the spring, and $r_{FW}$ is the radius of the wheel cylinder's cross-sectional area. Key parameters of the EV and powertrain are listed in Table 7.2.

### Data collection and processing
The CAN bus data are sampled with a frequency of 100 Hz Hz, and 6327 s of data are recorded in total. The raw data contains six standard NEDCs. The collected data of the vehicle speed and the brake pressure of the master cylinder are shown in Fig. 7.13 and further processed in Table 7.6.

Before training and testing the models through machine learning methods, the raw data is first smoothed and filtered as follows:

$$d_t = \frac{\sum_n^N d_{tn}}{N} \quad (7.76)$$

where $d_t$ is the value of a signal at time $t$, $d_{tn}$ is the $n$th sampled value of signal $d$ at time step $t$, and $N$ is the total amount of samples per second. Then the input

signals are scaled to the range from 0 to 1 in order to eliminate the influence brought by different units.

### Feature selection and model training
In this work, the braking correlated signals and important vehicle state variables are selected for the training of the braking intensity classifier and brake pressure estimation model. The brake pressure data is only used as a response signal for FFNN training, and it is not used during the testing process. When the EV is decelerating, the electric motor works at its regenerative brake mode, recovering the vehicle's kinetic energy. During this period, the value of battery current changes from positive to negative, indicating that the battery is charged by regenerative braking energy. Thus the signals of battery current and voltage are chosen as features. The signals used for model training in this work are listed in detail in Table 7.7.

As shown in Table 7.7, instead of using raw data from the CAN bus only, statistical information, including the mean value, max value, and STD of the original data in the past few seconds, are also adopted. In this work, the values of vehicle velocity during the past 5 s are stored and used to calculate the mean and STD values. Besides, the gradient values of the battery current and voltage are also utilized.

After the feature vector was determined, the supervised classification and regression models are trained. The overall sampled data are divided into two sets, namely, the training set and the testing set. The testing data set used for model training and validation contains 1400 sample points, which is randomly chosen from

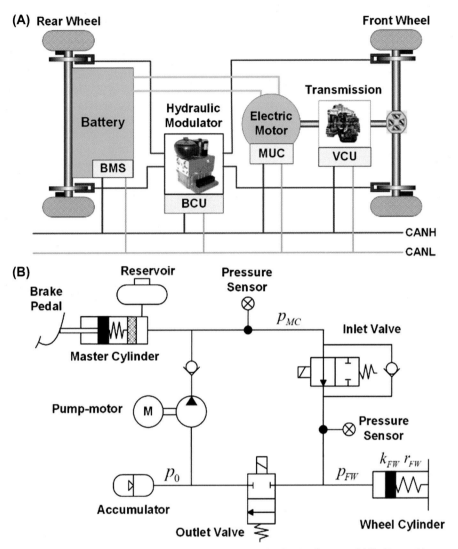

FIG. 7.14 The Structure of the Experimental Vehicle with the Brake System. *BMS*, Batter Management System; *BCU*, braking control unit; *MUC*, Motor Unit Controller; *VCU*, Vehicle Control Unit; *CANH*, CAN High; *CANL*, CAN Low signals.

the raw and training datasets. To modulate and evaluate the model performance, the K-fold cross-validation approach is used. K-fold cross-validation method randomly selects $K - 1$ folds from the training data to train the RF and FFNN models, and the rest fold is utilized for testing. The final assessment of the model performance is carried out according to the $K$ test results. In this work, the value of $K$ is set as 5. Considering the data quantity and the evidence that NN with one hidden layer is able to estimate most of the complex functions, the FFNN is constructed with a hidden layer. The FFNN

is then trained using a fast Levenberg-Marquardt algorithm with fivefold cross-validation.

## Experiment Results and Analysis

In this section, the experiment results of the abovementioned three machine learning tasks are described. First, labeling the results of the braking intensity level using GMM will be illustrated. Then the classification results of the braking intensity level with the RF approach are presented. Finally, the results braking pressure estimation using ANN will be shown. The algorithms are

**TABLE 7.6**
Statistics Data of Vehicle Velocity and Braking Pressure.

| Variables | Mean | Standard Deviation | Max | Min |
|---|---|---|---|---|
| Vehicle velocity (km/h) | 35.594 | 30.042 | 118.46 | 0 |
| Braking pressure (MPa) | 0.1857 | 0.4317 | 4.1728 | 0 |

**TABLE 7.7**
CAN Bus Signals Used for Model Training.

| No. | Signal | No. | Signal |
|---|---|---|---|
| 1 | Velocity (km/h) | 7 | Acceleration (m/s$^2$) |
| 2 | Accelerator pedal position | 8 | Mean velocity (km/h) |
| 3 | Battery current (A) | 9 | STD of velocity (km/h) |
| 4 | Battery voltage (V) | 10 | Max velocity (km/h) |
| 5 | Motor speed (rad/s) | 11 | Voltage gradient (V/s) |
| 6 | Motor torque (N· m) | 12 | Current gradient (A/s) |

STD, standard deviation.

implemented in the Intel Core i7 2.5GH computer with the MATLAB 2017a platform.

*Labeling result of braking intensity level using Gaussian mixture model*

The GMM takes the braking pressure as an input and outputs the Gaussian distribution of each cluster. In order to reach a good performance of the proper braking intensity identifier, a suitable GMM is achieved based on the proportion evaluation of each component. The final decision thresholds of the GMM can be given by

$$Label = \begin{cases} Low, & x < 0.05 \\ Middle, & x < 1.25, x > 0.05 \\ High, & x > 1.25 \end{cases} \quad (7.77)$$

where $x$ is the braking pressure.

Fig. 7.15 shows the labeling results with three different braking intensity levels. The first cluster, which is represented by the red dots, is of low-pressure area, representing the driver's low-intensity brake demand and the prebraking processes. In this area, the low-level pressure is used to eliminate the mechanical gaps of brake devices. The second cluster, which is represented by yellow dots with a pressure range from 0.15 to 1.1 MPa, indicates the driver's moderate braking intensity. The third cluster indicating intensive braking is illustrated by green dots, with pressure over 1.1 MPa. According to the results, the quantity of the cluster with low braking intensity is much larger than the other ones, taking up 85.3% of the overall data points. The other two clusters share the rest of the data with similar quantities of 7.7% and 7% each. The results show a good balance maintained between the proportions and the cluster's diversity.

Fig. 7.16 illustrates the learned Gaussian distribution of the GMM with the three components. As the low braking intensity group has many more data points and more significant characteristics, the classification confidence of this group is much higher than that of the other two groups. The moderate and the intensive braking levels have relatively wider Gaussian distributions and overlap with each other, showing that these two classes have similar characteristics and their data distributions can be roughly described by a single Gaussian distribution function.

*Random forest-based classification results of braking intensity level*

According to the abovementioned labeling, the classification results of the braking intensity level given by RF are analyzed as follows. The RF classifier is constructed with 50 decision trees, and the ANN is an FFNN with 50 neurons in the hidden layer. As mentioned before, the task is to solve the three-class classification problem with the labels given by GMM. The classification performance is assessed from three different aspects, namely, the classification accuracy, the execution time of model training, and the execution time of testing. The general accuracy is defined as follows:

$$ac = \frac{N_1 + N_2 + N_3}{N} \quad (7.78)$$

where $ac$ is the general accuracy of the classification, $N$ is the total amount of data, and $N_1, N_2, N_3$ are the number of correct classification cases, respectively.

Moreover, the RF classifier is also compared with other existing ones, including decision trees, support vector machine, K-nearest neighbor, and multilayer

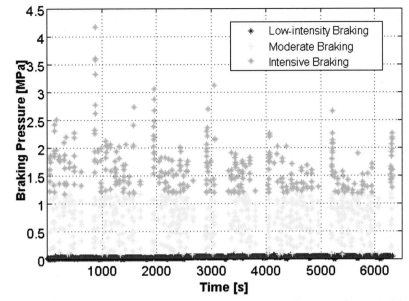

FIG. 7.15 Labeling results of braking intensity levels using Gaussian mixture model.

FFNN. The detailed classification performance and comparison results are illustrated in Table 7.8.

The detail classification performance of the RF is illustrated using a confusion matrix, as shown in Fig. 7.17. The result is generated from the preselected 1400 data samples, containing 1227 low-intensity samples, 106 moderate-intensity samples, and 67 intensive ones. According to the result, cluster 1, i.e., the low-intensity braking level, achieves results of 100% accuracy. The classification results of the second and third clusters are not as accurate as of those of the first one. These fault classifications can be explained by the

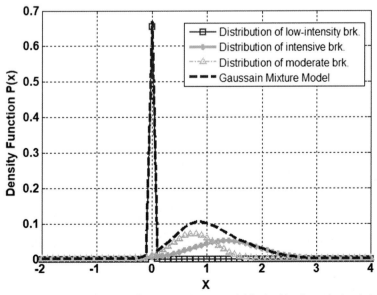

FIG. 7.16 Gaussian distribution of Gaussian mixture model for braking intensity level clustering.

TABLE 7.8
Comparison Results of the Braking Intensity
Classification Performance.

|  | General Accuracy | Training Time (s) | Testing Time (obs/s) |
|---|---|---|---|
| Decision tree | 0.969 | 4.061s | ~73000 |
| Quadratic SVM | 0.974 | 5.663s | ~59000 |
| Weighted KNN | 0.967 | 7.818s | ~12000 |
| ANN | 0.971 | 3.201s | ~30000 |
| Random forest | 0.977 | 6.086s | ~27000 |

ANN, artificial neural network; KNN, K-nearest neighbor; SVM, support vector machine.

overlap phenomena of the Gaussian distribution of the GMM. Although there are some misclassifications, their impacts on the overall system are limited, and this will be discussed in the following section.

As mentioned in the Braking intention classification using random forest part, the RF algorithm adopts the bootstrap method by randomly selecting training dataset and training predictors from raw data to construct various decision trees. In this process, about one-third of the data will not be used for the model training task, and this part of the data is called the out-of-bag (OOB) data [76]. The model testing error of the OOB data, which has a similar accuracy with real testing data, is sufficient to reflect the model generalization performance [80]. Therefore the OOB data is used as another source to assess the model performance and estimate the importance of the predictors [80,81].

Fig. 7.18 illustrates the estimation results of predictor importance with the OOB data set. It can be seen that the most important predictor in the feature vector is the battery current, following with vehicle velocity, acceleration, and the STD of velocity. The battery voltage, voltage variation rate, and current variation rate also exert impacts on the model classification performance.

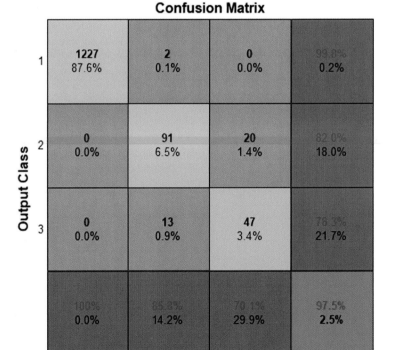

FIG. 7.17 Confusion matrix of the classification result given by random forest.

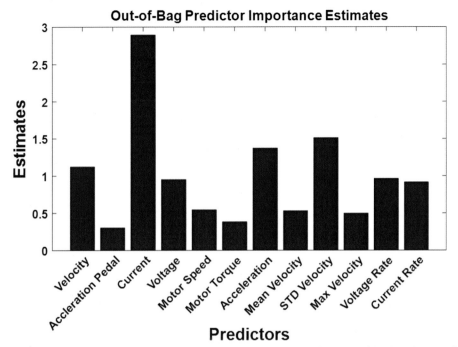

FIG. 7.18 The predictor importance estimation results of the random forest model using the out-of-bag dataset. *STD*, standard deviation.

### Estimation result of braking pressure based on artificial neural network

The ANN-based quantitative estimation result of braking pressure is analyzed as follows. The regression result is compared with other machine learning methods, including regression decision tree, support vector machine, Gaussian process model, and regression RF. The models are trained and tested with fivefold cross-validation methods. The ANN and RF have 50 neurons and trees, which are similar to the ones used in the classification task. The model performance is evaluated via four properties, namely, the coefficient of determination (denote as $R^2$), the root-mean-square error (RMSE), the training time, and the testing time, as shown in Table 7.9. $R^2$ and RMSE are common performance indexes that have been widely accepted to evaluate prediction accuracy.

The definitions of the $R^2$ and RMSE are presented as follows. Suppose the ground-truth dataset is $T = \{t_1 \cdots t_N\}$ and the predicted value is $Y = \{y_1 \cdots y_N\}$. The $R^2$ is calculated as

$$R^2 = 1 - \frac{E_{res}}{E_{tot}} \qquad (7.79)$$

where $E_{res}$ is the residual sum of the square and $E_{tot}$ is the total sum of the square. They are defined as

$$E_{res} = \sum_i (t_i - y_i)^2 \qquad (7.90)$$

$$E_{tot} = \sum_i (t_i - \overline{T})^2 \qquad (7.91)$$

where $\overline{T}$ is the mean value of the ground-truth data.

The RMSE can be calculated as

$$RMSE = \sqrt{\frac{\sum_i^N (y_i - t_i)}{N}} \qquad (7.92)$$

As shown in Table 7.9, the ANN algorithm yields the best performance of brake pressure estimation. The running time of training with ANN and RF is similar; however, the testing speed of ANN is much faster than that of the RF. Another interesting phenomenon is that the single decision tree algorithm has much shorter training time and a much faster testing speed than the other algorithms. In terms of real-time application, the regression decision tree could be a better candidate because of its simplicity and high computation efficiency.

The ANN model estimation result with testing data is shown in Fig. 7.19. The x-axis presents the 1400 data samples, and the y-axis shows the estimation results of the scaled pressure of the data samples. As the input and output data for model training is scaled to the range

**TABLE 7.9**
Comparison of Braking Pressure Estimation Performance.

| Methods | $R^2$ | RMSE (MPa) | Training Time (s) | Testing Time (obs/s) |
|---|---|---|---|---|
| Decision tree | 0.912 | 0.133 | 1.092 | ~240000 |
| Quadratic SVM | 0.867 | 0.188 | 141.93 | ~46000 |
| Gaussian process model | 0.921 | 0.125 | 156.89 | ~8100 |
| ANN | 0.935 | 0.101 | 3.42 | ~82000 |
| Random forest | 0.903 | 0.104 | 3.79 | ~36000 |

*ANN*, artificial neural network; *RMSE*, root-mean-square error; *SVM*, support vector machine.

of [0, 1], the model testing output falls in the range between 0 and 1 accordingly. Based on the results, the ANN model achieves high-precision regression performance in most cases and the overall RMSE is around 0.1 MPa, demonstrating the feasibility and effectiveness of the developed algorithm.

The two most accurate models, namely, the ANN and RF, are further studied in detail. The impacts of the different numbers of neurons and trees on the brake pressure prediction are analyzed. The neuron numbers of ANN and the ensemble tree numbers of RF range from 10 to 100. The prediction results of the algorithms are shown in Figs. 7.20 and 7.21. According to Fig. 7.20, the overall performance of ANN is better than that of the RF, and the best prediction performance yield is by FFNN, with the number of neurons at 70. Because of using the gradient-descent-based model learning method, compared to the RF with relatively stable results, the accuracy of different FFNNs varies

significantly. This shows that the prediction performance of RF changes little with the different number of ensemble trees, indicating its good robustness. Fig. 7.21 shows the linear regression results given by the most accurate FFNN model, with 70 neurons. It can be seen that the FFNN can accurately estimate the braking pressure information using vehicle status from the CAN bus.

To further validate the feasibility and effectiveness of the proposed approach, the vehicle road test is carried out with the real-time implementation of the algorithms. The test track is flat and it has a dry surface with a high adhesion coefficient. As shown in Fig. 7.22, the road test data with a duration of the 1500 s contains 10 individual normal deceleration processes. Among these decelerations, three typical intensities of the normal deceleration, namely, the intensive, medium, and mild ones, are covered. According to the experiment results, the proposed approach is

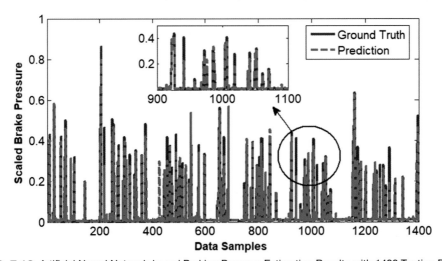

FIG. 7.19 Artificial Neural Network-based Braking Pressure Estimation Results with 1400 Testing Data.

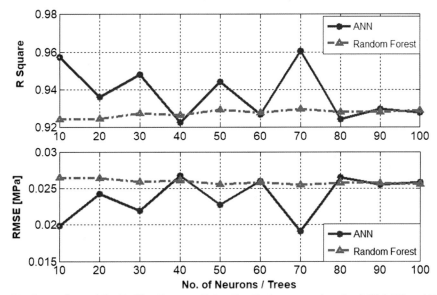

FIG. 7.20 Comparison of the Braking Pressure Estimation Performance given by Artificial Neural Network (ANN) and Random Forest (RF).

able to accurately predict the value of the brake pressure during deceleration. Comparing the prediction to the ground truth measured by the onboard hydraulic pressure sensor, the values of $R^2$ and RMSE are over 0.85 and 0.2 MPa, respectively, under such naturalistic

driving conditions, demonstrating the accuracy and robustness of the developed methods.

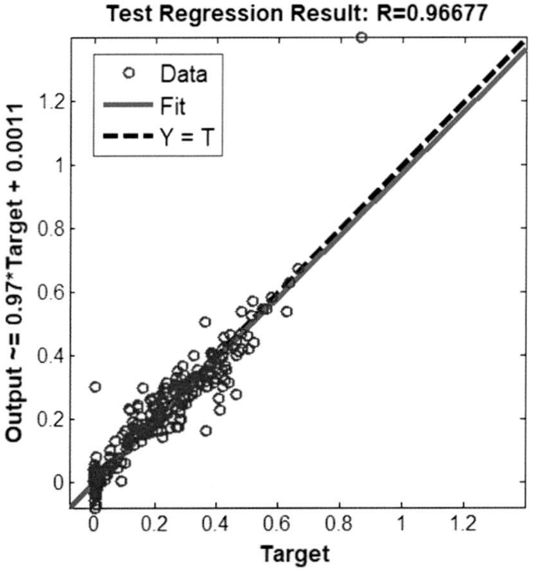

### Discussions

In this section, the fault classification of the intensive braking cases, which has been shown in the previous section, is further analyzed based on the classification and regression results. Moreover, the methodology for improving classification and prediction performance by using a smaller feature vector, which is obtained by the predictor importance estimation result of the RF, is investigated.

#### Fault classification of the intensive braking

According to the results in Fig. 7.17, the fault classifications of RF mainly occur in the moderate and intensive braking levels. There are 13 moderate braking points that are incorrectly classified into the intensive ones and 20 intensive cases are identified as moderate ones. From the perspective of braking safety, intensive braking being incorrectly identified as a moderate one is much more dangerous than vice versa. Thus the fault detection of intensive braking cases is analyzed as follows.

There are 67 intensive braking samples within the testing dataset, among which 20 cases are of fault classification and 47 are correct ones. Fig. 7.23 shows the braking pressure prediction results of the total 67 samples using the ANN algorithm. The upper and lower

**Test Regression Result: R=0.96677**

FIG. 7.21 Regression Performance of Feedforward Neural Network with 70 Neurons.

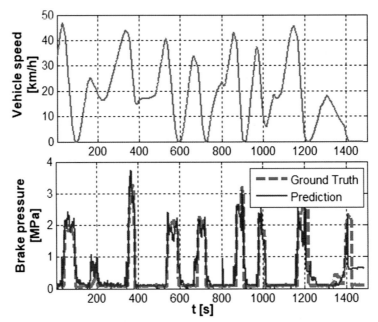

FIG. 7.22 Vehicle Road Test Results of the Proposed Approaches.

subplots present the fault and correct classification cases, respectively. As the upper subplot shows, the braking pressure of the fault classifications range from 0.3 to 0.4, corresponding to 1.2−1.6 MPa in real measurement. However, this pressure range is around the lower bound of the overall intensive braking samples,

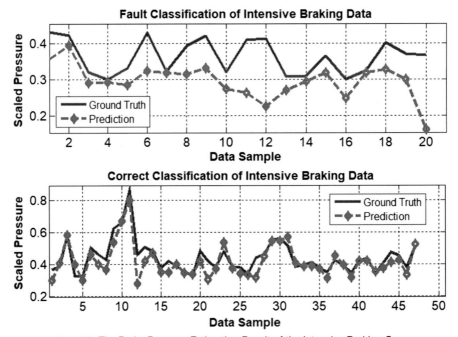

FIG. 7.23 The Brake Pressure Estimation Result of the Intensive Braking Cases.

so it can be accepted to be classified into the moderate level group. Besides, most of the estimated pressure points generated by the trained FFNN model can accurately follow the real value. Therefore we believe that the 20 fault classification samples have small effects on the whole intensive braking process.

## Performance With a Reduced Order Feature Vector

According to the analysis results illustrated in Fig. 7.18, the four most important predictors of the raw data for classification are the battery current, the vehicle speed, the acceleration, and the STD of the speed. In order to extend the proposed approach to conventional internal combustion engine (ICE) vehicles by removing signals related to electric powertrains, in this section, a new feature vector and a dataset containing only the aforementioned four signals are constructed and adapted to assess the classification and prediction performance. The confusion matrix of the classification and the predicted braking pressures with the new feature vector and dataset are illustrated in Figs. 7.24 and 7.25, respectively.

As shown in Fig. 7.18, the new RF model, which is trained with a low-dimensional dataset, generates more accurate classification results. The classifier is tested with 1400 randomly selected testing data points.

The accuracies of the moderate and intensive braking classification results are improved from 85.8% to 91.7% and from 70.1% to 74.1%, respectively.

The estimation results of the braking pressure with the FFNN model is shown in Fig. 7.25. Owing to the dimension reduction, the prediction accuracy of the FFNN slightly decreases. The $R^2$ and RMSE values of the testing dataset become 0.905 and 0.123 MPa, respectively. Although the braking pressure prediction performance reduces, the FFNN model is still able to estimate braking pressure with small errors in most of the situations, indicating that the proposed approach can be applied in conventional ICE vehicles.

## Conclusions

In this section, a hybrid-learning-based classification and quantitative recognition methodology of driver braking intensity are investigated. Three different braking intensity levels, namely, low-intensity, moderate, and intensive braking, are first identified using the GMM algorithm. Then an RF algorithm is proposed to classify the braking intensity level based on the output label of GMM and vehicle state variables from CAN bus. Finally, a continuous estimation algorithm for braking pressure observation using FFNN is proposed. High-accuracy results of the braking intensity classification and brake pressure prediction are achieved with the developed

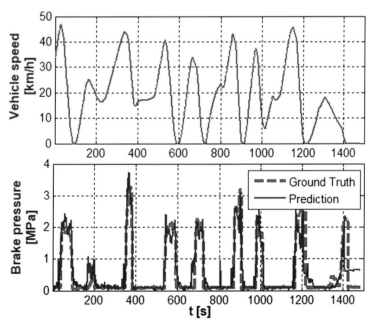

**FIG. 7.24** Braking Intensity Classification Results Given by Random Forest using a Low-dimensional Feature Vector with 50 Ensemble Trees.

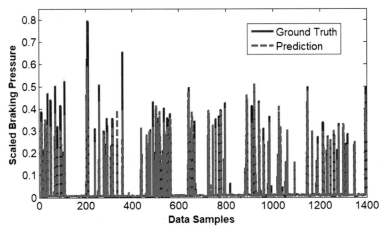

FIG. 7.25 Braking Pressure Prediction Result Given by Feedforward Neural Network Using a Reduced Dimensional Feature Vector with 50 Neurons.

hybrid machine learning methods. In order to validate the algorithms, experimental data are first collected from an EV under standard NEDCs. Then the hybrid learning methods are tested and improved with the collected dataset. The test results show that the proposed methods, which require simple modeling and parameter identification procedures, are able to accurately classify the braking intensity levels and predict the braking pressure correspondingly. It enables the sensorless technology and provides additional redundancy for the safety-critical braking system, and it can be widely utilized in energy management and active chassis control for various types of ground vehicles.

## REFERENCES

[1] Z. Bareket, P.S. Fancher, H. Peng, et al., Methodology for assessing adaptive cruise control behavior, IEEE Transactions on Intelligent Transportation Systems 4 (3) (2003) 123–131.

[2] H. Rong, J. Gong, Study on intelligent control of road condition perception and forward collision warning system, in: International Conference on Intelligent Computing and Integrated Systems (ICISS), 2010, IEEE, 2010, pp. 285–289.

[3] J.F. Liu, Y.F. Su, M.K. Ko, et al., Development of a vision-based driver assistance system with lane departure warning and forward collision warning functions, in: Digital Image Computing: Techniques and Applications (DICTA), 2008, IEEE, 2008, pp. 480–485.

[4] T. Qu, H. Chen, D. Cao, H. Guo, B. Gao, Switching-based stochastic model predictive control approach for modeling driver steering skill, IEEE Transactions on Intelligent Transportation Systems 16 (1) (2015) 365–375.

[5] D.A. Crolla, D. Cao, The impact of hybrid and electric powertrains on vehicle dynamics, control systems and energy regeneration, Vehicle System Dynamics 50 (Suppl. 1) (2012) 95–109.

[6] C.M. Martinez, X. Hu, D. Cao, et al., Energy management in plug-in hybrid electric vehicles: recent progress and a connected vehicles perspective, IEEE Transactions on Vehicular Technology 66 (6) (2016) 4534–4549.

[7] X. Hu, J. Jiang, B. Egardt, D. Cao, Advanced power-source integration in hybrid electric vehicles: multicriteria optimization approach, IEEE Transactions on Industrial Electronics 62 (12) (2015) 7847–7858.

[8] C. Lv, J. Zhang, Y. Li, Extended-Kalman-filter-based regenerative and friction blended braking control for electric vehicle equipped with axle motor considering damping and elastic properties of electric powertrain, Vehicle System Dynamics 52 (11) (2014) 1372–1388.

[9] C. Lv, J. Zhang, Y. Li, Y. Yuan, Mode-switching-based active control of powertrain system with nonlinear backlash and flexibility for electric vehicle during regenerative deceleration, Proceedings of the Institution of Mechanical Engineers - Part D: Journal of Automobile Engineering 229 (11) (2015) 1429–1442.

[10] Y. Zheng, S.E. Li, J. Wang, et al., Stability and scalability of homogeneous vehicular platoon: study on the influence of information flow topologies, IEEE Transactions on Intelligent Transportation Systems 17 (1) (2016) 14–26.

[11] J. Wang, L. Zhang, D. Zhang, et al., An adaptive longitudinal driving assistance system based on driver characteristics, IEEE Transactions on Intelligent Transportation Systems 14 (1) (2013) 1–12.

[12] N.M. Enache, S. Mammar, S. Glaser, et al., Driver assistance system for lane departure avoidance by steering and differential braking, IFAC Proceedings Volumes 43 (7) (2010) 471–476.

[13] H. Zheng, S. Ma, Development and Test of Braking Intention Recognition Strategies for Commercial Vehicle, SAE Technical Paper, 2015.

[14] A. Lopez, R. Sherony, S. Chien, et al., Analysis of the braking behaviour in pedestrian automatic emergency braking, in: 2015 IEEE 18th International Conference on Intelligent Transportation Systems, IEEE, 2015, pp. 1117–1122.

[15] G.S. Aoude, V.R. Desaraju, L.H. Stephens, et al., Driver behavior classification at intersections and validation on large naturalistic data set, IEEE Transactions on Intelligent Transportation Systems 13 (2) (2012) 724–736.

[16] J.C. McCall, M.M. Trivedi, Human behavior based predictive brake assistance, in: 2006 IEEE Intelligent Vehicles Symposium, IEEE, 2006, pp. 7–12.

[17] T. Wada, S. Doi, A. Nishiyama, et al., Analysis of braking behavior in car following and its application to driver assistance system, Proceedings of International Symposium on Advanced Vehicle Control 2 (2008) 577–583.

[18] S. Haufe, J.-W. Kim, I.-H. Kim, A. Sonnleitner, M. Schrauf, G. Curio, B. Blankertz, Electrophysiology-based detection of emergency braking intention in real-world driving, Journal of Neural Engineering 11 (5) (2014), 056011.

[19] S. Haufe, M.S. Treder, M.F. Gugler, M. Sagebaum, G. Curio, B. Blankertz, EEG potentials predict upcoming emergency brakings during simulated driving, Journal of Neural Engineering 8 (5) (2011), 056001.

[20] T.K. Moon, The expectation-maximization algorithm, IEEE Signal Processing Magazine 13 (6) (1996) 47–60.

[21] J. MacQueen, Some methods for classification and analysis of multivariate observations, Proceedings of the Fifth Berkeley Symposium on Mathematical Statistics and Probability 1 (14) (1967) 281–297.

[22] B.S. Everitt, Finite Mixture distributions, John Wiley & Sons, Ltd, 1981, ISBN 0-471-90763-4. Wiley.

[23] C.E. Rasmussen, The Infinite Gaussian Mixture Model [C]//NIPS vol. 12, 1999, pp. 554–560.

[24] X. Hu, J. Jiang, D. Cao, et al., Battery health prognosis for electric vehicles using sample entropy and sparse Bayesian predictive modeling, IEEE Transactions on Industrial Electronics 63 (4) (2016) 2645–2656.

[25] B. Md Zakirul Alam, J. Wu, G. Wang, J. Cao, Sensing and decision making in cyber-physical systems: the case of structural event monitoring, IEEE Transactions on Industrial Informatics 12 (6) (2016) 2103–2114.

[26] F.-Y. Wang, The emergence of intelligent enterprises: from CPS to CPSS, IEEE Intelligent Systems 25 (4) (2010) 85–88.

[27] H. Gong, R. Li, J. An, W. Chen, K. Li, Scheduling algorithms of flat semi-dormant multi-controllers for a cyber-physical system, IEEE Transactions on Industrial Informatics 13 (4) (2017) 1665–1680.

[28] F.-Y. Wang, Control 5.0: from Newton to Merton in popper's cyber-social-physical spaces, IEEE/CAA Journal of Automatica Sinica 3 (3) (2016) 233–234.

[29] Q. Zhou, W. Zhang, S. Cash, O. Olatunbosun, H. Xu, G. Lu, Intelligent sizing of a series hybrid electric power-train system based on Chaos-enhanced accelerated particle swarm optimization, Applied Energy 189 (2017) 587–601.

[30] Chen Lv, H. Wang, D. Cao, High-precision hydraulic pressure control based on linear pressure-drop modulation in

valve critical equilibrium state, IEEE Transactions on Industrial Electronics 64 (10) (2017) 7984–7993.

[31] M.C. Kisacikoglu, F. Erden, N. Erdogan, Distributed control of PEV charging based on energy demand forecast, IEEE Transactions on Industrial Informatics 14 (1) (2017) 332–341.

[32] Y. Qin, R. Langari, Z. Wang, C. Xiang, M. Dong, Road excitation classification for semi-active suspension system with deep neural networks, Journal of Intelligent & Fuzzy Systems Preprint (2017) 1–12.

[33] S. Wang, Z.Y. Dong, et al., Stochastic collaborative planning of electric vehicle charging stations and power distribution system, IEEE Transactions on Industrial Informatics 14 (1) (2017) 321–331.

[34] Chen Lv, Y. Liu, X. Hu, D. Cao, et al., Simultaneous Observation of Hybrid States for Cyber-Physical Systems: A Case Study of Electric Vehicle Powertrain, IEEE Transactions on Cybernetics 48 (8) (2017) 2357–2367.

[35] M.J. Mirzaei, A. Kazemi, O. Homaee, A probabilistic approach to determine optimal capacity and location of electric vehicles parking lots in distribution networks, IEEE Transactions on Industrial Informatics 12 (5) (2016) 1963–1972.

[36] C. Lv, H. Wang, B. Zhao, et al., Cyber-physical system based optimization framework for intelligent powertrain control, 2017-01-0426, SAE International Journal of Commercial Vehicles 10 (2017) 254–264.

[37] Y. Huang, A. Khajepour, et al., A supervisory energy-saving controller for a novel anti-idling system of service vehicles, IEEE 22 (2) (2017) 1037–1046.

[38] X. Hu, S.J. Moura, N. Murgovski, B. Egardt, D. Cao, Integrated optimization of battery sizing, charging, and power management in plug-in hybrid electric vehicles, IEEE Transactions on Control Systems Technology 24 (3) (2016) 1036–1043.

[39] G. Cena, I.C. Bertolotti, et al., CAN with eXtensible in-frame reply: protocol definition and prototype implementation, IEEE Transactions on Industrial Informatics 13 (5) (2017) 2436–2446.

[40] C. Lv, J. Zhang, Y. Li, Y. Yuan, Novel control algorithm of braking energy regeneration system for an electric vehicle during safety–critical driving maneuvers, Energy Conversion and Management 106 (2015) 520–529.

[41] L. Martins, T. Gorschek, Requirements engineering for safety-critical systems: overview and challenges, IEEE Software 34 (4) (2017) 49–57.

[42] A.D. Ames, X. Xu, J.W. Grizzle, P. Tabuada, Control barrier function based quadratic programs for safety critical systems, IEEE Transactions on Automatic Control 62 (8) (2016) 3861–3876.

[43] Y. Shoukry, P. Nuzzo, A. Puggelli, A.L. Sangiovanni-Vincentelli, et al., Secure state estimation for cyber physical systems under sensor attacks: a satisfiability modulo theory approach, IEEE Transactions on Automatic Control 62 (10) (2017) 4917–4932.

[44] N. Ding, X. Zhan, Model-based recursive least square algorithm for estimation of brake pressure and road friction, in: Proceedings of the FISITA 2012 World

Automotive Congress, Springer, Berlin, Heidelberg, 2013, pp. 137–145.

[45] G. Jiang, X. Miao, Y. Wang, et al., Real-time estimation of the pressure in the wheel cylinder with a hydraulic control unit in the vehicle braking control system based on the extended Kalman filter, Proceedings of the Institution of Mechanical Engineers – Part D: Journal of Automobile Engineering 231 (10) (2017) 1340–1352.

[46] L. Li, J. Song, Z. Han, Hydraulic model and inverse model for electronic stability program online control system, Chinese Journal of Mechanical Engineering 44 (2) (2008) 139.

[47] J. Zhang, C. Lv, J. Gou, D. Kong, Cooperative control of regenerative braking and hydraulic braking of an electrified passenger car, Proceedings of the Institution of Mechanical Engineers - Part D: Journal of Automobile Engineering 226 (10) (2012) 1289–1302.

[48] J. Yao, Y. Zhang, J. Wang, Research on algorithm of braking pressure estimating for anti-lock braking system of motorcycle, in: IEEE International Conference on Aircraft Utility Systems (AUS), IEEE, 2016, pp. 586–591.

[49] K. O'Dea, Anti-lock Braking Performance and Hydraulic Brake Pressure Estimation, SAE Technical Paper, 2005. No. 2005-01-1061.

[50] H.B. Demuth, M.H. Beale, O. De Jess, M.T. Hagan, Neural Network Design, 2014.

[51] A. Soualhi, M. Makdessi, R. German, et al., Heath monitoring of capacitors and supercapacitors using the neo-fuzzy neural approach, IEEE Transactions on Industrial Informatics 14 (1) (2017) 24–34.

[52] C.M. Bishop, Pattern Recognition and Machine Learning, springer, 2006.

[53] G. Dreyfus, Neural Networks: Methodology and Applications, Springer Science & Business Media, 2005.

[54] M.T. Hagan, M.B. Menhaj, Training feedforward networks with the Marquardt algorithm, IEEE Transactions on Neural Networks 5 (6) (1994) 989–993.

[55] C. Lv, D. Cao, Y. Zhao, D.J. Auger, et al., Analysis of autopilot disengagements occurring during autonomous vehicle testing, IEEE/CAA Journal of Automatica Sinica 5 (2018) 57–68.

[56] A. Mallik, W. Ding, A. Khaligh, A comprehensive design approach to an EMI filter for a 6-kW three-phase boost power factor correction rectifier in avionics vehicular systems, IEEE Transactions on Vehicular Technology 66 (4) (2017) 2942–2951.

[57] H. Zhang, J. Wang, Active steering actuator fault detection for an automatically-steered electric ground vehicle, IEEE Transactions on Vehicular Technology 66 (5) (2017) 3685–3702.

[58] L. Yu, H. Wang, A. Khaligh, A discontinuous conduction mode single-stage step-up rectifier for low-voltage energy harvesting applications, IEEE Transactions on Power Electronics 32 (8) (2017) 6161–6169.

[59] N. Li, T. Misu, F. Tao, Understand driver awareness through brake behavior analysis: reactive versus intended hard brake, in: Intelligent Vehicles Symposium (IV), IEEE, 2017, pp. 1523–1528. IEEE, 2017.

[60] X. Zeng, J. Wang, A stochastic driver pedal behavior model incorporating road information, IEEE Transactions on Human-Machine Systems 47 (5) (2017) 614–662.

[61] B. Lorenzo, F.J. Gonzalez-Castano, Y. Fang, A novel collaborative cognitive dynamic network architecture, IEEE Wireless Communications 24 (1) (2017) 74–81.

[62] C.M. Martinez, X. Hu, D. Cao, et al., Energy management in plug-in hybrid electric vehicles: recent progress and a connected vehicles perspective, IEEE Transactions on Vehicular Technology 66 (2017) 4534–4549.

[63] S. Schnelle, J. Wang, H. Su, R. Jagacinski, A driver steering model with personalized desired path generation, IEEE Transactions on Systems, Man, and Cybernetics: Systems 47 (1) (2017) 111–120.

[64] V. Ivanov, D. Savitski, B. Shyrokau, A survey of traction control and antilock braking systems of full electric vehicles with individually controlled electric motors, IEEE Transactions on Vehicular Technology 64 (9) (2015) 3877–3896.

[65] C. Lv, J. Zhang, Y. Li, Y. Yuan, Directional-stability-aware brake blending control synthesis for over-actuated electric vehicles during straight-line deceleration, Mechatronics 38 (2016) 121–131.

[66] Chen Lv, Y. Xing, J. Zhang, X. Na, Y. Li, T. Liu, et al., Levenberg–Marquardt backpropagation training of multilayer neural networks for state estimation of a safety-critical cyber-physical system, IEEE Transactions on Industrial Informatics 14 (8) (2017) 3436–3446.

[67] S. Haufe, J.-W. Kim, I.-H. Kim, A. Sonnleitner, M. Schrauf, G. Curio, B. Blankertz, Electrophysiology-based detection of emergency braking intention in real-world driving, Journal of Neural Engineering 11 (5) (2014), 056011.

[68] H. Suzuki, Prediction of driver's brake pedal operation in vehicle platoon system: model development and proposal, in: 7th International Conference on Intelligent Systems, Modelling and Simulation (ISMS), 2016, IEEE, 2016.

[69] J. Wang, et al., An adaptive longitudinal driving assistance system based on driver characteristics, IEEE Transactions on Intelligent Transportation Systems 14 (1) (2013) 1–12.

[70] C. Sankavaram, et al., Fault diagnosis in hybrid electric vehicle regenerative braking system, IEEE Access 2 (2014) 1225–1239.

[71] Y. Song, B. Wang, Analysis and experimental verification of a fault-tolerant HEV powertrain, IEEE Transactions on Power Electronics 28 (12) (2013) 5854–5864.

[72] C.E. Rasmussen, The infinite Gaussian mixture model, News in Physiological Sciences 12 (1999).

[73] G. Xuan, W. Zhang, P. Chai, EM algorithms of Gaussian mixture model and hidden Markov model, in: Image Processing, 2001. Proceedings. 2001 International Conference on, vol. 1, IEEE, 2001.

[74] M. Fernández-Delgado, et al., Do we need hundreds of classifiers to solve real world classification problems, Journal of Machine Learning Research 15 (1) (2014) 3133–3181.

[75] L. Breiman, Random forests, Machine Learning 45 (1) (2001) 5–32.

[76] S.R. Safavian, D. Landgrebe, A survey of decision tree classifier methodology, IEEE transactions on systems, man, and cybernetics 21 (3) (1991) 660−674.

[77] L. Breiman, Bagging predictors, Machine Learning 24 (2) (1996) 123−140.

[78] H.B. Demuth, et al., Perception learning rule, in: Neural Network Design. Martin Hagan, 2014.

[79] C. Lv, J. Zhang, Y. Li, Y. Yuan, Mechanism analysis and evaluation methodology of regenerative braking contribution to energy efficiency improvement of electrified vehicles, Energy Conversion and Management 92 (2015) 469−482.

[80] D.H. Wolpert, W.G. Macready, An efficient method to estimate bagging's generalization error, Machine Learning 35 (1) (1999) 41−55.

[81] W.-Y. Loh, Y.-S. Shih, Split selection methods for classification trees, Statistica Sinica (1997) 815−840.

# CHAPTER 8

# Driver Lane-Change Intention Inference

## HOST DRIVER INTENTION INFERENCE

### Introduction

Traffic accidents can give rise to millions of injuries and death each year worldwide. Most of the traffic accidents are caused by human misbehaviors, such as driver cognitive overload, judgment mistake, and operation errors [1−3]. As drivers are within the center of the Traffic-Driver-Vehicle (TDV) loop, a proper understanding of driver behaviors and driving-related intention can largely reduce the number of traffic accidents [4,5]. In the past 2 decades, a large number of advanced driver assistance system (ADAS) products have been implemented on commercial vehicles. Currently, most of the ADAS products, such as Lane Departure Warning (LDW) [6,7], Adaptive Cruise Control (ACC) [8], and Side Warning Assistant (SWA) [9], are designed to provide additional traffic context information to assist the driver during driving. Although these products can be treated as active safety systems, these functions interact with the human driver in a passive manner, which fails to monitor and understand the driver in real time. The active interaction between the human driver and the intelligent units are the major object for the next-generation ADAS products [10,11].

Therefore, in this section, a driver behavior monitoring system toward the lane-change intention inference is proposed. The reasons to understand human drivers and infer their intentions are summarized as follows. First, the main reason for driver lane-change intention inference (LCII) is to increase driving safety. Driver intention inference enables ADAS or intelligent vehicles to focus on the potential context as early as possible. Hence, the chance that the driver assistance system prevents the accident from happening would largely increase. Meanwhile, as reported in Ref. [12], the turn signals are only used in 66% lane changes and less than 50% of the turn indicator activation happens in the initial phase of the lane-change maneuver. Therefore the driver intention inference system would benefit the ADAS to efficiently interact with human drivers. As most of the ADAS tend to share vehicle control authority with the drivers, the LCII system is also helpful to decrease the conflicts between the driver and the vehicle. Finally, a hotpot of the intelligent and automated vehicle study is to design human-like decision-making and vehicle control algorithms. A better understanding of the human driver intention mechanism will contribute to forming a more naturalistic onboard decision system for automated vehicles. In addition, LCII has a close relationship with driver behavior imitation [13,14], which can accelerate the development of the parallel automated vehicle evaluation system [15].

Second, as most of the ADAS or more intelligent autonomy controllers tend to share the control authorities with the driver, the LCII system can decrease the conflicts between the driver and the intelligent vehicle. The mutual understanding between the human driver and the automation is an important task for the construction of highly intelligent shared control strategies. For example, Pentland and Liu describe the human into a device that has a large number of internal mental states and certain control behaviors. If a machine can anticipate human behavior, it can serve humans' needs better [65]. In Ref. [14], driver path planning intention is recognized based on the lateral offset and lateral velocity and is integrated into the shared obstacle-avoidance model predictive controller. However, current integration between the intention system and the shared controller mainly rely on the driver control command such as the steering angle and velocity, which cannot provide an early intention prediction and are unable to fully exploit the cooperate potential of the integrated shared control strategies [14,15]. The intelligent vehicles are believed to have great potential in comprehensive sensing and perception of the surrounding context, and understanding the driver's intention will help the vehicle to generate smart sensing and control assistance strategies for the driver; this can make the intelligent vehicle much easier to be accepted by the public.

Finally, understanding the driver intention mechanism is expected to contribute to a more naturalistic decision-making system for autonomous vehicles and can be used to design the human-like decision-making and behavior generation algorithms [16]. People may argue future autonomous vehicles will not maintain the driver in the loop; however, it is not to

Advanced Driver Intention Inference. https://doi.org/10.1016/B978-0-12-819113-2.00008-7

say the human-like decision-making strategies are not necessary. Learning how human driver generating intentions and making decisions is a long-term task for the automated driving vehicles, as the law-based decision-making methods cannot meet the various situations in the real world. The data-driven intention and decision-making models are expected to be more efficient in the future [55,64–66].

Driver intention inference was widely studied in the past 2 decades from different aspects. Liu and Pentland employed hidden Markov model (HMM) to predict several driver intentions on a car simulator [16]. At that moment, only the vehicle status data such as the steering angle, steering velocity, and the vehicle velocity were used. They reported an average of 88.3% detection rate was achieved after 0.5 s when the action was initiated. As discussed in the next section, the vehicle status data are hard to be used for intention prediction but can be used for intention recognition. After that, Oliver and Pentland proposed another intention prediction model using the coupled hidden Markov model (CHMM) [17]. The driver behavioral signals such as head pose and eye gaze were included along with the front lane and vehicle positions. The CHMM model gave a high prediction accuracy for the start and stop intention 1 s before any significant change in the car, while they achieved 29.4% and 6.3% detection rate for the lane-change left and right maneuver, respectively. The detection rates for lane changes are unreliable in real life at that moment. After these early studies, it was found that a precise intention inference system should rely on a holistic approach, which needs to fuse the multimodal data in the TDV loop together [18,33].

The signals from the TDV loop for lane-change intention come from different sensors. For the traffic context, the most widely used signals are lane markings, surrounding vehicle positions, digital maps, and GPS [19–21]. Vision-based LDW also provides the distance to the lane boundaries, vehicle position, vehicle yaw angle, and road curvature. The driver's behavioral signals consist of the head motion, eye gaze, body gestures, and even the electroencephalogram (EEG) [22–25]. The vehicle status signals are normally collected from the Controller Area Network (CAN), which contains vehicle speed, acceleration, steering wheel angle and velocity, turn signal, and pedal position [12,26,27].

Rafael introduced an approach to predict lane change on the highway. GPS/IMU sensors are used to collect the highway context data [28]. The GPS device can give the location and time information in complex weather conditions, which are more robust than the camera and Lidar devices. Salvucci introduced a four-step-based lane-change intention detection system, which contains data collection, model simulation, action tracking, and thought inference [27,29]. The steering wheel angle, accelerator, and front vehicle position were used as the model input. Schmidt and Beggiato proposed a lane-change intention recognition method based on performing an explicit steering wheel angle mathematical model [30]. In Ref. [12], early driver intention was detected by observing the easily accessible vehicle and traffic signals. The pedal positions and global vehicle position on a digital map were used to identify an intention at a very early stage after the lane change was started. Campbell and Bajcsy use multiple discriminative models to identify three kinds of driver intentions, which were lane keeping, preparation for lane changing, and lane changing [31]. Henning et al. proposed a lane-change intention recognition system, which focused on the analysis of the impact of the environmental indicator on the prediction of lane-change intention [32].

Driver behavioral signals such as the head and eye movement can give earlier clues about the driver's intention. Many studies have evaluated their influence on the intention prediction problem [34–36]. In Ref. [37], the authors used the pupil information as the cognitive signals to predict the lane-change intent. In Ref. [38], the authors proposed a head tracking system for driver lane-change intention inference based on the glance area estimation. Zhou et al. evaluated how driver behavior being affected by the cognitive distraction during lane-change preparation process through the analysis of driver eye movement [39]. They concluded that a secondary task could affect the intention of inference accuracy. In Ref. [40], the authors constructed a queuing network-based cognitive architecture to model driver behavior during normal/emergency lane-change maneuvers and lane keeping. Comparing with those methods based on the eye gaze and head moment estimation, this method can be easily extended into real-world vehicles at a low cost. In Ref. [41], the authors used an HMM as driver lane-change intention prediction. A new feature named CDI was introduced to represent the surrounding context of the host vehicle and the lane-change willing level. Li et al. proposed an integrated intention inference algorithm based on HMM and Bayesian Filter (BF) techniques [42]. A preliminary output from the HMM was further filtered using the BF method to make the final decision. The HMM-BF framework achieved recognition accuracy of 93.5% and 90.3% for the right and left lane change, respectively. In Ref. [43], the authors proposed a lane-change detection method based on the object-oriented Bayesian networks. The system was designed according to the modularity and reusability of the Bayesian network, which makes it easy to extend the system according to different requirements.

McCall and Trivedi designed a lane-change intent inference system using a sparse Bayesian learning (SBL) methodology [44]. The lane tracking, driver head motion, and vehicle CAN bus data are fused and fed into the SBL model. They found that the driver's head pose signals lead to early recognition of the lane-change intent. Following this system, Doshi and Trivedi evaluated the impact of eye gaze and head pose on the recognition of driver lane-change intent [34]. The eye-gaze locations in the front were divided into nine different areas. With the various combination of the input data, they found that the eye-gaze signals were not as informative as the head motion and did not significantly contribute to a precise intention inference. In Ref. [45], a discriminative Relevance Vector Machine (RVM) classifier was used to predict driver lane-change intent with a sliding-time window. The multimodal signals from ACC, SWA, LDW, and head motion were fused. It was found that the LDW system is more useful to predict the intent between 0 and 1.5 s before the lane change while the head motion is more informative between 2 and 3 s before the lane change. The surrounding vehicle position given by the SWA system does not affect the precise intention inference. At the moment of 2 s before the lane-change maneuver, the RVM-based intention inference system achieved 82% prediction accuracy.

Jain et al. [46,47] designed several intelligent algorithms to anticipate future maneuvers for the driver. The autoregressive input—output HMM and recurrent neural network (RNN)-long short-term memory (LSTM) algorithms were used to predict the intention and future maneuvers, respectively. The algorithms take the inside and outside video streams, vehicle dynamics, GPS, and street maps as input signals to anticipate the lane change, turn, and normal driving maneuvers. By comparing the predict results with different machine learning algorithms, they finally conclude that the RNN model leads to the best precision for 90.5% and can anticipate the maneuvers 3.5 s before they occur. However, the strategical plan about the route and the utilization of digital map reduced the system reliability at the unseen street.

## The Framework of Comprehensive Driver Intention Recognition

In this section, a driver lane-change intention framework is introduced. The framework describes the general procedure of lane-change intention and the relationship between each level. This framework also can be extended to describe any other intention inference problem.

### Driver intention inference framework

Michon pointed out that the cognitive structure of human behavior in the traffic environment is a four-level hierarchical architecture, which contains the road user, transportation consumer, social agent, and psychobiological organism [48]. Among these levels, the road user level is directly connected with drivers and can be further divided into three levels: strategy, tactical, and operational level (also known as control level) according to the time constant property. As shown in Fig. 2.9, the strategical level defines the high-level plans for each trip such as the route, destination, and risk assessment. The time constant of this level is much longer than that in the remaining two levels, which can last for minutes or hours. The tactical level tasks are the driving maneuvers that a driver can take during normal driving, such as lane change, lane keeping, brake, and acceleration. The time constant of these maneuvers takes several seconds. Finally, the control level describes the human control actions on the vehicles and stands for the willingness of the driver to remain safe and comfortable in the traffic situation. The time constant at this level is normally in milliseconds [26]. As shown in Fig. 2.9, each tactical intention and maneuver can consist of a series combination of different control actions. Different from the strategical level intention that has explicit original and destination, the tactical intention is more random. A driver has to choose different maneuvers according to the real-time traffic context. Therefore the study of the tactical level intention is more challenging than the higher-level intention.

A general lane-change intention framework is shown in Fig. 2.9. Fig. 2.9 indicates the relationship between the tactical level and the operational level. As discussed in Ref. [49], human intention is generated by some stimuli. For example, the reason to make a lane change normally can be summarized as following a traffic rule or facing an uncomfortable driving context. The traffic context is the key stimuli for most of the lane-change intentions. Once an intention occurs, it can be reflected by different driver dynamics, such as the variation of EEG waves, the head and eye motion for mirror checking, etc. These behavioral signals play a key role in the estimation of the intention. After the drivers determine to execute the intention, they will control the vehicle through the steering wheel and the acceleration/brake pedal. Finally, the vehicle response to these control actions with the variation of vehicle dynamics.

After the analysis of the intention framework, it can be found that before the lane-change maneuver is initiated, the driver has to execute a series of behaviors. The traffic context and driver behaviors are the two important clues for predicting the lane-change intention. The drivers will control the vehicle only after they have decided to finish their lane-change intention. Therefore the vehicle dynamic signals will have a limited contribution to the prediction of the intention. However, these signals are still useful for the recognition of the intention before the vehicle crosses the lane.

Fig. 8.1 illustrates some exemplary sensors systems that can be used for driver intention inference considering the critical moments of lane-change maneuver. The camera-based systems have a lower cost than the Lidar and Radar-based systems, which make them more popular in a real-world application. As illustrated in Fig. 2.9, driver intention mainly occurs based on outer traffic stimuli, which can be destination-oriented or surrounding vehicle influence. Therefore the object detection systems can be used to analyze the driver's intention during the whole process (intention occur—maneuver start—maneuver finish). Based on the cognitive process of driver intention given in Fig. 8.1, it can be found that driver behavior and physical dynamics such as electroencephalogram is believed to be more straightforward than the behavior monitoring as the EEG signals are closer to the real intention generation moment. Then, driver behavior monitoring systems (head tracking, gaze tracking, breath, heart rate, etc.) can be used as the primary systems to recognize the driving intention before the maneuver. Once the maneuver is initialized, the driver normally maintains their hands on the wheel without significant body motion. Hence, at this moment, the global GPS positioning, vehicle dynamics, and local lane position detection become important cues. One of the objectives of this section is to analyze the driver's intention based on

the features that can bring early cues. Therefore only the driver's head pose, eye gaze, and lane marking styles are used in this part.

### Lane-change intention formulation

Based on the intention inference framework in Fig. 2.9, this section will try to solve the following problems. First, to construct an efficient LCII system based on the fusion of multimodal sensors. The primary object of this section is to infer the lane-change intention before the driver initiates the maneuver. Therefore the steering wheel and pedal signals will not be used. Second, as the driver's intention is not an instant detection task, the inference model should be able to process the temporal information before the lane-change start and capture the temporal dependency between the features. Therefore the LSTM-based RNN model is adopted in this section. Next, it is important to understand the naturalistic driver behaviors and analyze the statistic roles within the lane-change maneuver. The statistic results for the mirror-checking moment before the lane change and the lane-change duration are analyzed. Finally, one assumption made in this part is that only the intended lane-change maneuvers are studied. The unintended and aborted lane changes are not considered.

## Methodologies in Driver Lane-Change Intention Inference
### Experimental setup and naturalistic highway data collection

In this part, the experimental data are collected from naturalistic driving on the highway. The vehicle testbed is a commercial sport utility vehicle, which is equipped with multiple sensors. The sensory platform includes three low-cost CMOS cameras and a VBOX. The three cameras are all mounted inside the vehicle cabin. One

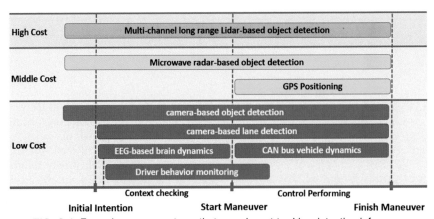

FIG. 8.1 Exemplar sensor systems that are relevant to driver intention inference.

is mounted in front of the driver to capture driver head motion and eye-gaze information. The one that faces the front outside traffic context is mounted on the top middle of the front window. The remaining one is mounted on the sunroof to monitor driver hand motion and only used for reference. A general system architecture is shown in Fig. 8.2.

All three cameras are synchronized with a timestamp and operate at 25 fps. The video streams are recoded at 640 × 480 resolution. Two VBOX antennas are fixed on the outside of the roof ceiling to measure the vehicle dynamics such as the velocity, acceleration, and heading. The sampling frequency for the VBOX data logger is 20 Hz. All the signals are collected using a laptop with an Intel Core i7 2.5 GHz CPU. Three adult drivers with varying ages and experiences have participated in the data collection process. They were asked to drive as usual without telling the real objective of the experiment. Each driver drove the vehicle on the highway for about 1 h and a total of 150 miles of naturalistic data are collected.

### Traffic context and vehicle dynamic features

The road-facing camera captures the front traffic context, focusing on lane detection. The lane detection system is constructed with an edge detector and a Hough transform. This method has been successfully used for lane detection in many studies [6]. Then, the detected lanes are tracked using a Kalman filter. After the lane positions are detected, a line style detector based on the lane sampling and voting scheme is proposed. Three lane styles are known as the solid, double solid, and dashed are recognized. First, an odd number of sampling points are generated in the detected lane position. Then, the line style for the left and right lanes can be recognized by extending the sampling points to short sampling segments. The scanning is proposed on the edge image given by the Sobel edge detector and the lane style is determined by counting the rising edge along each sampling line. The conclusion is made by the voting results of the sampling lines. A detailed description can be found in Chapter 4.

Two vehicular signals are known as the vehicle speed and heading angle are collected using the VBOX. Then, the total feature vector for the outside traffic context and vehicular dynamics time $t$ can be formed as a four-dimensional vector.

$$O_t = [L_r \ L_l \ V \ H] \tag{8.1}$$

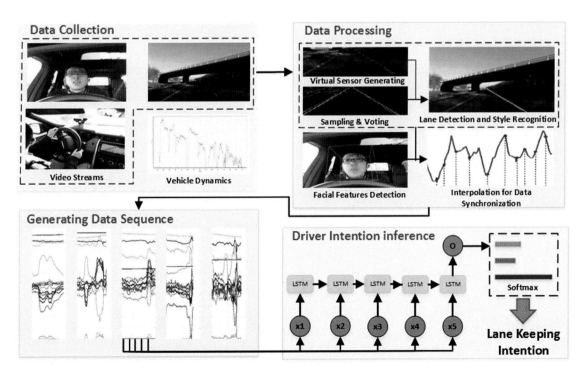

FIG. 8.2 The system architecture of lane-change intention inference system.

where $L_r$ and $L_l$ are the lane style for right and left lanes, and $L \in [-1, 0, 1]$ represents the three different lane styles.

### Driver behavioral features

The inside features for driver head motion and eye gaze are detected using an open-source system, namely, Openface [50]. The driver head position and facial landmarks are detected using the conditional local neural field (CLNF) approach [51]. The CLNFs estimate the 3D head pose angles by projecting the 3D representation of the facial marks to the image plane using orthographic camera projection. The pupil locations for both eyes are detected with the deformable shape registration approach. The gaze directions are estimated according to the pupil locations and the 3D eyeball center [52]. The inside feature vector at each time constant can be formed as follows.

$$I_t = [G_r \; G_l \; G_a \; H_t \; H_r] \tag{8.2}$$

$$\text{Feature Vector} = [O_t, I_t] \tag{8.3}$$

where $G_a$ is the 2D gaze angle in $x$ and $y$ coordinates, $G_r$, $G_l$, $H_t$, and $H_r$ are the 3D gaze direction for each eye, head pose translation vector, and head pose direction vector, respectively. This gives a 14-dimensional vector inside the feature vector. A comparison of the head yaw angles rotation is illustrated in Fig. 8.3.

### Algorithms in Driver Lane-Change Intention Inference

#### Support vector machine

The support vector machine (SVM) belongs to the supervised learning algorithm and has been widely used in data analyzing, pattern recognition, classification, and regression. SVM also belongs to the discriminative model that is designed for binary classification at the beginning and has been developed to be able to solve multiple classes. SVM was first introduced by Corrina and Vapnik in 1955. The theory fundamental of SVM is the Vapnik-Chervonenkis dimension and structural risk minimization, which using optimal theory to determine the best combination of complexity and learning ability. It has its unique advantage in solving pattern recognition problems with small samples, nonlinear, and high dimensional characteristics. Given training data and their label, the SVM model is trained to separate the categories with a maximized gap. When a new

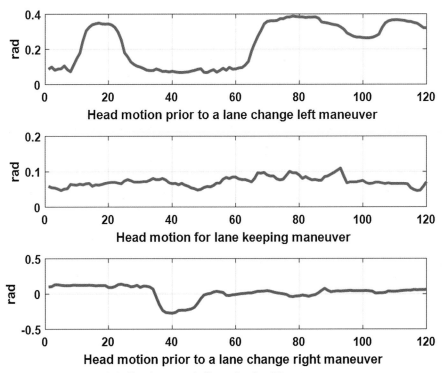

FIG. 8.3 Head yaw angle illustration for different maneuvers.

dataset is fed into the SVM model, it will map the data to its class. SVM tries to find a plane called a hyperplane. In terms of a binary classification problem, SVM determines whether the candidate belongs to $+1$ class or $-1$ class. The standard SVM uses a linear decision boundary like $w^T x_{new} + b$, this line boundary separates the two classes and classify the object after given new data. The decision function can hence be performed like

$$t_{new} = sign\left(w^T x_{new} + b\right) \qquad (8.4)$$

where *sign* is a sign function whose output is either $+1$ or $-1$.

Given a set of training data, the learning task for SVM is to find the best $w$ and $b$ that lead to the maximum margin between the two classes. A good training model has a plane with the largest functional margin as a larger margin will lead to a smaller generalization error. The margin is defined as the distance from the closest points to the decision boundary, as shown in Fig. 8.4.

The margin can be calculated as

$$2\gamma = \frac{1}{\|w\|} w^T (x_1 - x_2) \qquad (8.5)$$

$w$ and $b$ are designed to satisfy

$$w^T x + b = \pm 1 \qquad (8.6)$$

Then,

$$2\gamma = \frac{1}{\|w\|} w^T (x_1 - x_2) = \frac{1}{\|w\|} \left(w^T x_1 + b - w^T x_2 - b\right)$$

$$= \frac{1}{\|w\|} (1+1) \qquad (8.7)$$

To train SVM and find the maximum margin, constrained optimization with Lagrange multipliers method is used:

$$\underset{w}{\mathrm{argmin}} \frac{1}{2}\|w\|^2 \qquad (8.8)$$

Subject to $t_n(w^T x_n + b) \geq 1$, for all $n$.

To solving this problem, a Lagrange multiplier $\alpha$ is introduced to the objective function:

$$\underset{w,\alpha}{\mathrm{argmin}} \frac{1}{2} w^T w - \sum_{n=1}^{N} \alpha_n (t_n (w^T x_n + b) - 1) \qquad (8.9)$$

Subject to $\alpha_n \geq 0$, for all $n$. Detailed information shall see Ref. [67].

In multiple classes, multiple classification planes will be studied from the training data and used to classify these classes. For those multiple classes and two classes that are not easy to classify with linear classification in lower-dimensional, SVM maps the data into higher dimensions by using a kernel function to perform a nonlinear classification and add some terms to $x$ and extend $w$. Then, the optimization and decision function that includes kernel function can be written as follows:

$$\underset{w}{\mathrm{argmax}} \sum_{n=1}^{N} a_n - \frac{1}{2} \sum_{n=1}^{N} a_n a_m t_n t_m k(x_n, x_m) \qquad (8.10)$$

Subject to $\sum_{n=1}^{N} a_n t_n = 0$ and $0 \leq a_n \leq C$ for all $n$.

The new parameter C controls to what extent do we wish the points to sit on the wrong side.

**(A)**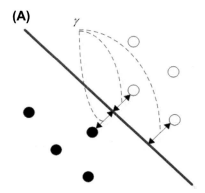

An optimal decision boundary with maximum margin

**(B)**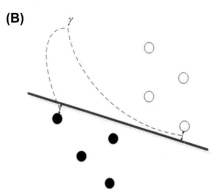

A non-optimal decision boundary

FIG. 8.4 The classification margin defined as the perpendicular distance from the decision boundary to the nearest points in both sides [67].

Three most popular kernel functions are as follows:

$$\text{Linear}: k(x_n, x_m) = x_n^T x_n \qquad (8.11)$$

$$\text{Gaussian}: k(x_n, x_m) = \exp\left\{-\gamma(x_n - x_m)^T(x_n - x_m)\right\} \qquad (8.12)$$

$$\text{Polynomial}: k(x_n, x_m) = \left(1 + x_n^T x_m\right)^{\gamma} \qquad (8.13)$$

SVM has a rich and mature mathematical background. It is also famous for its robustness and ability in accurately dealing with small sample problems. Therefore it has been adopted in much prior research as a powerful way to classify driver intentions.

**A case study of SVM in driver workload estimation.** Measuring driver workload is of great significance for improving the understanding of driver behaviors and supporting the improvement of advanced driver assistance systems technologies. In this part, a hybrid method for measuring driver workload estimation for real-world driving data is proposed based on the support vector regressor to estimate the driver workload in real time. An error reduction ratio causality, a nonlinear causality detection approach, is proposed to assess the correlation of each measured variable to the variation of workload. A full model describing the relationship between the workload and the selected important measurements is then trained via a support vector regression model. Real driving data of 10 participants, comprising 15 measured physiological and vehicle-state variables, are used for the purpose of validation. Test results show that the developed error reduction ratio causality method can effectively identify the important variables that relate to the variation of driver workload, and the support vector regression-based model can successfully and robustly estimate workload.

Intelligent vehicles have been gaining increasing attention from both academia and industrial sectors [70]. The field of intelligent vehicles exhibits a multidisciplinary nature, involving transportation systems, automotive engineering, information technology, energy, and security [71–78]. Intelligent vehicles have increased their capabilities in high, and even fully, automated driving. However, unresolved problems do arise due to strong uncertainties surrounding driving experience and complex driver–vehicle interactions. Before transitioning to fully autonomous driving, driver behavior should be better understood and integrated to enhance vehicle performance and traffic efficiency [79,80].

Measuring driver workload is of great significance for improving the understanding of driver behaviors, and

worthwhile investigating for the purposes of enhancing driver–vehicle interactions [81,82]. The workload indicates the proportion of an operator's limited capacity that is needed to conduct a specific task [83]. Driving tasks also require drivers to allocate certain amounts of physical and cognitive workload. A driver's workload is dynamically varied with their different driving behaviors, including straight-line driving, cornering, U-turns, rapid acceleration and deceleration, shifting gears, and changing lanes. Furthermore, the level and the variation of drivers' workload that are affected by the earlier behaviors could be also influenced by many subjective and objective factors, including driving skills, driving styles, trip objectives, personal tendencies, gender, road conditions, traffic conditions, and so on.

A lot of studies focusing on the measurement and estimation of drivers' workloads have been conducted via different methodologies in recent years. In Ref. [84], a method of quantifying driver's workload with five discrete levels was proposed by the subjective measurement of vehicle data. In Ref. [85], the correlation between distraction condition and drivers' mental load was investigated. The results showed that three variables, namely the driver's left-pupil size, skin conductance, and pulse-to-pulse interval could be used for efficiently identifying a driver's distraction. In Ref. [86], the driver's subjective mental workload and the multiple task performance were modeled through a proposed queuing network method. In Ref. [87], drivers' workloads under lane-change maneuvers were investigated through driving simulations. During simulations, the drivers were required to verbally rate the level of their workloads. In Ref. [87], the drivers' physiological information, including the electrocardiogram (ECG) signals, eye blinking, pupil diameters, and head rotational angles, was measured and used to estimate the workload in a driving simulation scenario. Although the earlier studies provided multiple potential means to estimate drivers' workload quantitatively by using various physiological signals of drivers, all these studies were performed in the simulation environment, which could not reflect the real driving situations and therefore replicate and measure the impact of the potential uncertainties.

Outside of the simulation environment, driver workload estimation with data from the real driving environment has also been studied. Analysis of drivers' workload was conducted using the driving data of a real vehicle [88]. EEG data were collected by a sensor mounted on the driver's head during actual driving conditions. The experimental data showed that the EEG signals increased when the vehicle speed went over a threshold limit. Moreover, with respect to the driving scenarios, the

EEG signals tended to rise with left cornering and downhill. However, the EEG measurement is very sensitive to external disturbances [89,90]. In this analysis, the original EEG data were used without filtering noise signals caused by vehicle vibrations and other factors, which may affect the reliability of results. Nevertheless, the existing research in driver workload estimation is mainly in the stage of driving simulations, and the methodology of measuring workload with real vehicle data is still very challenging, it is still worthwhile improving this.

In this section, the SVM will be used to estimate the continuous driver workload value, which plays a role as the regression model. In terms of using SVM to do regression tasks, Vapnik [68] pointed out that the goal for the support vector regression model ($\varepsilon - SVR$) is to have at most $\varepsilon$ deviation between the actual target data and the predicted value. Meanwhile, the linear decision boundary $f(x) = w, x + b$ should be as flat as possible. Therefore by applying a soft margin, the loss function for Support Vector Regression (SVR) can be represented as follows:

$$\min \frac{1}{2}\|w\|^2 + C \sum_{i=1}^{l} (\xi_i + \xi_i^*) \qquad (8.14)$$

$$\text{s.t.} \begin{cases} y_i - \langle w, x_i \rangle - b \leq \varepsilon + \xi_i \\ \langle w, x_i \rangle + b - y_i \leq \varepsilon + \xi_i^* \\ \xi_i, \xi_i^* \geq 0 \end{cases} \qquad (8.15)$$

where $\xi_i$ and $\xi_i^*$ are the slack variables in the soft margin loss function that allows error cases larger than $\varepsilon$. $C > 0$ controls the trade-off between the flatness of function $f(x)$ and the tolerate ratio for the cases that have larger deviations than $\varepsilon$. To efficiently solve the optimization problem, the linear decision function can be further nonlinearized with kernel function, which has the following form.

$$K(x_i, x_j) = \varnothing(x_i)^T \varnothing(x_j) \qquad (8.16)$$

where $\varnothing(x)$ is the mapping function that maps the raw data points into higher dimensions [69]. In this part, a specific kernel function named radial basis function is adopted and represented as follows.

$$K(x_i, x_j) = (-\gamma \|x_i - x_j\|^2), \quad \gamma > 0 \qquad (8.17)$$

According to the aforementioned concept, the penalty term $C$ and the kernel function parameter $\gamma$ are two of the most important parameters that determine the model parameters training of the SVM. Therefore in the following, a genetic algorithm will be used to iteratively searching the optimal $C$ and $\gamma$ that can maximize the classification margin.

A genetic algorithm is a specific form of evolutionary algorithm inspired by the process of natural population selection. Genetic Algorithm (GA) has been widely applied for optimization and path searching in previous studies. The major parts of GA contain encoding, selection, crossover, and mutation. Specifically, encoding is a process of genetic representation that describes the SVM parameters using multibit binary values like genes. Each individual in the population group represents one possible value of the SVM parameter pair. The selection process selects individuals with good fitness and passes their genes to the next generation. To increase the gene diversity, crossover aims to make the genes better by randomly combining two well-fitted individuals and the mutation prevent the optimization results from being blocked in the local minimum according to the mutation probability. As the fitting function is hard to be described using a single mathematic equation in this part, the training error is adopted in this part to stand for the fitness. Specifically, the candidate $C$ and $\gamma$ are decoded and fed into the SVM, the training error, which is the average difference between the predicted workload values and the ground truth values will be used as the fitness to evaluate whether the parameter pairs are optimal or not. Some important parameters for GA are illustrated in Table 8.1.

The procedure of model estimation is summarized in Table 8.2.

Assessing the drivers' workload in a simulator study is hardly possible because drivers always know that they are navigating through a virtual world. Using the proposed method, this section analyzed a public dataset [29] collected through a real-world driving study. The physiological states of the participants, including skin conductance response (SCR), hand temperature, and heart rate using ECG, were recorded. The GPS position, brightness level, and acceleration were also recorded. Two cameras were used to record the driver's view onto the road, and the participants were asked to rate the workload offline based on these videos to provide

| TABLE 8.1 Parameters for GA-Based SVM Model. | |
|---|---|
| Maximum Generation | 100 |
| Size of Population | 30 |
| Crossover Probability | 0.4 |
| Mutation Probability | 0.01 |
| C Boundary | [0.1 100] |
| $\gamma$ Boundary | [0.01 1000] |

**TABLE 8.2**
**The Procedure for Combining SVR and GA.**

| | |
|---|---|
| 1. | Input: train data with label |
| 2. | Initialize GA parameters and generate the first population |
| 3. | for $i = 1, 2, \ldots$, Max generation |
| 4. | Decoding chromosomes |
| 5. | Computing fitness using SVR for all population |
| 6. | Select population according to individual fitness |
| 7. | Crossover and mutation to create offspring |
| 8. | end for |
| 9. | Find the best model parameters |
| 10. | Train SVR with optimized parameters |
| 11. | Test SVR regression model |
| 12. | Output: performance index and optimized parameters for SVR |

the baseline. The video rating of workload is in the range of 0−1000. A value of 1000 indicates a maximum workload. Ten participants (3 females, 7 males) aged between 23 and 57 years took part in these experiments.

A data preprocessing step was undertaken before applying the developed method. As the sampling frequency of each measure is different, the data have been resampled at the frequency of 1 Hz. There are 1515 (25 min and 15 s) data points for each variable. An example of the selected measures and the video rating for participant 1 is shown in Fig. 8.5.

SVR models for 10 participants were trained based on the selected seven indicators that are latitude GPS, longitude GPS, altitude GPS, speed GPS, ECG, SCR, and temperature. The boundary of the parameter C was in the range of [0.1, 100] and $\gamma$ was in the range of [0.01, 1000]. The maximum generation was set as 100, the size of the population was set as 30, crossover probability was set as 0.4, and mutation probability was set as 0.01. The notion of these parameters can be seen in Ref. [28]. For each group of data, 600 data points were randomly selected for the testing purpose and the remaining points were utilized for training. By using the genetic algorithm, the optimal C and $\gamma$ for 10 participants were estimated and are given in Table 8.3.

Fig. 8.6 illustrates the estimation of workload based on SVR for all the 10 participants. Note 600 randomly

selected data points (about 1/3 of the total data) for each participant were used to test the trained SVR model. As shown in the figure, SVR generates precise estimations of workload for all participants, demonstrating that SVR is a robust and reliable estimator for observing the workload of different subjects.

To evaluate the performance of different input combinations on the workload estimation, three different scenarios were tested and compared: (a) human body features only (ECG, SCR, and temperature); (b) GPS signals only (latitude, longitude, altitude, and speed); and (c) GPS and human body features. The model performance of each participant, represented by the value of R2 (Pearson correlation coefficient) between the model prediction and recorded data, is shown in Table 8.4.

According to the results in Table 8.4, the average value of R2 with the workload estimation based on the selected three human body features only is 0.70, while the average value of R2 under the estimation based on the four GPS signals reaches 0.83. This leads to an interesting conclusion that vehicle state measurements are more relevant to driver workload estimation than those physiological signals. It should be noted that some GPS features (e.g., latitude, longitude, altitude) describe the position of the vehicle. In other words, they reflect the road condition, which is subjective. The participants rate the workload based on the video, which captures the road condition and traffic condition. It is therefore not surprising to observe that GPS features perform well in prediction. The final target of this research is to use the human body features to estimate the workload. Therefore this part is interested in the difference of R2 between human body features and GPS & human body features. It is observed that the combination of human body features and vehicle's GPS information would construct more relevant feature vectors for estimating driver workload, with an overall improvement of R2 of 0.2. It has also been observed that this difference varies between different participants. For participants 2 and 10, the difference is smaller than 0.05, which indicates that the selected human body features are sufficient to describe the workload. For participants 4 and 5, the difference is larger than 0.3, which indicates that the selected human body features are not sufficient and more features (e.g., motion, eye gaze) should be considered to better estimate the workload.

### Hidden Markov model

HMM is a very popular statistic tool for modeling time-series probability problems. It is a kind of dynamic Bayesian network, which is first introduced by Andrei

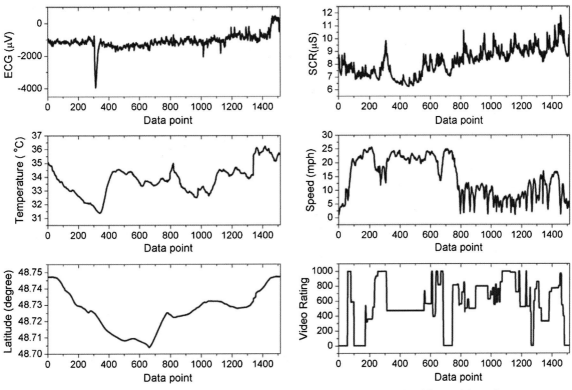

FIG. 8.5 Selected measures and the video rating representing the workload for participant 1.

## TABLE 8.3
### The Identified Parameters for the SVR Models.

| Participants | C | $\gamma$ |
|---|---|---|
| 1 | 97.18 | 48.75 |
| 2 | 84.60 | 64.54 |
| 3 | 76.22 | 73.57 |
| 4 | 69.59 | 130.36 |
| 5 | 93.94 | 73.03 |
| 6 | 90.09 | 133.23 |
| 7 | 85.79 | 74.56 |
| 8 | 85.82 | 145.90 |
| 9 | 14.80 | 70.63 |
| 10 | 84.42 | 65.78 |

Markov in his mathematical theory of Markov process in the early 20th century. Then, in the 1960s, Baum and his colleagues developed the Markov process theory and created HMM. After then, HMM has become a powerful theory in the area of signal processing, speech recognition, behavior recognition, and fault diagnosis. In HMM, an observation sequence is assumed to be directly influenced by a hidden state at each moment and the hidden states are distributed based on a Markov process. In terms of the basic HMM structure, observation at the current time step is only assumed to depend on its hidden state without any relationship with the past observation. However, for some complex or extended HMM, this assumption is not necessary. A simple HMM structure is shown in Fig. 8.7:

The shaded nodes are observations while the unshaded nodes are hidden states. As can be seen from Fig. 8.7, the most important characteristic of HMM is it is a time-series process and it is defined to meet the requirement of first-order Markov model in which the current state only depends on the nearest past state:

$$P(X_4|X_1, X_2, X_3) = P(X_4|X_3) \qquad (8.18)$$

Moreover, the Joint probability of HMM is

$$P(\{X_t, Y_t\}) = P(X_1)P(Y_1|X_1)\prod_{t=2}^{T}P(X_t|X_{t-1})P(Y_t|X_t) \qquad (8.19)$$

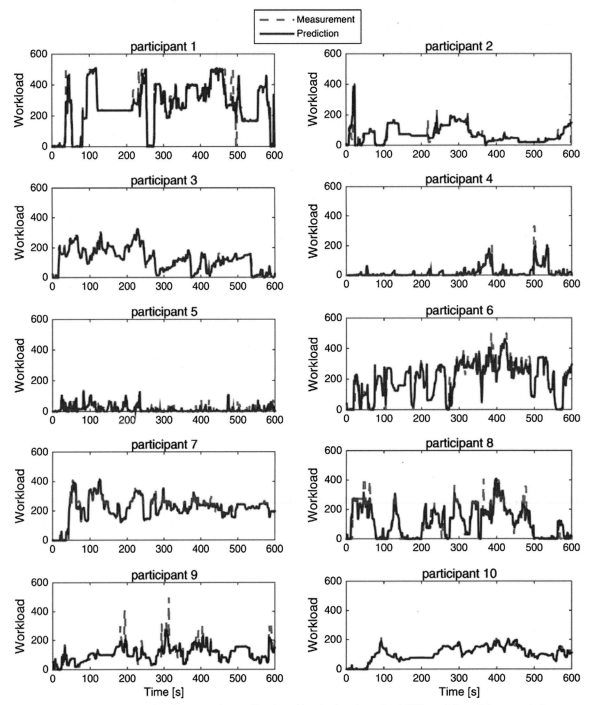

FIG. 8.6 The comparison between the predicted workload using the trained SVR models and the recorded video rating for 10 participants.

**TABLE 8.4**
Comparison of Model Performance With Different Input Combinations.

| | | $R^2$ | |
|---|---|---|---|
| Participant | Human Body Features | GPS Features | GPS and Human Body Features |
| 1 | 0.68 | 0.85 | 0.91 |
| 2 | 0.87 | 0.67 | 0.92 |
| 3 | 0.81 | 0.91 | 0.97 |
| 4 | 0.54 | 0.86 | 0.88 |
| 5 | 0.38 | 0.69 | 0.74 |
| 6 | 0.72 | 0.89 | 0.94 |
| 7 | 0.72 | 0.83 | 0.94 |
| 8 | 0.75 | 0.86 | 0.90 |
| 9 | 0.62 | 0.77 | 0.83 |
| 10 | 0.95 | 0.96 | 0.97 |
| Average | 0.70 | 0.83 | 0.90 |

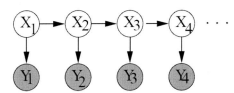

FIG. 8.7 A simple illustration of HMM architecture [91].

The formal equation definition of HMM is

$$h = (A, B, \pi) \tag{8.20}$$

where A is the state transition matrix that defines the transition probability between the states. B is the observation matrix, which defines the probability of the observation given a certain hidden state. $\pi$ is the initial probability matrix that describes the initial state and used for initializing HMM. Given a group of hidden states, observations, A, B, and $\pi$, an HMM can be constructed [92].

A very common problem for an HMM is to infer the current hidden state according to the observations. This is known as a decoding process of HMM and a well-known algorithm called the Viterbi algorithm is particularly suitable for this problem. The general process of utilizing HMM is first trained in the HMM according to the data both in a supervised way or unsupervised way (Baum–Welch algorithm). HMM can be used to

solve the decoding and evaluation problem (compute the probability of a certain set of observations) after the best architecture and parameters are determined.

As mentioned earlier, by connecting the observation with hidden states, HMM describes the process of human behavior to some extent. Human intention or their mental thoughts can be viewed as the hidden states and their behavior is observed. According to this, HMM has been widely used in many human intent recognition studies. Pentland and Liu [16] first apply HMM in reasoning driver intent according to the observed behavior. HMM can give an acceptable result in driver intention inference, however, some authors point out that HMM is not the perfect way to fit the driver model and more complex algorithms should be applied [93].

### Recurrent neural network

In this section, the recurrent neural network basics and the long short-term memory cell are introduced. The RNN model is used to learn the temporal dependency between the input data, and the LSTM efficiently increases the performance of the RNN model by capturing the long-term context dependencies.

As driver intention inference is not an instance detection task, previous driver behavioral data need to be taken into consideration. Therefore the recurrent neural network is applied in this section to process the sequential inputs. RNN allows exhibiting the dynamic temporal behavior of a sequence by forming a directed connection between previous states and the current state [53–55]. Fig. 8.8 illustrates a basic structure of the RNN model, and the right-side model in the graph is the unfolded version of the left circuit diagram. Current state $s_t$ can be viewed as the memory of the RNN network; it stores the information that happened in all the previous time steps.

The RNN model can be described as follows.

$$s_t = f(W_x x_t + H_s s_{t-1} + b_x) \tag{8.21}$$

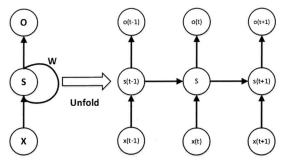

FIG. 8.8 A simplified recurrent neural network architecture.

$$o_t = softmax(W_o s_t + b_o) \tag{8.22}$$

where $f$ is the activation function of the hidden states, which normally is selected as *tanh* and *sigmoid* function. $X$, $S$, and $O$ represent the input, hidden states, and the output of the RNN, respectively. $W_x$, $H_s$, and $W_o$ are the model parameter matrix for the input, hidden states, and output. $b_x$ and $b_o$ are the bias vector.

To reduce the total number of parameters, the RNN share the same $W_x$, $H_s$, and $W_o$ at each time step. This means that the RNN executes the same task with different input values at each step. Although the number of training parameters is reduced with the parameter-sharing scheme, RNN still suffers another serious problem, namely, the gradient vanishing or exploding [56]. Therefore the simplified RNN structure has limited memory ability to remember long-term dependency.

*Long short-term memory*
Hochreiter and Schmidhuber developed the LSTM cell to overcome the drawbacks of RNN [57]. The LSTM-RNN solves the long-term dependency problem by introducing three extra gates, known as the input gate, forget gate, and output gate. The central idea behind LSTM is that the gates in the LSTM cell cooperate with each other to control how much information should be remained and forgot. Fig. 8.9 shows the LSTM architecture. The LSTM RNN still follows the chain-like structure as shown in Fig. 8.8 [58]. The difference is that LSTM RNN replaces the hidden unit (normally a *sigmoid* or *tanh* activation function) with an LSTM cell.

The LSTM cell has the following mathematic representation. First, the forget gate controls what

information to throw away. Then, the input gate chooses new information to be updated and stored. The output gate controls the candidate layer output.

$$f_t = \sigma(U_f x_t + W_f s_{t-1} + b_f) \tag{8.23}$$

$$i_t = \sigma(U_i x_t + W_i s_{t-1} + b_i) \tag{8.24}$$

$$o_t = \sigma(U_o x_t + W_o s_{t-1} + b_o) \tag{8.25}$$

The value $\tilde{c}_t$ is the candidate cell state that can be represented as

$$\tilde{c}_t = tanh(U_c x_t + W_c s_{t-1} + b_c) \tag{8.26}$$

The $c_t$ in the center is the internal memory cell state of the LSTM unit, which is the combination of previous $c_{t-1}$ and current candidate states.

$$C_t = f_t * c_{t-1} + i_t * \tilde{c}_t \tag{8.27}$$

Finally, the layer output is the products of the cell state $C_t$ and the candidate output from the output gate.

$$s_t = o_t * tanh(C_t) \tag{8.28}$$

where $\sigma$ in the above equations represents the *sigmoid* function. $*$ is element-wise production. $x_t$ and $s_{t-1}$ are the current input vector and the previous layer output. $f$, $i$, and $o$ are the forget gate, input gate, and output gate. $U$, $W$, and $b$ are the corresponding model parameters.

The input $x_t$ for the LSTM-RNN at each time step is the concatenation of the inside and outside feature vector $x_t = [I_t, O_t]$, which is an 18-dimensional vector in this part. Hence, each training sample for the LSTM-RNN is an 18-D temporal sequence and the total dataset can be formed as $\left\{ (x_{t-n}, x_{t-n+1}, \cdots, x_t)_j, y_j \right\}_{j=1}^{N}$, where $N$ is the total number of the training sample. $y_j$ is the intention label for this sequence. In this part, the target values are manually labeled according to the video stream, which has three candidate values as {*lane change left, lane change right, lane keeping*}. Moreover, 135 lane-change maneuvers are detected according to the experiment video with 65 lane-change left cases and 70 lane-change right cases. Then, 66 normal driving sequences are randomly picked. The LSTM-RNN model is trained with MATLAB deep learning toolbox.

## Performance Evaluation
In this section, statistical analysis is proposed based on the critical moment of lane-change maneuvers. Next, lane-change intention inference results using different algorithms are described.

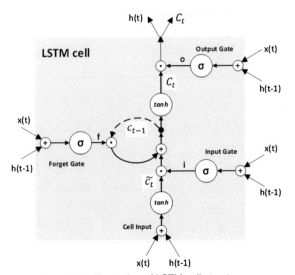

FIG. 8.9 Illustration of LSTM cell structure.

## Driver lane-change maneuver analysis

As shown in Fig. 2.14, four critical moments for one lane-change maneuver can be identified, which are the intention occur point (denoted as T1), maneuver start point (T2), lane crossing point (T3), and maneuver finishing point (T4), respectively. At T1, the driver generates a lane-change intention according to the traffic context stimuli. Most of the time, the specific moment when the driver arises an intention is undetectable. Therefore T1 is replaced by the first mirror-checking moment in the following analysis. At T2, the driver uses the turn signal to indicate their lane-change intention and then turn the steering wheel. Finally, T3 and T4 are the moments that the vehicle just crosses the lane and finishes the lane-change maneuver, respectively.

In this section, the statistical analysis of the gap between T1 and T2, T2 and T3 for all the lane-change maneuvers is proposed. The time interval between T1 and T2 (denoted as T1−T2) measures time cost for the lane-change preparation, and the interval between T2 and T3 (denoted as T2−T3) measures how long it takes to cross the lane after the driver starts the maneuver.

In Fig. 8.10, the time distributions of T1−T2 and T2−T3 are fitted with the lognormal distribution function. According to the statistical results shown in

| **TABLE 8.5** Descriptive Statistics of the Lane Change Preparation and Execution. | | |
|---|---|---|
| **Statistics** | **T1−T2** | **T2−T3** |
| Mean | 6.085 s | 1.881 s |
| SD | 3.033 s | 0.530 s |
| Variance | 9.200 s$^2$ | 0.281 s$^2$ |
| Median | 5.437 s | 1.805 s |
| Mode | 3.060 s | 1.890 s |
| Maximum | 22.423 s | 3.680 s |
| Minimum | 1.843 s | 0.930 s |

*SD is short for standard deviation.

Table 8.5, the drivers in the dataset would perform their first mirror-checking behavior about 6 s before they start the lane-change maneuver. The average time cost between initiate the lane change and just crossing the lane is about 1.88 s. Therefore to predict the lane-change intention before the driver turns on the signal or the steering wheel, a 6 s window is sufficient to cover the important driver behavior features. After the driver

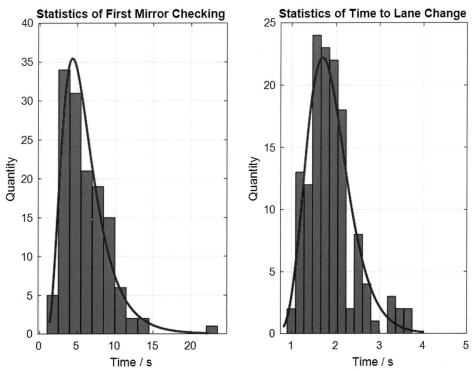

FIG. 8.10 Statistic results for critical time intervals of lane-change maneuver.

starts the lane-change maneuver, the detector has nearly 2 s to recognize an intended lane change. From the dataset it can be found that mirror-checking behaviors can occur during the lane-keeping maneuver; however, the duration for this mirror-checking behaviors is much shorter than that during the lane-change preparation process. Meanwhile, drivers tend to perform single-side mirror-checking behaviors multiple times during the lane-change preparation step instead of performing both sides checking behaviors as they do during the normal lane-keeping process.

### Lane-change intention inference results

In this section, we examine the performance of the intention inference model from two aspects, which are the detection accuracy and the prediction horizon. The prediction horizon measures the different inference results along with the time gap before the maneuver. In Ref. [46], the intention prediction results given by various machine learning models have been richly studied. The authors found that the performance of the RNN model overweighs the other models such as the IOHMM, HMM, and SVM. Therefore in the following

parts, only an feedforward neural network (FFNN) model that trained in a different manner compared to the method in Ref. [46] is used for comparison purposes. To predict the lane-change intention before the maneuver, the temporal sequence that is up to 6 s before the maneuver is selected. As all the drivers are asked to drive as usual, some lane-change maneuvers are not indicated by the turn signal. Therefore the start point of the maneuver is marked as either turn on the signal or the first moment of turning the steering wheel. To train the LSTM-RNN model, 70% of the data are randomly selected for model training and the test data are used for testing. Fig. 8.10 illustrates the confusion matrix for the intention of inference results using LSTM-RNN with complete features.

As shown in Fig. 8.11, the general inference accuracy for the three maneuvers is 96.7%. The left lane-change intentions are all correctly predicted before the driver initiates the lane-change maneuver. For the right lane change and straight driving intention in the testing dataset, each one has one misclassified sample. Different from [47], which trains the nontemporal discriminative model by concatenating the feature vector at each step

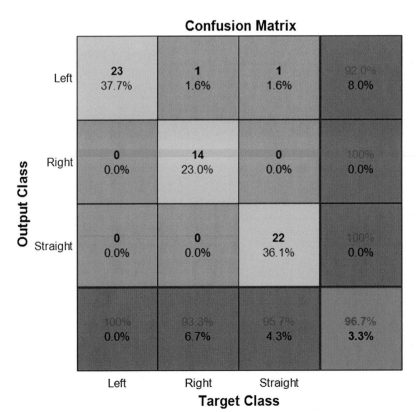

FIG. 8.11 Confusion matrix for lane-change intention inference using LSTM-RNN.

into a large feature set, in this section, the FFNN is trained with a much smaller feature vector. For each training sample, the 18-D temporal sequence data are transformed into a $72 \times 1$ feature vector. The mean, standard deviation, maximum value, and minimum value for the 18 channels are computed. The 72-dimensional feature vector is much smaller compared with the 3840-dimensional vector in Ref. [47]. The intention prediction results using the FFNN is shown in Fig. 8.12. The general performance of the FFNN model is 91.8%. Although the detection rate using the discriminative model is lower than the LSTM-RNN model, the proposed FFNN still shows an acceptable ability on the intention prediction task and achieves a better result compared with the other discriminative models in Ref. [46]. In addition, as shown in Fig. 8.12, the inference accuracy for the left lane-change intention is still 100%, which means that for these two different models, the left lane-change intention is the easiest to be detected. One of the most important reason is that when preparing the left lane-change maneuver (right-side driving vehicle), the driver has to check the far-end left-side mirror or the rear mirror by rotating their head and check multiple times

instead of only gaze searching as they did in many right lane-change maneuvers, which makes the system easier to detect the corresponding behaviors and the inner intention.

The model prediction performance for the RNN model and FFNN model is evaluated using the sliding window method. Specifically, the testing sequence is shifted back for every 0.5 s, which means that the intention is predicted every 0.5 s before the maneuver. As shown in Figs. 8.13 and .8.14, an ensemble learning-based method, which combines three bi-directional LSTM models [62] achieved the most accurate results among several baseline methods. The LSTM-based approach is more accurate than the FFNN model when inferring the intention 1 s early. In contrast, the FFNN model gives a more robust detection when the intention is inferred more than 1 s.

This may be due to the fact that as the testing sequence move earlier, more irrelevant information is involved in the temporal sequence, which would confuse the RNN model on the intention of inference tasks. On the other hand, the FFNN model is trained with the statistic features of the sequence. Therefore the FFNN may still be able to capture the significant

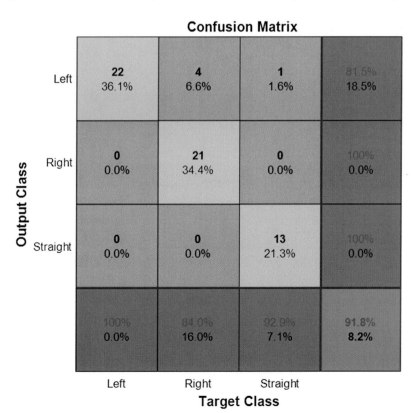

FIG. 8.12 Confusion matrix for lane-change intention inference using FFNN.

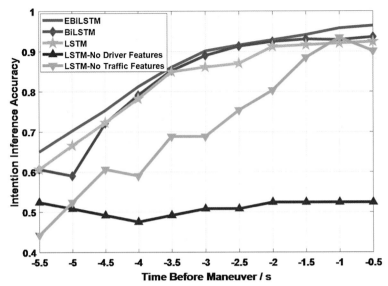

FIG. 8.13 Prediction performance versus time-to-maneuver for lane change with the LSTM-RNN model.

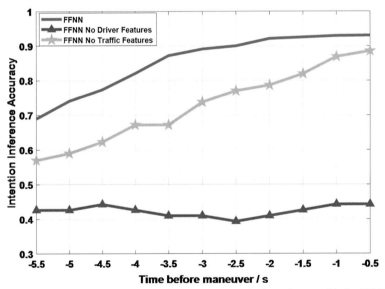

FIG. 8.14 Prediction performance versus time-to-maneuver for lane change with the FFNN model.

features within the temporal sequence. The impacts of the driver behavioral data, as well as the traffic context data on the intention inference, are also evaluated. Based on the results from the two different methods, the traffic context features do not contribute to accurate detection of the driver's intention before the maneuver. On the contrary, the model based on the driver's behavioral features achieved slightly lower accuracy than the model with a full feature vector, which means the

driver's intention can be roughly detected according to the driver's intended checking behaviors.

Fig. 8.15 illustrates the intention of inference accuracy with respect to the prediction time. In this part, the models are trained with completed sequence data, while, at the testing step, the testing data are cropped. The results indicate the intention prediction results by observing partial temporal context. As the testing sequence getting smaller, more important features

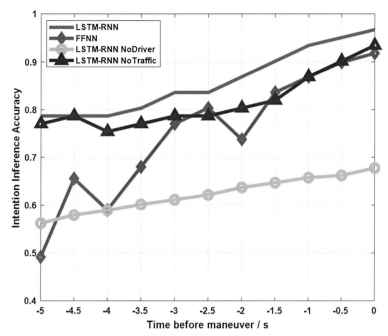

**FIG. 8.15** Prediction performance versus time-to-maneuver for lane change with the incomplete sequence.

along with the lane-change preparation duration will lose. The purple line and the yellow line in Fig. 8.15 represent the results by using driver behavior features only and using traffic context and the vehicle dynamic features only. As shown in Fig. 8.15, the driver behavioral features are more important than the traffic context features. The red line indicates the prediction results given by the FFNN model. Different from the temporal sequence-based methods, the FFNN accuracy drops more rapidly as the testing sample getting smaller. This means that the FFNN model is more sensitive to the partially observed dataset.

Fig. 8.16 illustrates the real-time intention inference results with the ensemble learning-based LSTM model as shown in Fig. 8.13. The testing dataset was expanded about 5 s before and after a lane-change maneuver to simulate a whole inference cycle for the lane-change intention. The normal lane-keeping intent is labeled with number three while number two and number one for right lane change and left lane change, respectively. The first three rows represent the left lane change, and the middle three rows are the right lane change. The bottom three rows indicate the lane-keeping processing. Here the lane-change initiate point is manually selected as the first mirror-checking behavior that before a lane-change maneuver and the finished point is determined as the moment when the vehicle is just crossing the lane. As shown in Fig. 8.16, the LSTM-based LCII system

can efficiently detect the lane-change intention in a very early stage after the driver generates the mirror-checking behaviors. As the RNN models take the sequence data as inputs, the lane-change intention signal can still be generated even after the vehicle has crossed the lane. However, this should be a significant concern in the real world as the LCII system can be integrated with the LDW system so that to increase recognition accuracy. Moreover, as the intention inference tasks are more important than this status recovery issue, the main objectives for this section are to concentrate on the fast and early inference for the lane-change intention.

### Discussions and Perspectives

In this section, the driver intention inference (DII) system is proposed based on the naturalistic driving data on the highway. It has been found that it is possible to infer human driver intention before they initiate the lane-change maneuver. However, as far as we concern, a few works are expected before the driver's intention inference can be implemented into the real-world system.

First, the intention inference system in this section assumes the human drivers always concentrate on the driving tasks during the experiment. The experiment duration for each participant is about 1 h, which can make sure the drivers are not overloaded. However, in the real world, the DII system should not be working

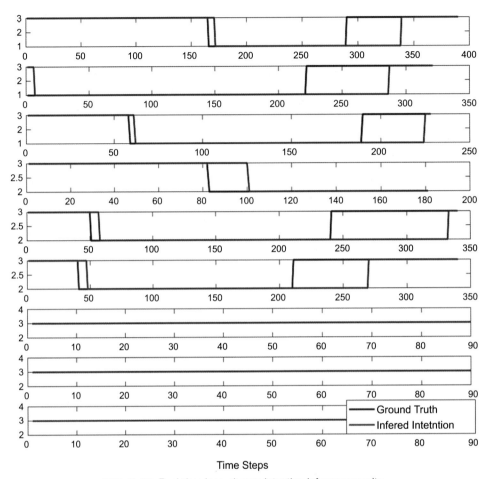

FIG. 8.16 Real-time lane-change intention inference results.

as an isolated function. The system needs to cooperate with other driver assistance systems to make an accurate prediction result. The most relevant system is the driver workload estimation module. It is found that when the drivers are overloaded, their behaviors change significantly, and the driver workload is highly related to the intention inference system [59]. Therefore an integrated system that combines the driver workload estimation and intention inference system is needed for real-time driver assistance.

Second, as the vehicle intelligence increasing significantly nowadays, more and more driver assistance system will share the control authority with the human driver. What happens if the drivers are performing the secondary tasks and how to correctly infer their intention? The DII system has to clarify the driver status before estimating the true intention. By considering driver distraction and intention has a whole, the control

conflicts between the driver and the vehicle can be minimized. Meanwhile, for those partially automated vehicles, the estimation of secondary tasks will help the decision system to determine the driver's intention and whether they are capable to take over the control on some emergency tasks [60,61].

Finally, more challenge work is to have a more comprehensive understanding of the cognitive intention generation process according to the traffic context and human behaviors. By analyzing the driver's intention with respect to the current traffic context, a human-like intention generation system can be proposed. This knowledge will benefit the design of the decision-making system of automated vehicles.

## Conclusions

In this section, a driver lane-change intention inference is proposed based on the LSTM-RNN model. The experimental data are collected on the highway environment in the real world. A general framework for intention inference framework is designed. According to the framework, the outside traffic context is viewed as the stimuli for driver intention, while the driver behaviors and the vehicle dynamics are the corresponding responses to the intention. Based on the statistics of the experiment data, it is found that the driver tends to perform the first mirror-checking behavior 6 s before the lane-change maneuver and it takes around 2 s for the vehicle to cross the lane. An LSTM-RNN and FFNN model is used for intention inference. The results indicate that the RNN model achieved an average of 96.7% inference accuracy before the maneuver being initiated. The LSTM-RNN model is proved more accurate and robust with partially observed data than the FFNN model. Future works will concentrate on the comprehensive analysis of the driver intention and its integration with other driver status detection system.

## LEADING VEHICLE INTENTION INFERENCE-TRAJECTORY PREDICTION

Forecasting the motion of the leading vehicle and the corresponding driving intention is a critical task for connected autonomous vehicles as it provides an efficient way to model the leading–following vehicle behavior and analyze the interactions. In this section, a personalized time-series modeling approach for leading vehicle trajectory prediction considering different driving styles is proposed. The method enables a precise, personalized trajectory prediction for leading vehicles with limited intervehicle communication signals, such as vehicle speed, acceleration, space headway, and time headway of the front vehicles. Based on the learning nature of human that human always tries to solve problems based on grouping and similar experience, three different driving styles are first recognized based on an unsupervised clustering with a Gaussian mixture model (GMM). The GMM generates a specific driving style for each vehicle based on the speed, acceleration, jerk, time, and space headway features of the leading vehicle Then, a personalized joint time-series modeling (JTSM) method based on the long short-term memory (LSTM) recurrent neural network model (RNN) is proposed to predict the front vehicle trajectories. The JTSM contains a common LSTM layer and different fully connected regression layers for different driving styles. The proposed method is tested with the next-generation simulation (NGSIM) data on US101 and I-80. The JTSM is tested for making predictions 1 s ahead.

Results indicate that the proposed personalized JTSM approach shows a significant advantage over the other algorithm.

### Introduction

Surrounding driver intention inference is a difficult task as early cues like driver behavior signals are not available to the ego driver. Hence, the driving intention of the surrounding driver cannot be estimated before the maneuver. However, it is still possible to estimate the driver's intention based on the observation of time-series driving patterns and trajectory. For example, the following driver will understand the lane-change maneuver of the leading driver by estimate the short-term driving trajectory. The driving behaviors of the surrounding vehicles can be estimated by the vision-based perception module on the vehicle, also, it can be obtained based on the vehicle-to-vehicle communication technique. In this part, a surrounding driver driving intention recognition module is designed, which focuses on using the easily acceptable driving behavior data to recognize the leading driver's intention.

Connected vehicles and vehicle-to-vehicle (V2V) communication techniques are designed for safer and more efficient transportation, which plays a key role in constructing next-generation road mobilities and transportations [94,95]. Furthermore, the V2V technique will significantly influence the design of future intelligent and automated driving vehicles [96,97]. The final vision of V2V is to benefit all road entities such as buses, vehicles, motorcycles, and pedestrians to safely use the public road and transport based on short-range and long-range broadcast [98].

Although V2V techniques are expected to largely increase road safety and traffic efficiency in the next decades, how to maximize the influence of V2V techniques based on the limited short-range signal transfer capability and the quantities of signals to enhance the scene understanding ability of the intelligent vehicles are still an open question [99]. For example, as shown in Fig. 8.17, the following vehicles are interested in predicting the future trajectories of the leading vehicles so that to understand the future motion of the leading vehicles better. Conventional methods to predict the leading vehicle trajectory have to estimate its motion based on vehicle perception methods. However, this approach is more complex and less efficient than the V2V technique that the leading–following pairs can share brief status information.

Therefore this section aims to provide an efficient, personalized traffic context understanding algorithm

FIG. 8.17 Leading–following pair communication example.

based on the joint estimation of driving styles and future vehicle trajectory for the connected vehicles. As the nature of the human learning process is to always rely on grouping similar tasks and solving these tasks with similar experience, in this section, we developed a similar human-like reasoning scheme for the connected vehicles toward high precise motion prediction. Different driving styles are unsupervised learned in the first; then, it can be used to group vehicles with similar behaviors together. Separate prediction networks with a shared temporal feature extraction layer can be used to improve the final prediction accuracy. Based on the proposed system, the connected vehicles will obtain a more informative understanding of the nearby vehicles based on the limited transferable vehicle data through the V2V network. Moreover, the prediction accuracy for the future trajectory prediction of the leading vehicle can also be further improved.

The high-level system architecture is shown in Fig. 8.18. In the bottom layer, the leading vehicle data, as well as ego-vehicle data, are collected and processed. Then, an unsupervised clustering module based on GMM is applied to extract three different driving styles for the leading vehicle according to the collected vehicle information. Once driving styles are determined, the total training data will be used to train a common temporal pattern extraction layer based on the LSTM. Then, for each group, a personalized fully connected regression layer is trained to estimate the vehicle trajectory in each group.

In Ref. [100], Lefèvre et al. proposed a well-organized survey study about motion modeling and prediction approaches, which can be separated into three different groups, namely, physics-based motion models [101,102], maneuver-based models [103,104], and interaction-aware motion models [105,106]. Specifically, physics-based motion prediction models represent the vehicle according to the laws of physics and make

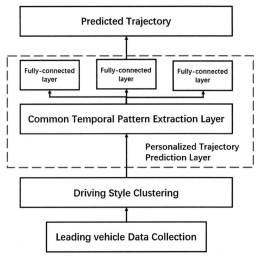

FIG. 8.18 High-level system architecture.

trajectory prediction based on dynamic and kinematic models. Maneuver-based models, on the other hand, represent the vehicles as independent entities and the motion consists of a series of discrete maneuvers. Lastly, the interaction-aware models represent the vehicles as maneuver entities, which can interact with each other according to the traffic laws. Considering the dependence between different vehicles and the traffic context enables a longer-term and more precise motion prediction of the surrounding vehicles compared with the physics-based and maneuver-based methods. Therefore most current motion and trajectory prediction studies focus on the interaction-aware based approaches.

Most of the early interaction-aware algorithms use dynamic Bayesian network (DBN) to learn the mutual dependencies between the multiple entities. In Ref. [107], an integrated physics-based and maneuver-based trajectory prediction method was proposed based on the Bayesian network. A generalized time to critical probability was used to validate arbitrary uncertainty multiagents driving context and long-term prediction. In Ref. [108], a hierarchical DBN model, which consists of three dynamic layers, namely the observation layer, dynamic layer, and context layer, was proposed. For each vehicle, the dynamic layer stores the low-level information such as the pose, orientation, and velocity based on the onboard sensors. Then, the context layer captures the high-level semantic information such as the relative location concerning nearby vehicles. In Ref. [109], an interaction-aware intention and maneuver prediction framework were proposed for complex traffic scene understanding. The motion intentions of surrounding vehicles were estimated iteratively based on the game-theoretic algorithm, and spatiotemporal cost maps are used for predicting surrounding vehicle intention considering the traffic context. In Ref. [110], three types of predictive features were designed for the DBN to recognize future vehicle maneuvers, which were road structure-based feature (existence of lane and lane curvatures), interaction-aware feature (state of adjacent and leading vehicles), and physics-based features (vehicle dynamics). In Ref. [111], an agent-sensitive motion prediction method, namely the general agent motion prediction model for autonomous driving (GAMMA) was developed based on Bayesian inference learning. The GAMMA predicts the multi-agent's motion by integrating the different kinematic and geometry features and generate hypotheses trajectories by inferring the human intention with velocity and acceleration characteristics.

Some of the existing studies have analyzed the effects of LSTM to the surrounding vehicle motion prediction. Inspired by the social LSTM architecture given by Alexandre [112], many studies have successfully transferred this method into the vehicle motion prediction task. For example, in Ref. [113], an LSTM encoder–decoder model was proposed for surrounding vehicle maneuver and trajectory prediction. An improved convolutional social pooling scheme was developed for the maneuver-based decoder. The model generated the multi-modal distributions for the future motion of the surrounding vehicles based on the predicted maneuvers. In Ref. [114], a multiagent tensor fusion network was proposed to model the social interaction behavior between a varying number of vehicles on the highway.

Although some interesting studies have been proposed in the past for vehicle trajectory prediction, few studies consider the impact of different driving styles on the trajectory prediction. Hence, in this section, a personalized track prediction algorithm based on the modeling of driving styles is proposed. The contribution of this section can be summarized as follows.

First, an unsupervised driving style recognition method for connected vehicles on the highway is proposed. This section focuses on the analysis of the driving style of the leading vehicle with V2V communication, which uses the limited and easy to access signals to measure the driving styles. Then, a joint deep LSTM-based RNN network is proposed for the trajectory prediction of different types of vehicles. Different fully connected regressions networks are proposed for different styles with a shared LSTM layer. The joint framework can decrease model volume and increase prediction accuracy. The personalized JTSM method is comprehensively evaluated and tested with different prediction horizons and multiple metrics.

The remainder of this section is organized as follows. Section II introduces the GMM-based driving style generation. Section III introduces the deep learning scheme based on the RNN. In Section IV, the experimental design and results are proposed. Finally, this section is concluded in Section V.

### Driving Style Recognition Based on GMM

In this part, the GMM unsupervised clustering method is applied to generate the most distinctive driving styles for the connected vehicles. GMM can be represented by a weighted sum of sub-Gaussian components [115]. Each cluster is modeled according to different Gaussian distribution function.

The GMM can be described as the following merged distribution function:

$$p(x_i|\theta) = \sum_{K=1}^{K} \pi_k N\left(x_i|\mu_k, \sum_k\right) \quad (8.29)$$

where $x_i$ is the data point and can be multiple dimensions, $\theta$ is the parameters of the GMM, which can be represented as $\theta = \{\pi_k, \mu_k, \sum_k\}$, $\pi_k$ are the weight of each component Gaussian distribution function and the sum of $\pi_k$ equals to one. $\mu$ and $\sum$ are the mean and covariance parameter of multivariate Gaussian function, $K$ is the total number of components in the model, and $N\left(x_i|\mu_k, \sum_k\right)$ is the univariate Gaussian distribution function in this case with the following form.

$$N\left(x|\mu, \sum\right) = \frac{1}{(2\pi)^{D/2}\left|\sum\right|^{1/2}} \exp\left[-\frac{1}{2}(x-\mu)^T \sum{}^{-1}(x-\mu)\right]$$

$$(8.30)$$

The GMM model can be trained with the expectation-maximization maximum likelihood estimation algorithm, which computes the maximization of the cost function iteratively.

Most of the existing driving styles recognition systems were designed to recognize the host vehicles in the past [116], while, in this section, the impact of the GMM on the recognition of surrounding driving styles based on V2V communication is developed. The GMM-based driving style recognition system can learn the leading or the surrounding vehicle styles in an unsupervised manner. Based on the publicly available data that are allowed to be transferred through vehicles, vehicle velocity, acceleration, space headway, and time headway of learning vehicles were extracted. Moreover, the jerk information, which is the derivative of acceleration along time, is also adopted. The vehicle jerk is calculated as follows.

$$Jerk(t) = \frac{acc(t + \Delta t) - acc(t)}{\Delta t} \quad (8.31)$$

To increase the diversity as well as the robustness of the driving style recognition algorithm, the statistical measurement such as the standard deviation (STD), mean, maximum, and minimum for these features are used. The whole feature vector for the GMM is a 16-dimensional vector, as shown in Table 8.6.

The rear vehicle will take 3 s to observe the dynamics of its leading vehicle and generate the corresponding style label based on the GMM. Therefore the GMM is trained with the statistical values of the above feature vector in the length of 3 s. An assumption made here

is that the driving style for each vehicle will not change during the whole driving cycle. This is true in the real world as the driving style is the reflection of driving habits, which can hardly be changed [117]. When the predicted trajectories are significantly different from the observed position of the learning vehicle, the algorithm can reestimate the driving style and predict the vehicle track based on a more proper subnetwork.

Based on the driving style recognition analysis, the different driving styles are visualized. The visualization for different driving styles is shown in Figs. 8.19 and 8.20, respectively. As shown in the two figures, the three driving styles show different significant distributions on the most representative feature space.

### Joint Feature Learning and Personalized Trajectory Prediction

In this section, the recurrent neural network and the long short-term memory unit are introduced. The RNN model is used to learn the temporal dependency within the time sequence data, while the LSTM can increase the performance of the RNN model by introducing the memory and forget gate unit to control the information flow.

#### Recurrent neural network and LSTM

The LSTM-based RNN is applied in this section to process the sequential trajectory data and generate a future prediction. RNN exhibits the dynamic temporal pattern of a sequence of data by forming a directed connection between previous states and the current state [54,118]. Current state stores the information of the previous hidden states and the past inputs. A major problem of vanilla RNN is the so-called gradient vanishing or exploding, which makes RNN unable to capture the long-term dependency and information for the sequential data [56]. Hochreiter and Schmidhuber developed the LSTM cell to overcome the drawbacks of RNN [57]. The LSTM-RNN solves the long-term dependency problem by introducing three extra gates, known as the input gate, forget gate, and output gate. The gates cooperate to control the information flow [57]. The chain-like LSTM RNN model replaces the hidden states with the LSTM cell.

#### Joint time-series model construction

The JTSM model is trained based on the following—leading pairs extracted from the NGSIM dataset. The data contain the trajectory data from lane one to lane six of the US-101 and I-80 freeway. There are 5574 following—leading pairs used in this section. The sequential data contain the six following vehicle

**TABLE 8.6**
**Statistics of the Features Used for Driving Style Recognition.**

| Features | STATISTIC VALUES | |
|---|---|---|
| Velocity | Velocity STD | Mean velocity |
| Acceleration | Acceleration STD | Mean acceleration |
| Jerk | Max jerk<br>Jerk STD | Min jerk<br>Mean jerk |
| Space headway | Max space headway<br>Time headway STD | Min space headway<br>Mean time headway |
| Time headway | Max time headway<br>Time headway STD | Min time headway<br>Mean time headway |

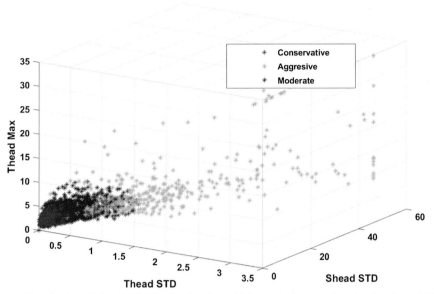

FIG. 8.19 Visualization of the distribution for the three driving styles concerning time headway STD, time headway Max, and space headway STD.

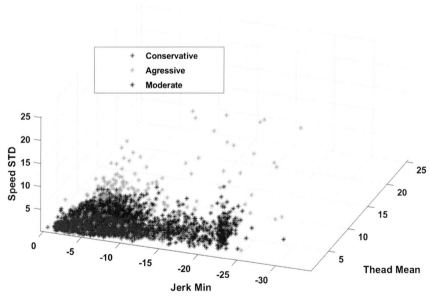

FIG. 8.20 Visualization of the distribution for the three driving styles concerning time headway means, speed STD, and Jerk Min.

features and four leading vehicle features. The features used in this section are shown in Table 8.7.

The sequential feature extraction layer of JTSM is constructed with LSTM cells, while the personalized

trajectory prediction layer contains three different fully connected regression neural networks. The training objective of the personalized JTSM is to estimate the leading vehicle trajectories based on the easily

**TABLE 8.7**
**The Feature Vector for the Construction of JTSM.**

| Vehicle | Features |
|---|---|
| Following | Lateral position, longitudinal position, speed, acceleration, Space headway, time headway |
| Leading | Lateral position, longitudinal position, Speed, acceleration |

accessible vehicle states of the leading vehicle through real-time communication. The common temporal pattern extraction layer based on the LSTM is first trained with the whole set of training data to increase the diversity of the state. Then, according to the different driving styles, different fully connected networks are trained and fine-tuned separately with the common LSTM layer based on transfer learning. The whole leading–following pairs are used to generate the driving style categories.

There are 80% of the data that are randomly selected as the training data, and the rest is used for model testing. Among the 4458 training data, 362 cases are classified into the conservative part, 1644 cases belong to the moderate group, and the rest 2429 samples are

classified into the aggressive part. For the conservative group model training, the 362 samples are too few to train a precise model. Therefore data augmentation is applied to the third part, which uses a random sampling method with a random sampling rate selected between 0.75 and 0.9 for each iteration. For each sequence, the sampling unit randomly generates a number within the range of the sampling rate and randomly extracts the data, which is a subsequence of the original sequence. The augmentation is taken four times, and the final data volume of the third group is four times of the original data. The whole networks are developed with MATLAB 2019a. A detailed training and testing process is illustrated in Fig. 8.21. The testing data are only used for model testing and evaluation to prevent data leakage issues.

### Experimental Results

In this section, the statistical analysis of the trajectory prediction is made based on different evaluation metrics and different prediction horizons.

#### Evaluation metrics and baselines

The results of the model are compared based on multiple metrics. Specifically, the Euclidean distances between the predicted and the ground truth trajectories are evaluated with the root of the mean square error (RMSE) format. The worst 5% and 1% of cases reported for the tracks are also analyzed. In addition, the modified Hausdorff distance (MHD) method is also adopted

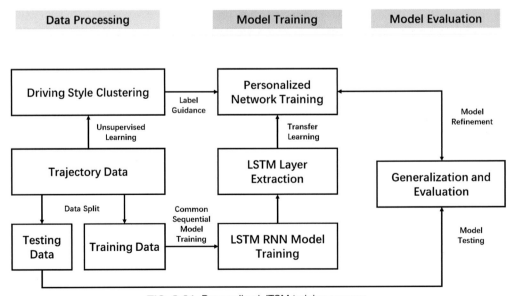

FIG. 8.21 Personalized JTSM training process.

[119], following with the worst 5% and 1% records for the MHD scores. The general MHD is the mean of the MHD for the lateral position and longitudinal position. The general RMSE measures for the tracks are calculated as

$$RMSE = \sqrt{\frac{1}{N}\sum_{i=1,2,\cdots N}\left((\widehat{x}_i - x_i)^2 + (\widehat{y}_i - y_i)^2\right)} \quad (8.32)$$

where $n$ is the total number of steps, $\widehat{x}_i$ and $\widehat{y}_i$ are the lateral and longitudinal values of the predicted trajectory.

Several baselines are used to make a comparison between the existing algorithms and proposed methods. The baselines include the following:

**1. Constant Kalman (CK) Filter.** The path of the leading vehicle is predicted with a constant velocity Kalman filter model.

**2. LSTM.** A single LSTM-RNN model is trained to predict future tracks without considering the driving styles.

**3. Multiple LSTM (MLSTM) for multiple classes.** Three different LSTM-RNN models are trained based on the driving styles without using the pretrained common LSTM layer. The training data for each group is used separately to train the models.

**4. JTSM.** A joint time-series model based on the shared LSTM layer and further fully connected regression networks for the different driving styles. First, a common LSTM temporal pattern network is trained. Then, for each driving style, the concatenate networks are fine-tuned.

*Performance evaluation*

In this part, the one-second ahead track prediction is proposed and compared within the algorithms above. The track prediction results for different driving styles are proposed with 1 s prediction horizon. The track prediction performance of different models is first quantitively analyzed in Table 8.8, then the exemplar track illustration is visualized in Fig. 8.22.

Fig. 8.22 indicates the trajectory prediction on some exemplary cases for the three different driving styles with a one-second prediction horizon. It is shown that the proposed JTSM method can generate a good trajectory estimation for different kinds of drivers. It is also important to understand whether different driving styles have a different impact on the estimation precision of future trajectories. Therefore track prediction results for different driving styles are compared. As shown in Table 8.8, the proposed JTSM method achieved the most accurate results than the other methods. It is shown that the joint temporal pattern extraction layer

**TABLE 8.8**

**Comparison of the Lateral Position Prediction Results for Different Methods.**

| Evaluation Metrics | CK | LSTM | MLSTM | JTSM |
|---|---|---|---|---|
| Eu RMSE | 1.419 | 0.764 | 0.682 | 0.568 |
| Eu 5% worst | 3.604 | 1.783 | 1.950 | 1.661 |
| Eu 1% worst | 5.595 | 3.150 | 3.443 | 3.029 |
| MHP mean | 0.058 | 0.176 | 0.189 | 0.132 |
| MHP 5% worst | 0.281 | 0.400 | 0.514 | 0.322 |
| MHP 1% worst | 0.505 | 0.514 | 0.718 | 0.469 |

along with the personalized prediction layer achieved more precise prediction than the separate LSTM networks.

**Conclusions**

In this section, a personalized leading vehicle trajectory prediction based on joint time series modeling is proposed for connected vehicles. By analyzing the leading vehicle status through vehicle-to-vehicle communication and driving style recognition, future trajectories can be estimated. The personalized JTSM algorithm generates three different driving styles for the leading vehicle; then, a joint LSTM temporal pattern extraction layer is used by the three different track prediction networks. With the transfer-learning-based JTSM algorithm, the track prediction achieved better results than the separate multiple networks and single LSTM-based methods. The track prediction for the leading vehicle achieved a 0.568-m RMSE result, which shows the efficiency of the proposed personalized JTSM method. The final results indicate that predicting future vehicle status based on the different groups is more efficient than predicting the status with a uniform model.

# MUTUAL UNDERSTANDING-BASED DRIVER–VEHICLE COLLABORATION

## Introduction

Driver intention inference plays a critical role in driver–vehicle collaboration as it enables the automation to analyze driver behaviors and understand his/her intention so that the automation can better interact with drivers based on this mutual understanding method. However, knowing driver intention along cannot design an intelligent and efficient driver–vehicle collaboration and driver–vehicle team system as there are many other

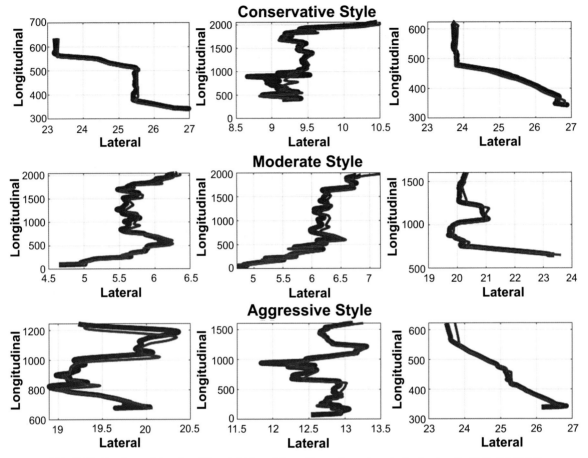

FIG. 8.22 Track prediction for different driving styles with x-axis and y-axis for the longitudinal and lateral position in the meter. The *red lines* are the target tracks, while the *blue lines* are the predicted tracks.

aspects that need to be considered. Hence, in this chapter, a brief literature review is proposed toward the basics of driver–vehicle collaboration and multiagents' teamwork. Then, based on the human–agents teamwork scheme, two collaboration manners are analyzed, which are "hard" collaboration, namely take-over control, and "soft" collaboration, namely shared control. The "hard" take-over collaboration enables the vehicle control authority switch so that vehicle automation can do repeat works for a human driver. On the contrary, the "soft" shared control allows the driver and vehicle to collaborate in a dynamic fashion so that the more complex vehicle control can be achieved. This part is simply a literature review based to show a preliminary discussion on how driver intention inference can benefit the driver–vehicle collaboration system and future autonomous driving technology.

### Literature Review

In Ref. [120], the author reviewed the intent of communication between human and autonomous agents. It is shown that intent communication plays a critical role in human–agent teamwork and collaboration. Moreover, the shared situation awareness through transparency user interface improves the efficiency collaboration for the human and highly automated and intelligent agents. Besides, the misperception of the human member can increase the possibility of trust degradation and unnecessary intervention as intent are not well shared. It is pointed out that the development of shared situation awareness for user and nonuser framework is the most important and difficult task for the efficient human–agents collaboration and teamwork. In Ref. [121], Sycara and Sukthankar proposed an initial review about the human–agents collaboration teamwork model.

The teamwork styles were roughly classified into three categories, which are human-only, agent-only, and human−agent teamwork. Cohen [122] defined the agent teamwork as "a set of agents having a shared objective and a shared mental state." In contrast, the human teams have a similar but slightly more complex definition according to Salas [123], which is "a distinguishable set of two or more people who interact dynamically, interdependently, and adaptively toward a common and valued goal/objective/mission."

The human teamwork can be further divided into three dimensions according to Cannon-Bowers [124], which are cognitions, skills, and attitudes. Specifically, cognition represents the information and understanding of the team's mission, tasks, objectives, norms, and resources. Team skills include adjustability, adaptability, communication pattern, leadership, and interpersonal coordination. Last, the attitude represents the percipient's feelings about the team, task, team beliefs, mutual trust, team cohesion, and the importance of teamwork. According to Ref. [121], the distinct agents' roles can be identified based on the functionality of the agents within the team, which are agents supporting individual team members to complete their own task, agents supporting the team as a whole, and agents working as an equal team member with the human operators. They identified the trust concept as the fundamental building block of effective human−agent teamwork and the shared knowledge between the human operators and agents promote mutual understanding between them. When the agents working in the mode of supporting an individual team member, there are normally no other humans involved in the task, and no direct communication is made with other humans. When the agents support the whole team, the agents usually assist the whole team members in the group by aiding communication, coordination, and focus of attention. Last, if the agents working as a team member individually and can assume equal roles with other team members, it is critically a challenge to obtain a comparable competency with human operators. Finally, regarding the human−agent interaction, three major aspects are identified, which are the mutual predictability of teammates, team knowledge and shared understanding, and direct ability and mutual adaptation. Team knowledge cannot be represented by a unitary concept, instead, it refers to a series of knowledge (shared understanding, collective cognition, shared cognition, shared mental models, etc.) that need to be communicated between the team members. Similarly, mutual predictability also enables the teammate to communicate and interact with each other about their ongoing tasks and motions. Detectability means to allocate roles and responsibilities to the team members and mutual adaptation refers to the manner that team members announce and adjust their roles to satisfy the requirements of the team object.

In Ref. [125], Chen and Barnes proposed a comprehensive review and discussion on the human−agent (H-A) teaming for multirobot control. It shows that within the H-A teams, the agents are always acting as a subordinate member, who can be controlled to work autonomously only under specific conditions and environment. Then, five different types of intelligent agents were discussed, which are teaming agents, hierarchical agents, adaptive automation, adjustable automation, and mixed-initiative systems. The adaptive systems assign tasks under specified conditions for the human operator and the agents before the mission. A trigger is needed to indicate the transition of the control authority between the automation and the human driver. In contrast, the adjustable automation defines several automation levels for the mission and can select the level of autonomy based on the types of the subtask. Lastly, the mixed-initiative system enables a joint action and decision-making between the human operator and the agents for the mission. The authors also reviewed the trust mechanism of the H-A team. It shows that the agent's predictability is a critical aspect of trust development and maintenance. The trust is mostly influenced by the agent's performance and human−agent interaction manner. Moreover, the trust among the H-A was also analyzed from the human side, it shows that human-related factors such as the evaluation errors and intent errors can also affect the trust of the agents. Another important factor that significantly influences the effectiveness of the H-A team is human situation awareness (SA). As for the H-A teamwork, the human operator can perform secondary tasks if the agent can handle the primary task well, SA recovery and improvement function for the human is needed when designing the interface. Finally, the authors developed a framework for the guidance of the H-A team interface design. Several major functions for the interface are identified as flexible human−agent interaction, maintaining operators' ultimate decision authority, support operators' multitasking performance, automation transparency, visualization, and training techniques to enhance H-A collaboration, and human individual differences must be part of the human−agent design process.

In Ref. [126], the authors reviewed the fundamental issues of the human−autonomy teaming scheme and a particular focus on the flight deck operations. Based on the experiences from human—human teaming, the authors identified seven key aspects for the human—autonomy interaction, which are team communication, shared mental models, mutual performance monitoring, backup behavior, team adaptability, team leadership, and team orientation and mutual trust. Team communication is reviewed as an information-sharing process rather than simply a matter of pushing information. An information-sharing scheme should be constructed in the case to let human pilot pull information from the automation, as well as provide information feedback to the automation. The compatible mental models within the human and automation about the understanding of the teamwork and operation environment is strong guidance for effective team performance. It is always critical and effective if the human operator and the agents can provide their own understanding when significant changes and actions are made. The automation should provide monitoring about the inconsistent input from the human operator and suggest and implement alternatives on critical events. In addition, the automation is required to have the ability to explain its behavior and reasons for this intervention. The automation should be able to assist the human operator when being required. In addition, it has to respond adaptively to changing tasks according to the variation of environmental conditions.

In Ref. [127], Bradshaw et al. argued that it is not proper to view the H-A collaboration as a process of allocating tasks, instead, the primary concern should be how the task can best be shared by both the human and the agent in concert. Hence, according to the study of Fitts in 1951 [128], the human surpasses machine in the ability of detection, perception, judgment, induction, improvisation, and long-term memory. On the contrary, the machines surpass the human in the ability of speed, power, computation, replication, simultaneous operation, and short-term memory. In Ref. [127], the current H-A team system can be categorized into two major groups. The first one is software agents networked multiagent systems that help address data-to-decision problems based on an intelligent user interface for monitoring, analyzing, and management. The second group is robots and autonomous machines with software agents that are embedded in specialized hardware for a specific usage (search, rescue, space, etc.). Bradshaw et al. further emphasized that the interdependency in joint H-A activities should not only focus on the task allocation for the human and agents. The interdependency among the task and the team members is a really important issue that determines

the team interactions, such as predictability, observability, and directability. In Ref. [129], the authors showed that when designing an H-A team system, it is always more important to let the automation recognize the interdependency between the teammates and the coactivity of the team as a whole than improving the autonomy of the agent system.

In Ref. [130], driver situation awareness study was proposed based on the transition of four different levels of autonomy (fully autonomous, autonomous steering, autonomous speed control, and no automation). Five emergent scenarios, which contain two car cut-off events, a car stop event, and two pedestrian incursion events were used to simulate the driving performance. Driver situation awareness was measured by a time to initiation an evasive measure in response to a potential accident. Moreover, the trust of the autonomous car and comfort was measured based on a questionnaire approach. Specifically, it was found that the reliability, feeling of safe, level of confidence, desire for control, worry, and discomfort were the highly representative features for trust measurement. Besides, the passengers' comfort feeling to the car can be measured by the comfort, enjoyability, annoying, feelings of comfort, calmness, and relaxation while the car drove. The final results indicate that the 48 participants show significant high trust and comfort on the fully autonomous car when encountering the emergent events, while drivers with autonomous speed control perform the worst reaction time from autonomous to manual driving. Moreover, most of the participants submit a low score on the trust and confidence of the autonomous speed control, which shows that the autonomous driving or autonomous steering makes the driver feel that the car can control the critical events well.

Based on the 10 levels of autonomy defined in Ref. [131], and [132], the authors in Ref. [133] emphasized that it is the primary step to define a criterion that allows the selection of the participant for specific subtasks when two or more agents have formed a team. Three main aspects to study the H-A cooperation domain are proposed, which are the identification of the capabilities of each agent to be involved in the task, the identification of the capabilities of each agent to share the knowledge and communicate information with each other, and the identification of the criteria to define the work allocation and split. Considering the different levels of operation and task control, three cooperation levels can be further identified, which are cooperation in activity level, cooperation in plan level, and meta-cooperation at the meta-level. The authors in Ref. [134] reviewed several classical areas for multiagents' interaction. It is shown that a primary application issue for multirobot system is the interdependency communication between the agents. It is a challenging task that needs to be solved toward an efficient multirobot system. Another challenge for such a system is the energy consumption issue that needs to be optimized to maximize the usage efficiency of the overall system energy consumption.

In Ref. [135], the authors evaluated the individual differences in H-A teaming and particular focus on the attention control differences. The eye movements dynamics such as fixation count, average fixation duration, blink rate, saccade amplitude, and pupil diameter were used to evaluate the relationship between eye movement and situation awareness as well as the workload. The eye movements of each participant on three different autonomy tasks (manual, semiautonomous, fully autonomous) were analyzed. It was shown that there is a linear relationship between the workload and level of autonomy. In addition, there is no significant difference in the eye movement between the semiautonomous mode and the fully autonomous mode. However, correlation analysis between SA and eye movement indicates a significant positive correlation exists between the fixation count (FC) and SA in the manual and fully autonomous condition, while an obvious negative correlation occurs between the FC and SA in the semiautonomous condition. Finally, it shows that the participants who have higher attentional control scores generate fewer fixations across all conditions than those with low attentional control scores.

In Ref. [136], the authors claimed that human–machine interaction cannot be regarded as the same notion as collaboration. Human–machine collaboration requires a mutual understanding between the team members, preemptive task management, and shared progress tracking. In Ref. [137], Bratman analyzed the shared cooperative activity and summarized three key features to measure whether an activity can be said as shared and cooperative, which are mutual responsiveness, commitment to the joint activity, and commitment to mutual support. The mutual responsiveness ensures that the actions of different members can be synchronized. The commitment to the joint activity assures that the teammates will not drop his side at some point. Finally, the commitment to the mutual support provides backup plans or actions when one teammate has the inability to finish his part of the plan.

In Ref. [138], Hoffman and Breazeal proposed a collaborative framework for the human–robot team that primarily based on the joint intention theory. An original work for the development of H-A collaboration was developed with an interactive robot, namely, "Leonardo," which can work as a partner with the operator. The team members can communicate with each other to maintain a shared belief about the task and coordinate their actions toward the shared plan. In addition, the team members should be able to demonstrate their commitment to finish their own tasks, as well as realizing the partner's task, providing mutual supports, and gaining mutual beliefs about the task. It shows that to design a collaboration framework for the teamwork, the robots should work jointly with the human operators rather than acting upon others, which need the robot to have the ability of reason human intention, beliefs, desires, and goals in the case to perform proper actions. Hence, when introducing a new agent to the teamwork, several key questions need to be answered and solved. For example, the agent should recognize the human operators' actions and intent before planning what to do next; mutual support must be provided to make sure that the task can be finished, and an efficient and clear communication channel should be maintained to establish mutual beliefs and common ground. Finally, the authors claimed that it is a critical and efficient solution for H-A collaboration if the robot can not only understand the ongoing actions of the human operator but also being capable of reasoning the operator's intention or goals under the specific behaviors.

In Ref. [139], Klein et al. emphasize the joint activity for H-A collaboration should involve four basic commitments, which are a team agreement, namely basic compact for the team members, the mutual predictable in the actions, mutual directable, and maintain common ground. Specifically, basic compact refers to a tacit agreement for collaboration and work under shared goals. Mutual predictability requires the team members to infer and predict others' actions when planning their own actions based on shared knowledge and action coordination. Mutual detectability requires the team members to be able to deliberatively assessing and modifying other members' actions if the situation and environment of the task changed. Lastly, common ground is the fundamentals of effective coordination, which contains pertinent knowledge, beliefs, and assumptions toward the task. Based on the four basic aspects of joint activity for the teamwork, 10 major challenges toward efficient and intelligent H-A collaboration were further highlighted, which were fulfill the requirement of basic compact, mutual predictable, directable, capable of revealing the status and intentions to the teammates, observer and interpret pertinent signals of status and intention, able to engage in goal negotiation, support collaborative planning, participate in managing attention, and all teammates must help to control the costs of coordinate effects for global optimal energy consumption.

In Ref. [140], the authors reviewed the driver—vehicle collaboration research and a particular focus on vehicle autonomy control and shared control aspects. The authors summarized four key problems that exist in the driver—vehicle collaboration research, which are personalized collaboration approach that can deal with the significant differences in the driving styles and diversity, how to precisely estimate and recognize different driver states and driver intention, robust driver—vehicle collaboration control toward complex driving context and situation, and how to efficiently test and evaluate collaboration system. Then, based on the literature of existing driver—vehicle collaboration research, they were divided into three types of collaboration methods, which are the intelligent driving assistant system for driver perception ability enhancement, vehicle control authority allocation under a specific condition, and dynamic control authority split for shared control.

In Ref. [141], the authors developed a so-called "driver-more" approach for the highly automated vehicle to construct a driver—vehicle collaboration scheme. The human-to-automation collaboration manner between the driver and the vehicle was studied to assess the driver's workload and acceptance of the highly automated vehicle. The proposed AutoMate Eu project was designed to analyze the mutual support in perception and in action between the driver and automation. Hence, before fully autonomous vehicles are accepted, the partially automated vehicle can still compensate its limitations in decision-making and complex perception by involving the driver in the loop. By introducing the driver—vehicle collaboration scheme, the strength of the driver and automation can be further consolidated, and the limitations of each part can be overcome and compensated. The authors agree with the idea that the success of highly automated driving vehicle depends on how well the vehicle interact, communicate, and cooperate with humans. Meanwhile, the quality of cooperation and communication will largely influence the user's trust and acceptance of the automation. Based on a preliminary experiment, it is found that the human-to-automation interaction (automation requests support to the human) is well understood and accepted by the participants. Moreover, it seems that the automation request for perception assistant was less demanding than the request for action assistant to the drivers when they were not driving and were asked to give support.

In Ref. [142], an anticipatory interface was designed for driver—automation collaboration on the highly automated vehicle. Two different interfaces were designed to announce the driver the future steering action of the vehicles that are based on physical anticipatory steering to show the rotation of the steering wheel and a visual anticipatory steering method based on the rotation of the LED light that was mounted on the wheel. The steering wheel and the road wheel are decoupled so that the steering wheel can move 1 s early before the actual vehicle steering in the case to announce the driver the future action of the automated vehicle. The 60 participants were told to keep the vehicle within the lane when disturbance events occur due to the perceived failure of the automation. It shows that the physical anticipatory steering shows great significance over the visual anticipatory steering method as it makes the driver easier to understand the vehicle plan and intervene in the false decision and control of the automation much earlier.

Task shifting from autonomous driving to manual driving requires considerable time for the physical and cognitive preparation of the human driver [143]. Hence, an efficient Human-Machine-Interface (HMI) is required to the L3 automated driving vehicle to recover the driver situation awareness from the secondary tasks and prepare proper control actions to the emerging and challenging situations. To prevent the sudden increase of driver workload and improving the quality of control actions such as steering and pedal, a multimodal driver—vehicle interface system was designed for efficient driver automation collaboration. Instead of letting driving to take over the vehicle completely and transferring to manual driving directly, a maneuver-based tactical level input (TLI) for collaborative control was proposed based on the multimodal HMI (equipped with gesture interface, touchscreen, and haptic interface). It shows that the TLI take-over approach can significantly lower the driver reaction time from the secondary tasks and decrease the workload during the take-over control process.

A preliminary discussion on autonomous driving and cooperative driving was proposed in Ref. [144]. The authors claimed that the three major motivations for developing autonomous vehicles improve the efficiency of the utilization of roadway infrastructure to reduce traffic congestion, improve safety and reduce traffic crash, and reduce energy consumptions and emissions. It shows that in the United States, the maximum highway capacity per lane is about 2200 vehicles per hour and the minimum headway to maintain safety and comfortability is about 45.5 m, which means the vehicle only occupies 11% of the lane and remaining 89% unused (for a normal car with 5 m length). Hence, developing autonomous driving and platooning can significantly improve the efficiency of road usage. It shows that about 90% of crashes are caused by human misbehavior. The development of autonomous driving and replacing the human with a highly autonomous agent can improve safety and reduce crash. Finally, the energy consumption and pollution can be reduced by regulating the vehicle at a highly efficient speed, minimizing stop-and-go maneuvers, and operating the vehicle at high speed with platooning to reduce aerodynamic drag. Regarding cooperative driving for autonomous vehicles, the cooperative approach with vehicle communication can provide additional information beyond the vehicle sensors, avoid misunderstanding about the intended maneuver, enable direct negotiations, enable infrastructure-based intelligence with global knowledge, and enable the replacement of expensive vehicle perception sensors with low-cost communication devices.

In Ref. [145], a cognitive vehicle framework was discussed toward the intelligent ADAS. It shows that there can be two distinct directions of developing the cognitive ADAS for intelligent vehicles, specifically, the stimuli-decisions–actions loop that focuses on the driver side and the perception–enhancement–action–suggestion–function–delegation loops that focus on the vehicle side. The driver-side cognitive ADAS design focuses on the stimuli and drivers' positions/actions recognition, drivers' physiological/psychological state recognition, and driver decision–action pattern analysis. Similarly, three key aspects for the vehicle side studies were estimated, which are driver perception enhancement, action–suggestions and human–driver interface, and advanced vehicle motion control delegation.

In Ref. [146], safety, reliance, and adaptability are viewed as three key drivers for human–automation collaboration. There are three stages that need to be developed toward an efficient human–machine collaboration scheme, which are mutual explanation, mutual understanding, and mutual trust. Besides, the affective computing for human emotion recognition will significantly influence the development and design for the human-centered automation as mutual trust is not only influenced by the performance of the agents but also determined by the mutual understanding (such as the intention and emotion). Moreover, artificial agents should be able to understand how their human counterpart is emotionally influenced by their decision and need to be capable of explaining their actions.

In Ref. [147], a new Swedish initiative, namely, AIR, was introduced to exploit the impact of mutual action and intention recognition between the human and autonomous vehicles on the development of successfully autonomous vehicles. Two key research topics were highlighted for the 4 years project, which is how to predict the action/intention of the autonomous vehicle and how the autonomous vehicle and human behaviors be described within the traffic system; and what information need to be revealed to communicate the action/intention of the automated driving vehicles and how to be presented. To describe the action/intention behaviors of the autonomous vehicles within the human embodied cognitive scope, two distinct branches of cognitive theory can be used. The first branch of the view, the behavior of the autonomous vehicle with an extended and distributed view, which explains the autonomous system is the extension of the human mind while cognition is distributed and should be reasoned according to the interaction manner with the material and social world. Another branch of the view describes the social cognition (interaction) behaviors that multiagents coregulating their coupling with the effect that their autonomy will not be destroyed and their dynamics acquire autonomy of their own.

In Ref. [148], a fictive driver activity parameter was introduced into the shared control system for cooperative vehicle lateral control between the driver and the lane-keeping assistance system. A time-varying control was proposed based on Takagi–Sugeno fuzzy approach to adaptively adjust the assistance torque of the steering assistance system with the real-time driving activity of the driver. The shared control strategy iteratively estimates the driver activity based on the driver torque and driving state monitoring system to decide how much assistance torque was required by the Lane Keeping Assistant Systems (LKAS). By introducing the measured weighting parameter into the shared control system, adaptive assistance splitting can be optimized and the conflict between the driver and automation can be effectively managed.

In Ref. [149], Inagaki proposed a smart collaboration framework for human–automation interaction based on mutual understanding, which was the outcome of one of the national projects in Japan, namely advanced safety vehicle. From the human point of view, the machine must be designed understandable so that it is easy for the human to understand the machine's capability, can give directions to the machine, monitor the ongoing tasks of the machine, and intervene in the machine control when needed. In contrast, from the machine's point of view to perform smart, the human can be monitored and reasoned from the situation awareness, intention and action understanding, the properness of the interpretation of the traffic environment, and whether the driver is inactive psychologically or physiologically. Based on the mutual understanding motivation, six categories of assistance from the machine to the driver were highlighted, which are perception enhancement, workload reduction, presentation of information, arousing driver's attention, providing warnings, and accident-avoidance control.

The human-centered automation was analyzed in Ref. [150] according to some preliminary studies. Specifically, several principles of human-centered automation were identified such as human must be in command; human must be involved and informed through effective communication; functions must be automated only if a good reason is on hold; automation must be understandable, action explainable, predictable, and directable; automation must be capable of understanding its human partner; mutual understanding about the intention, SA, action, emotions should be built between the teammates; and automation must be designed in a simple manner to learn and operate. Within such a collaboration system, several negative consequences to the human who work together with the automation can be estimated such as out-of-loop performance, loss of situation awareness, complacency or lack of vigilance, over trust, and automation surprises. Accordingly, two different types of assistance from the automation can be developed, which are soft protection and hard protection. Soft protection assists the driver to avoid any collision by warning the driver or assisting the driver's misbehaviors within the safety scope. However, if the assistance is canceled by the human driver who has improper SA, the system cannot prevent the driver from dangerous situations. The hard protection, on the other hand, mainly prevents the human from doing the dangerous action with machine-initiated decisions and controls. Unsurprisingly, hard protection may cause automation surprises to some operators who are not familiar with the system. Therefore a transparency human–machine interface is needed for the human to share the SA with the automation, recognize the decision of the automation and understand its judgment, and estimate the functional limitation of the automation [151].

In Ref. [152], a hazard-anticipatory shared braking control system was proposed to provide braking assistance to elderly people when encountering suddenly approaching pedestrians. The collision-avoidance driving trajectory and maneuver from expert drivers were analyzed and applied for risk predictive reference driver model design. The proposed system shows the effectiveness of elder driver assistance on the driving simulator. According to Ref. [153] sharing control between human driver and automation can be classified into four types, which are extension (extend driver capability), protection (prevent inappropriate behavior), relief (reduce driver workload), and partitioning (divide driving tasks into different parts and provide separate automation).

In Ref. [154], a shared collision-avoidance path following approach was proposed based on the evaluation of driver intention, situation assessment, and performance evaluation. Specifically, a linear time-varying model predictive controller was designed to follow the obstacle-avoidance path that was estimated by the artificial potential method. The lateral offset and lateral velocity were used to estimate the driver lane-change intention. The situation assessment of the risk was indicated by time to collision, and the path following performance was evaluated based on the lateral deviation between the planned path and the executed path. The shared control strategy for control authority allocation was proposed based on the fuzzy controller, which determines the control authority according to jointly evaluate the driver's intention, driver's situation assessment, and the lateral deviation performance.

In Ref. [155], the impact of drowsiness on the takeover time request and quality was studied based on the 31 participants on a real AUDI highly automated driving vehicle. It was found that the drowsiness is quite interindividual and show significant distribution pattern among the participants. In addition, the manipulation of automation is less suitable to investigate the drowsiness as some drivers can never reach a high drowsiness level on the HAD vehicle during the 3 h test. Experiment on simple take-over scenarios indicates that the drowsiness does not show a significant influence on the reaction time and request to intervene (RtI) quality. However, drivers with high-level drowsiness may generate a startling sound and react to the events even faster, which means drowsiness can lead to a startled or surprised reaction.

In Ref. [156], Zhang et al. reviewed the take-over time (TOT) of drivers on the automated driving vehicles based on 129 studies. The within-study analysis, between-study analysis, and linear mixed-effects model were used to investigate the take-over time features from different studies. Several interesting conclusions were made toward the take-over time characteristic. First, the urgency of the situation has a significant impact on the TOT as a driver tends to spend more time to prepare and resume the takeover if the situation is not emergent and the time budget is sufficiently long. Similarly, the authors in Ref. [157] argued that it is not always necessary to react as soon as possible because drivers may take time to assess the situation and perform proper control. Second, it shows that performing a handheld device for nondriving tasks significantly increase TOT. Third, drivers on higher-level automated driving vehicles show longer TOT than that on the low-level automated driving vehicle. Fourth, prior experience with takeover will reduce the TOT as the repeated trial will contribute to shorter response time. Fifth, visual-only time to request (TOR) is less efficient than auditory or vibrotactile TOR in the decrease of TOT. Sixth, the age factor has no obvious influence on the reduction of TOT. Last, the moderate effect of the surrounding traffic context exists on the TOT.

In Ref. [157], the experimental results on 49 participants showed that the more time the participants have to prepare for takeover, the more steering wheel will be used and the less braking will be performed. If the drivers have no sufficient time to make a lane change, the braking pedal will be used to increase situational awareness to make proper decisions or lane-change maneuvers. Moreover, it was suggested that the longer the TOR, the later the driver will react due to sufficient context checks. On the contrary, if the TOR time is short and insufficient, the driver will react to the request faster, but generate worse control quality.

In Ref. [158], the authors evaluated the situation awareness ability of a human driver. It was found that a 7 s period is enough for a driver to percept the position of surrounding vehicles, which is a consistent finding compared with existing studies listed in Ref. [156]. However, regarding awareness of the relative speed of the surrounding vehicle to host vehicle, it needs 20 s to gain the SA. In Ref. [158], the authors compared the SA levels between hazardous and nonhazardous situations with different time budgets. It was found that no significant difference for SA exists between these two situations as the precepted speed and the distance error of surrounding vehicles are similar to each other. However, the decision accuracy for drivers under the two situations is obviously different. Specifically, the decision accuracy toward driving safety for nonhazardous situation is much higher than that for the hazard situation (95% vs. 80%) because, in the hazard situation, the driver has to determine whether to take over and select the right decisions among the possible maneuvers (lane change right or left and brake).

In Ref. [159], driver behaviors experiment on Tesla Model S with autopilot 7.x shows that driver on L2

automated driving vehicles tend to perform nondriving-related tasks and can be difficult to monitor the traffic context for a long period of time due to the complacency and over-trust to the automation when the automated driving model was initiated.

In Ref. [160], Banks and Stanton identified three different roles that a driver can perform with the partially automated driving vehicle. Specifically, a driver can perform a driving role, monitoring role, and nondriving role. With the increase of the automation level, the role that the driver can perform will transfer from driving to monitoring and nondriving roles. However, transferring to the middle monitoring state can generate new considerations, such as a decrease of SA, erratic changes to mental workload, driving skill degradation, trust issues, and complacency. Based on the construction of task networks, social networks, and information networks, it was found that the monitoring mode does not reduce the driver workload and increase comfort. In contrast, transfer to the monitoring mode means an extra node, namely the automation will be introduced to the networks, which can significantly improve the network complexity by generating more links and edges. Hence, with the increase in the level of automation, transferring to the passive monitor mode will significantly improve the driver workload. Moreover, it is much easier to shift to the nondriving mode from the monitoring mode than that from the active driving mode. The sudden response to the take-over request in the nondriving mode can cause serious mental workload increment, SA reduction, and startle. Besides, it is impossible to build any networks for the driver under the nondriving mode, as s/he can perform any activity. Therefore more future efforts are expected in terms of maintaining driver monitoring role on high automation vehicles, and how to construct the efficient role transition strategy for a human driver to maintain proper workload, SA, trust, and skills [161].

Similarly, in Ref. [162], Saffarian et al. identified six potential challenges to the human driver when interacting with high-level automated driving vehicles, which are overreliance, behavioral adaptation, erratic mental workload, skill degradation, reduced SA, and inadequate mental model of automation functioning. Then, several solutions for efficient driver-automation interaction were estimated, such as shared control, adaptive automation, use of information portal for mutual communication and automation surprise avoidance, and new training methods for new drivers. A similar study on the analysis of human factors issues for autonomous vehicles can be found in Ref. [163]. In addition to the estimated human factors given in Ref. [162], motion sickness was also reviewed as an important human factor issue in the realm of autonomous driving. As shown in Ref. [163], more than 10% of American drivers are expected to suffer motion sickness regularly in automated driving vehicles. Accordingly, the authors in Ref. [163] further pointed out five future research that can be a potential solution for efficient human—automation cooperation, which are reengaging the driver, updating the user interface and communication, driver states monitoring, personalization of automation, and considering acceptance issues of the human driver to the autonomous vehicle.

# REFERENCES

[1] L. Yang, F.-Y. Wang, Driving into intelligent spaces with pervasive communications, IEEE Intelligent Systems 22 (2007) 1.

[2] E. Bellis, J. Page, National Motor Vehicle Crash Causation Survey (NMVCCS) SAS Analytical Users Manual, 2008. No. HS-811 053.

[3] C.M. Martinez, et al., Driving style recognition for intelligent vehicle control and advanced driver assistance: a survey, IEEE Transactions on Intelligent Transportation Systems 19 (3) (2017) 666—676.

[4] C. Lv, et al., Characterization of driver neuromuscular dynamics for human-automation collaboration design of automated vehicles, IEEE/ASME Transactions on Mechatronics 23 (6) (2018) 2558—2567.

[5] E. Ohn-Bar, M.M. Trivedi, Looking at humans in the age of self-driving and highly automated vehicles, IEEE Transactions on Intelligent Vehicles 1 (1) (2016) 90—104.

[6] V. Gaikwad, S. Lokhande, Lane departure identification for advanced driver assistance, IEEE Transactions on Intelligent Transportation Systems 16 (2) (2015) 910—918.

[7] Y. Saito, M. Itoh, T. Inagaki, Driver assistance system with a dual control scheme: effectiveness of identifying driver drowsiness and preventing lane departure accidents, IEEE Transactions on Human-Machine Systems 46 (5) (2016) 660—671.

[8] V. Milanés, et al., Cooperative adaptive cruise control in real traffic situations, IEEE Transactions on Intelligent Transportation Systems 15 (1) (2014) 296—305.

[9] A. Mukhtar, L. Xia, T.B. Tang, Vehicle detection techniques for collision avoidance systems: a review, IEEE Transactions on Intelligent Transportation Systems 16 (5) (2015) 2318—2338.

[10] A. Tawari, et al., Looking-in and looking-out vision for urban intelligent assistance: estimation of driver attentive state and dynamic surround for safe merging and braking, in: Intelligent Vehicles Symposium Proceedings, 2014 IEEE, IEEE, 2014.

[11] Y. Xing, et al., Identification and analysis of driver postures for in-vehicle driving activities and secondary tasks recognition, IEEE Transactions on Computational Social Systems 5 (1) (2018) 95—108.

[12] H. Berndt, J. Emmert, K. Dietmayer, Continuous driver intention recognition with hidden markov models, in: Intelligent Transportation Systems, 2008. ITSC 2008. 11th International IEEE Conference on, IEEE, 2008.

[13] A. Kuefler, et al., Imitating driver behavior with generative adversarial networks, in: Intelligent Vehicles Symposium (IV), 2017 IEEE, IEEE, 2017.

[14] J. Morton, M.J. Kochenderfer, Simultaneous Policy Learning and Latent State Inference for Imitating Driver Behavior, arXiv preprint arXiv:1704.05566, 2017.

[15] L. Li, et al., Intelligence testing for autonomous vehicles: a new approach, IEEE Transactions on Intelligent Vehicles 1 (2) (2016) 158—166.

[16] A. Liu, A. Pentland, Towards real-time recognition of driver intentions, in: Intelligent Transportation System, 1997. ITSC'97., IEEE Conference on, IEEE, 1997.

[17] N. Oliver, A.P. Pentland, Driver behavior recognition and prediction in a SmartCar, Proceedings of SPIE-The International Society for Optical Engineering 4023 (2000).

[18] M. Liebner, F. Klanner, Driver intent inference and risk assessment, in: Handbook of Driver Assistance Systems: Basic Information, Components and Systems for Active Safety and Comfort, 2014, pp. 1—20.

[19] G.-P. Gwon, et al., Generation of a precise and efficient lane-level road map for intelligent vehicle systems, IEEE Transactions on Vehicular Technology 66 (6) (2017) 4518—4533.

[20] W. Balid, H. Tafish, H.H. Refai, Intelligent vehicle counting and classification sensor for real-time traffic surveillance, IEEE Transactions on Intelligent Transportation Systems 19 (6) (2017) 1784—1794.

[21] F.-Y. Wang, Artificial intelligence and intelligent transportation: driving into the 3rd axial age with its, IEEE Intelligent Transportation Systems Magazine 9 (4) (2017) 6—9.

[22] C. Gou, et al., A joint cascaded framework for simultaneous eye detection and eye state estimation, Pattern Recognition 67 (2017) 23—31.

[23] E. Murphy-Chutorian, M.M. Trivedi, Head pose estimation in computer vision: a survey, IEEE Transactions on Pattern Analysis and Machine Intelligence 31 (4) (2009) 608—626.

[24] J. Chen, Q. Ji, A probabilistic approach to online eye gaze tracking without explicit personal calibration, IEEE Transactions on Image Processing 24 (3) (2015) 1076—1086.

[25] S. Haufe, et al., EEG potentials predict upcoming emergency brakings during simulated driving, Journal of Neural Engineering 8 (5) (2011) 056001.

[26] A. Doshi, M.M. Trivedi, Tactical driver behavior prediction and intent inference: a review, in: Intelligent Transportation Systems (ITSC), 2011 14th International IEEE Conference on, IEEE, 2011.

[27] D.D. Salvucci, et al., Lane-change detection using a computational driver model, Human Factors 49 (3) (2007) 532—542.

[28] R. Toledo-Moreo, M.A. Zamora-Izquierdo, IMM-based lane-change prediction in highways with low-cost GPS/INS, IEEE Transactions on Intelligent Transportation Systems 10 (1) (2009) 180−185.

[29] D.D. Salvucci, A. Liu, The time course of a lane change: driver control and eye-movement behavior, Transportation Research Part F: Traffic Psychology and Behaviour 5 (2) (2002) 123−132.

[30] K. Schmidt, et al., A mathematical model for predicting lane changes using the steering wheel angle, Journal of Safety Research 49 (2014). 85-e1.

[31] K. Driggs-Campbell, R. Bajcsy, Identifying modes of intent from driver behaviors in dynamic environments, in: Intelligent Transportation Systems (ITSC), 2015 IEEE 18th International Conference on, IEEE, 2015.

[32] M.J. Henning, et al., Modelling driver behaviour in order to infer the intention to change lanes, Proceedings of European Conference on Human Centred Design for Intelligent Transport Systems 113 (2008).

[33] S.S. Beauchemin, et al., Portable and scalable vision-based vehicular instrumentation for the analysis of driver intentionality, IEEE Transactions on Instrumentation and Measurement 61 (2) (2012) 391−401.

[34] A. Doshi, M.M. Trivedi, On the roles of eye gaze and head dynamics in predicting driver's intent to change lanes, IEEE Transactions on Intelligent Transportation Systems 10 (3) (2009) 453−462.

[35] F. Lethaus, et al., A comparison of selected simple supervised learning algorithms to predict driver intent based on gaze data, Neurocomputing 121 (2013) 108−130.

[36] E. Murphy-Chutorian, M.M. Trivedi, Head pose estimation and augmented reality tracking: an integrated system and evaluation for monitoring driver awareness, IEEE Transactions on Intelligent Transportation Systems 11 (2) (2010) 300−311.

[37] Y.-M. Jang, R. Mallipeddi, M. Lee, Identification of human implicit visual search intention based on eye movement and pupillary analysis, User Modeling and User-Adapted Interaction 24 (4) (2014) 315−344.

[38] T. Pech, P. Lindner, G. Wanielik, Head tracking based glance area estimation for driver behaviour modelling during lane change execution, in: Intelligent Transportation Systems (ITSC), 2014 IEEE 17th International Conference on, IEEE, 2014.

[39] H. Zhou, M. Itoh, T. Inagaki, Influence of cognitively distracting activity on driver's eye movement during preparation of changing lanes, in: SICE Annual Conference, 2008, IEEE, 2008.

[40] L. Bi, et al., Detecting driver normal and emergency lane-changing intentions with queuing network-based driver models, International Journal of Human-Computer Interaction 31 (2) (2015) 139−145.

[41] J. Ding, et al., Driver intention recognition method based on comprehensive lane-change environment assessment, in: Intelligent Vehicles Symposium Proceedings, 2014 IEEE, IEEE, 2014.

[42] K. Li, et al., Lane changing intention recognition based on speech recognition models, Transportation Research Part C: Emerging Technologies 69 (2016) 498−514.

[43] D. Kasper, et al., Object-oriented Bayesian networks for detection of lane change manoeuvres, IEEE Intelligent Transportation Systems Magazine 4 (3) (2012) 19−31.

[44] J.C. McCall, et al., Lane change intent analysis using robust operators and sparse bayesian learning, IEEE Transactions on Intelligent Transportation Systems 8 (3) (2007) 431−440.

[45] A. Doshi, B. Morris, M. Trivedi, On-road prediction of driver's intent with multimodal sensory cues, IEEE Pervasive Computing 10 (3) (2011) 22−34.

[46] A. Jain, et al., Brain4cars: Car that Knows Before You Do via Sensory-Fusion Deep Learning Architecture, arXiv preprint arXiv:1601.00740, 2016.

[47] A. Jain, H.S. Koppula, B. Raghavan, S. Soh, A. Saxena, Car that knows before you do: anticipating manoeuvres via learning temporal driving models, Proceedings of the IEEE International Conference on Computer Vision (2015) 3182−3190.

[48] J.A. Michon, A critical view of driver behavior models: what do we know, what should we do?, in: Human Behaviour and Traffic Safety Springer, Boston, MA, 1985, pp. 485−524.

[49] S.-J. Youn, K.-W. Oh, Intention recognition using a graph representation, World Academy of Science, Engineering and Technology 25 (2007) 13−18.

[50] T. Baltrušaitis, P. Robinson, L.-P. Morency, Openface: an open source facial behavior analysis toolkit, in: Applications of Computer Vision (WACV), 2016 IEEE Winter Conference on, IEEE, 2016.

[51] T. Baltrusaitis, P. Robinson, L.-P. Morency, Constrained local neural fields for robust facial landmark detection in the wild, in: Computer Vision Workshops (ICCVW), 2013 IEEE International Conference on, IEEE, 2013.

[52] E. Wood, T. Baltrusaitis, X. Zhang, Y. Sugano, P. Robinson, A. Bulling, Rendering of eyes for eye-shape registration and gaze estimation, Proceedings of the IEEE International Conference on Computer Vision (2015) 3756−3764.

[53] A. Zyner, S. Worrall, E. Nebot, A recurrent neural network solution for predicting driver intention at unsignalized intersections, IEEE Robotics and Automation Letters 3 (3) (2018) 1759−1764.

[54] J. Morton, T.A. Wheeler, M.J. Kochenderfer, Analysis of recurrent neural networks for probabilistic modeling of driver behavior, IEEE Transactions on Intelligent Transportation Systems 18 (5) (2017) 1289−1298.

[55] O. Olabiyi, et al., Driver Action Prediction Using Deep (Bidirectional) Recurrent Neural Network, arXiv preprint arXiv:1706.02257, 2017.

[56] Y. Bengio, P. Simard, P. Frasconi, Learning long-term dependencies with gradient descent is difficult, IEEE Transactions on Neural Networks 5 (2) (1994) 158−166.

[57] S. Hochreiter, J. Schmidhuber, Long short-term memory, Neural Computation 9 (8) (1997) 1735−1780.

[58] K. Greff, et al., LSTM: a search space odyssey, IEEE transactions on neural networks and learning systems 28 (10) (2017) 2222–2232.

[59] Y. Xing, et al., Driver workload estimation using a novel hybrid method of error reduction ratio causality and support vector machine, Measurement 114 (2018) 390–397.

[60] H.-Y.W. Chen, et al., Self-reported engagement in driver distraction: an application of the theory of planned behaviour, Transportation Research Part F: Traffic Psychology and Behaviour 38 (2016) 151–163.

[61] A. Eriksson, N.A. Stanton, Takeover time in highly automated vehicles: noncritical transitions to and from manual control, Human Factors 59 (4) (2017) 689–705.

[62] M. Schuster, K.K. Paliwal, Bidirectional Recurrent Neural Networks, IEEE Press, 1997.

[63] Deleted in review

[64] D. Yagil, Reasoned action and irrational motives: a prediction of drivers' intention to violate traffic laws 1, Journal of Applied Social Psychology 31 (4) (2001) 720–739.

[65] A. Pentland, A. Liu, Modeling and prediction of human behavior, Neural Computation 11 (1) (1999) 229–242.

[66] L. Fridman, Human-centered Autonomous Vehicle Systems: Principles of Effective Shared Autonomy, arXiv preprint arXiv:1810.01835, 2018.

[67] N. Cristianini, J. Shawe-Taylor, An Introduction to Support Vector Machines and Other Kernel-Based Learning Methods, Cambridge University Press, 2000.

[68] V. Vapnik, The Nature of Statistical Learning Theory, Springer Science & Business Media, 2013.

[69] A.J. Smola, B. Schölkopf, A tutorial on support vector regression, Statistics and Computing 14 (3) (2004) 199–222.

[70] J.L. Gabbard, G.M. Fitch, H. Kim, Behind the glass: driver challenges and opportunities for AR automotive applications, Proceedings of the IEEE 102 (2) (2014) 124–136.

[71] W. Hernandez, Optimal estimation of the relevant information coming from a rollover sensor placed in a car under performance tests, Measurement 41 (1) (2008) 20–31.

[72] J.C. Castellanos, F. Fruett, Embedded system to evaluate the passenger comfort in public transportation based on dynamical vehicle behavior with user's feedback, Measurement 47 (2014) 442–451.

[73] G. Andria, F. Attivissimo, A. Di Nisio, A.M.L. Lanzolla, A. Pellegrino, Development of an automotive data acquisition platform for analysis of driving behavior, Measurement 93 (2016) 278–287.

[74] Q. Zhang, Q. Wu, Y. Zhou, X. Wu, Y. Ou, H. Zhou, Webcam-based, non-contact, real-time measurement for the physiological parameters of drivers, Measurement 100 (2017) 311–321.

[75] M.A. Sotelo, Electrical, connected, and automated transportation editor's column, IEEE Intelligent Transportation Systems Magazine 8 (2) (2016) 2.

[76] V. Faure, R. Lobjois, N. Benguigui, The effects of driving environment complexity and dual tasking on drivers' mental workload and eye blink behavior, Transportation Research Part F: Traffic Psychology and Behaviour 40 (2016) 78–90.

[77] C.M. Martinez, X. Hu, D. Cao, E. Velenis, B. Gao, M. Wellers, Energy management in plug-in hybrid electric vehicles: recent progress and a connected vehicles perspective, IEEE Transactions on Vehicular Technology 66 (6) (2016) 4534–4549.

[78] D. Pecchini, R. Roncella, G. Forlani, F. Giuliani, Measuring driving workload of heavy vehicles at roundabouts, Transportation Research Part F: Traffic Psychology and Behaviour 45 (2017) 27–42.

[79] P. Choudhary, N.R. Velaga, Analysis of vehicle-based lateral performance measures during distracted driving due to phone use, Transportation Research Part F: Traffic Psychology and Behaviour 44 (2017) 120–133.

[80] M. Niezgoda, A. Tarnowski, M. Kruszewski, T. Kamiński, Towards testing auditor-y–vocal interfaces and detecting distraction while driving: a comparison of eye-movement measures in the assessment of cognitive workload, Transportation Research Part F: Traffic Psychology and Behaviour 32 (2015) 23–34.

[81] M. Chan, S. Nyazika, A. Singhal, Effects of a front-seat passenger on driver attention: an electrophysiological approach, Transportation Research Part F: Traffic Psychology and Behaviour 43 (2016) 67–79.

[82] B. Okumura, M.R. James, Y. Kanzawa, M. Derry, K. Sakai, T. Nishi, D. Prokhorov, Challenges in perception and decision making for intelligent automotive vehicles: a case study, IEEE Transactions on Intelligent Vehicles 1 (1) (2016) 20–32.

[83] R.D. O'Donnell, F.T. Eggemeie, Workload assessment methodology, in: K.R. Boff, L. Kaufman, J.P. Thomas (Eds.), Cognitive Processes and Performance, John Wiley and Sons, 1986.

[84] S. Sega, H. Iwasaki, H. Hiraishi, F. Mizoguchi, Verification of driving workload using vehicle signal data for distraction-minimized systems on ITS, in: 18th ITS World Congress, 2011.

[85] Y. Liang, M.L. Reyes, J.D. Lee, Real-time detection of driver cognitive distraction using support vector machines, IEEE Transactions on Intelligent Transportation Systems 8 (2) (2007) 340–350.

[86] W. Changxu, Y. Liu, Queuing network modeling of driver workload and performance, IEEE Transactions on Intelligent Transportation Systems 8 (3) (2007) 528–537.

[87] E.T.T. Teh, S. Jamson, O. Carsten, How does a lane change performed by a neighbouring vehicle affect driver workload?, in: 19th ITS World Congress, 2012.

[88] J.-B. Lim, S.-B. Lee, K.-H. Kim, S.-Y. Kim, J.-S. Choi, A study of the relationship between driver's anxiety EEG and driving speed in motorway sections, Journal of the Korean Surgical Society 27 (3) (2012) 167–175.

[89] Y. Zhao, S.A. Billings, H.-L.W. Wei, P.G. Sarrigiannis, A parametric method to measure time-varying linear

and nonlinear causality with applications to EEG data, IEEE Transactions on Biomedical Engineering 60 (11) (2013) 3141−3148.

[90] Y. Zhao, S.A. Billings, H. Wei, F. He, P.G. Sarrigiannis, A new NARX-based Granger linear and nonlinear casual influence detection method with applications to EEG data, Journal of Neuroscience Methods 212 (1) (2013) 79−86.

[91] L.E. Baum, T. Petrie, Statistical inference for probabilistic functions of finite state Markov chains, The Annals of Mathematical Statistics 37 (6) (1966) 1554−1563.

[92] D. Barber, Bayesian Reasoning and Machine Learning, Cambridge University Press, 2012.

[93] D. Polling, M. Mulder, M.M. van Paassen, Q.P. Chu, "Inferring the driver's lane change intention using context-based dynamic Bayesian networks," 2005, IEEE International Conference on Systems, Man and Cybernetics 1 (2005) 853−858.

[94] J. Harding, et al., Vehicle-to-vehicle Communications: Readiness of V2V Technology for Application, No. DOT HS 812 014, National Highway Traffic Safety Administration, United States, 2014.

[95] S. Biswas, T. Raymond, F. Dion, Vehicle-to-vehicle wireless communication protocols for enhancing highway traffic safety, IEEE Communications Magazine 44 (1) (2006) 74−82.

[96] H. Guo, et al., Simultaneous trajectory planning and tracking using an MPC method for cyber-physical systems: a case study of obstacle avoidance for an intelligent vehicle, IEEE Transactions on Industrial Informatics 14 (9) (2018) 4273−4283.

[97] J.E. Siegel, D.C. Erb, S.E. Sarma, A survey of the connected vehicle landscape—architectures, enabling technologies, applications, and development areas, IEEE Transactions on Intelligent Transportation Systems 19 (8) (2017) 2391−2406.

[98] S.R.K. Narla, The evolution of connected vehicle technology: from smart drivers to smart cars to… self-driving cars, ITE Journal 83 (7) (2013) 22−26.

[99] N. Lu, et al., Connected vehicles: solutions and challenges, IEEE Internet of Things Journal 1 (4) (2014) 289−299.

[100] S. Lefèvre, D. Vasquez, C. Laugier, A survey on motion prediction and risk assessment for intelligent vehicles, ROBOMECH Journal 1 (1) (2014) 1.

[101] J. Hillenbrand, M.S. Andreas, K. Kristian, A multilevel collision mitigation approach—its situation assessment, decision making, and performance tradeoffs, IEEE Transactions on Intelligent Transportation Systems 7 (4) (2006) 528−540.

[102] A. Polychronopoulos, et al., Sensor fusion for predicting vehicles' path for collision avoidance systems, IEEE Transactions on Intelligent Transportation Systems 8 (3) (2007) 549−562.

[103] M.G. Ortiz, et al., Behavior prediction at multiple time-scales in inner-city scenarios, in: 2011 IEEE Intelligent Vehicles Symposium (IV), IEEE, 2011.

[104] B. Morris, A. Doshi, M. Trivedi, Lane change intent prediction for driver assistance: on-road design and evaluation, in: 2011 IEEE Intelligent Vehicles Symposium (IV), IEEE, 2011.

[105] M. Bahram, et al., A combined model-and learning-based framework for interaction-aware maneuver prediction, IEEE Transactions on Intelligent Transportation Systems 17 (6) (2016) 1538−1550.

[106] N. Deo, M.M. Trivedi, Multi-modal trajectory prediction of surrounding vehicles with maneuver based lstms, in: 2018 IEEE Intelligent Vehicles Symposium (IV), IEEE, 2018.

[107] M. Schreier, V. Willert, J. Adamy, An integrated approach to maneuver-based trajectory prediction and criticality assessment in arbitrary road environments, IEEE Transactions on Intelligent Transportation Systems 17 (10) (2016) 2751−2766.

[108] G. Agamennoni, J.I. Nieto, E.M. Nebot, Estimation of multivehicle dynamics by considering contextual information, IEEE Transactions on Robotics 28 (4) (2012) 855−870.

[109] M. Bahram, et al., A game-theoretic approach to replanning-aware interactive scene prediction and planning, IEEE Transactions on Vehicular Technology 65 (6) (2015) 3981−3992.

[110] J. Li, et al., A dynamic bayesian network for vehicle maneuver prediction in highway driving scenarios: framework and verification, Electronics 8 (1) (2019) 40.

[111] Y. Luo, P. Cai, GAMMA: A General Agent Motion Prediction Model for Autonomous Driving, arXiv preprint arXiv:1906.01566, 2019.

[112] A. Alahi, et al., Social lstm: human trajectory prediction in crowded spaces, in: Proceedings of the IEEE Conference on Computer Vision and Pattern Recognition, 2016.

[113] N. Deo, M.M. Trivedi, Convolutional social pooling for vehicle trajectory prediction, in: Proceedings of the IEEE Conference on Computer Vision and Pattern Recognition Workshops, 2018.

[114] T. Zhao, et al., Multi-agent tensor fusion for contextual trajectory prediction, in: Proceedings of the IEEE Conference on Computer Vision and Pattern Recognition, 2019.

[115] C.E. Rasmussen, The infinite Gaussian mixture model, in: Advances in Neural Information Processing Systems, 2000.

[116] M. Van Ly, S. Martin, M. Mohan, Trivedi, Driver classification and driving style recognition using inertial sensors, in: 2013 IEEE Intelligent Vehicles Symposium (IV), IEEE, 2013.

[117] C.M. Martinez, et al., Driving style recognition for intelligent vehicle control and advanced driver assistance: a survey, IEEE Transactions on Intelligent Transportation Systems 19 (3) (2017) 666−676.

[118] Y. Xing, et al., Driver lane change intention inference for intelligent vehicles: framework, survey, and challenges, IEEE Transactions on Vehicular Technology 68 (5) (2019) 4378−4390.

[119] M.-P. Dubuisson, A.K. Jain, A modified Hausdorff distance for object matching, in: Proceedings of 12th International Conference on Pattern Recognition, vol. 1, IEEE, 1994.

[120] K.E. Schaefer, et al., Communicating intent to develop shared situation awareness and engender trust in human-agent teams, Cognitive Systems Research 46 (2017) 26−39.

[121] K. Sycara, G. Sukthankar, Literature Review of Teamwork Models, vol. 31, Robotics Institute, Carnegie Mellon University, 2006, p. 31.

[122] P. Cohen, H. Levesque, I. Smith, On team formation, in: G. Holmstrom-Hintikka, R. Tuomela (Eds.), Contemporary Action Theory, Kluwer Academic, 1997.

[123] E. Salas, T. Dickinson, S. Converse, S. Tannenbaum, Towards an understanding of team performance and training, in: R. Swezey, E. Salas (Eds.), Teams: Their Training and Performance, Norwood, 1992.

[124] J. Cannon-Bowers, S. Tannenbaum, E. Salas, C. Volpe, Defining team competencies: implications for training requirements and strategies, in: R. Guzzo, E. Salas (Eds.), Team Effectiveness and Decision Making in Organizations, Jossey Bass, 1995.

[125] J.Y.C. Chen, M.J. Barnes, Human−agent teaming for multirobot control: a review of human factors issues, IEEE Transactions on Human-Machine Systems 44 (1) (2014) 13−29.

[126] K.L. Mosier, et al., Autonomous, Context-Sensitive, Task Management Systems and Decision Support Tools I: Human-Autonomy Teaming Fundamentals and State of the Art, 2017.

[127] J.M. Bradshaw, et al., Human-agent-robot teamwork, IEEE Intelligent Systems 27 (2) (2012) 8−13.

[128] P.M. Fitts, Human Engineering for an Effective Air-Navigation and Traffic-Control System, 1951.

[129] M.A. Goodrich, et al., Incorporating a robot into an autism therapy team, IEEE Intelligent Systems 2 (2012) 52−59.

[130] D. Miller, A. Sun, W. Ju, Situation awareness with different levels of automation, in: 2014 IEEE International Conference on Systems, Man, and Cybernetics (SMC), IEEE, 2014.

[131] D.B. Kaber, M.R. Endsley, The effects of level of automation and adaptive automation on human performance, situation awareness and workload in a dynamic control task, Theoretical Issues in Ergonomics Science 5 (2) (2004) 113−153.

[132] T.B. Sheridan, Telerobotics, Automation, and Human Supervisory Control, MIT press, 1992.

[133] M.-P. Pacaux, et al., Levels of automation and human-machine cooperation: application to human-robot interaction, IFAC Proceedings Volumes 44 (1) (2011) 6484−6492.

[134] R.N. Darmanin, M.K. Bugeja, A review on multi-robot systems categorised by application domain, in: 2017 25th Mediterranean Conference on Control and Automation (MED), IEEE, 2017.

[135] J.L. Wright, et al., Individual differences in human-agent teaming: an analysis of workload and situation awareness through eye movements, in: Proceedings of the Human Factors and Ergonomics Society Annual Meeting vol. 58 (1), SAGE Publications, Los Angeles, CA, 2014. Sage CA.

[136] R.K.E. Bellamy, et al., Human-agent collaboration: can an agent be a partner?, in: Proceedings of the 2017 CHI Conference Extended Abstracts on Human Factors in Computing Systems ACM, 2017.

[137] M.E. Bratman, Shared cooperative activity, Philosophical Review 101 (2) (1992) 327−341.

[138] G. Hoffman, C. Breazeal, Collaboration in human-robot teams, in: AIAA 1st Intelligent Systems Technical Conference, 2004.

[139] G. Klien, et al., Ten challenges for making automation a" team player" in joint human-agent activity, IEEE Intelligent Systems 19 (6) (2004) 91−95.

[140] Y. Hu, et al., Human-machine cooperative control of intelligent vehicle: recent developments and future perspectives, Acta Automatica Sinica 45 (7) (2019) 1261−1280.

[141] A. Castellano, S. Fruttaldo, E. Landini, R. Montanari, Andreas Luedtke, A "driver-more" approach to vehicle automation, in: Proceedings of the 6th Humanist Conference, 2018.

[142] M. Johns, et al., Looking ahead: anticipatory interfaces for driver-automation collaboration, in: 2017 IEEE 20th International Conference on Intelligent Transportation Systems (ITSC), IEEE, 2017.

[143] U.E. Manawadu, et al., Tactical-level input with multimodal feedback for unscheduled takeover situations in human-centered automated vehicles, in: 2018 IEEE/ASME International Conference on Advanced Intelligent Mechatronics (AIM), IEEE, 2018.

[144] S.E. Shladover, Cooperative (rather than autonomous) vehicle-highway automation systems, IEEE Intelligent Transportation Systems Magazine 1 (1) (2009) 10−19.

[145] L. Li, et al., Cognitive cars: a new frontier for ADAS research, IEEE Transactions on Intelligent Transportation Systems 13 (1) (2011) 395−407.

[146] Azevedo, R.B. Carlos, K. Raizer, R. Souza, A vision for human-machine mutual understanding, trust establishment, and collaboration, in: 2017 IEEE Conference on Cognitive and Computational Aspects of Situation Management (CogSIMA), IEEE, 2017.

[147] M. Nilsson, S. Thill, Z. Tom, Action and intention recognition in human interaction with autonomous vehicles, in: Experiencing Autonomous Vehicles: Crossing the Boundaries between a Drive and a Ride" Workshop in Conjunction with CHI2015, 2015.

[148] A.-T. Nguyen, C. Sentouh, J.-C. Popieul, Driver-automation cooperative approach for shared steering control under multiple system constraints: design and experiments, IEEE Transactions on Industrial Electronics 64 (5) (2016) 3819−3830.

[149] T. Inagaki, Smart collaboration between humans and machines based on mutual understanding, Annual Reviews in Control 32 (2) (2008) 253–261.

[150] C.E. Billings, Aviation Automation: The Search for a Human-Centered Approach, CRC Press, 2018.

[151] T. Inagaki, Design of human–machine interactions in light of domain-dependence of human-centered automation, Cognition, Technology and Work 8 (3) (2006) 161–167.

[152] Y. Saito, P. Raksincharoensak, Shared control in risk predictive braking maneuver for preventing collisions with pedestrians, IEEE Transactions on Intelligent Vehicles 1 (4) (2016) 314–324.

[153] T. Inagaki, Adaptive automation: Sharing and trading of control, in: E. Hollnagel (Ed.), Handbook of Cognitive Task Design, Lawrence Erlbaum, Mahwah, NJ, USA, 2003, pp. 147–169 (Chapter 8).

[154] M. Li, et al., Shared control driver assistance system based on driving intention and situation assessment, IEEE Transactions on Industrial Informatics 14 (11) (2018) 4982–4994.

[155] V. Weinbeer, et al., Highly automated driving: how to get the driver drowsy and how does drowsiness influence various take-over-aspects?, in: 8. Tagung Fahrerassistenz, 2017.

[156] B. Zhang, et al., Determinants of take-over time from automated driving: a meta-analysis of 129 studies, Transportation Research Part F: Traffic Psychology and Behaviour 64 (2019) 285–307.

[157] C. Gold, et al., "Take over!" How long does it take to get the driver back into the loop?, in: Proceedings of the Human Factors and Ergonomics Society Annual Meeting vol. 57 (1), SAGE Publications, Los Angeles, CA, 2013. Sage CA.

[158] Z. Lu, X. Coster, J. de Winter, How much time do drivers need to obtain situation awareness? A laboratory-based study of automated driving, Applied Ergonomics 60 (2017) 293–304.

[159] V.A. Banks, et al., Is partially automated driving a bad idea? Observations from an on-road study, Applied Ergonomics 68 (2018) 138–145.

[160] V.A. Banks, N.A. Stanton, Analysis of driver roles: modelling the changing role of the driver in automated driving systems using EAST, Theoretical Issues in Ergonomics Science 20 (3) (2019) 284–300.

[161] D.D. Heikoop, et al., Psychological constructs in driving automation: a consensus model and critical comment on construct proliferation, Theoretical Issues in Ergonomics Science 17 (3) (2016) 284–303.

[162] M. Saffarian, J.CF de Winter, R. Happee, Automated driving: human-factors issues and design solutions, in: Proceedings of the Human Factors and Ergonomics Society Annual Meeting vol. 56 (1), Sage Publications, Los Angeles, CA, 2012. Sage CA.

[163] M. Sivak, B. Schoettle, Motion Sickness in Self-Driving Vehicles, University of Michigan, Transportation Research Institute (UMTRI), Michigan, USA, 2015.

# CHAPTER 9

# Conclusions, Discussions, and Directions for Future Work

At the end of this book, the main motivation can be summarized as an advanced driver intention inference system designed based on the integration of multimodal data from both traffic context and driver behaviors. The major components of this book can be separated into the following aspects, which are integrated road perception algorithms and evaluation system introduction; the recognition of driver activities and secondary tasks toward a comprehensive driver understanding with either conventional feature engineering methods and end-to-end deep learning methods; driver lane change maneuver and intention analysis based on the timescale analysis and driver behaviors; and driver braking intention recognition toward a safety cyber-physical system and driver braking intensity estimation. Moreover, the proposed lane change and braking intention inference system are based on naturalistic data that are collected from the highway or standard driving cycles. The lane change intention inference system enables the prediction of the lane change intention a few seconds before the maneuver happens. Furthermore, several extensions for the driver's mental status studies such as the utilization of support vector machine on the driver workload estimation are based on multimodal signals. These contributions are elaborated in the following so as to discuss the limitations as well as future development and work directions.

The organization and summarization of this book are provided in Fig. 9.1.

## INTEGRATED ROAD DETECTION TOWARD ROBUST TRAFFIC CONTEXT PERCEPTION

Road and lane detection on a structured road is one of the most fundamental traffic context awareness approaches to assist the driving task. It not only is important to the conventional Advanced Driver Assistance Systems but also plays a critical role in the human-centered driver-automation collaboration. Understanding ego-lane position and the styles play a critical role in the driver intention estimation because a large amount of lane change maneuvers are generated according to the specific traffic laws and the ego-lane states. In this section, an integrated lane detection framework is proposed to overcome the robustness issue of the lane detection system. A comprehensive literature survey as well as the algorithm design and evaluation are discussed. The main contribution of this section can be summarized as follows. The construction of a deep sensor fusion network, which combines the lidar and camera for road detection, is discussed. According to the technical requirement, a state-of-the-art literature review on lane and road detection, integration, and evaluation is proposed in the beginning. In most of the existing studies, there is no such comprehensive analysis for the integration and evaluation methods of the lane detection system. Second, by analyzing the limitations of existing studies, an algorithm level integrated lane detection system is developed. The system is able to detect lane positions and evaluate the detection performance. The proposed integrated system satisfies the real-time application requirement and robustness to various road challenges (lighting, shadows, occlusion, etc.). Finally, a general parallel lane detection system framework is designed. By mapping the real-world problem into the virtual parallel world, unlimited experimental data can be collected. The parallel lane detection system is a powerful way for lane detection system construction and evaluation.

### Algorithm Limitation

Currently, the proposed lane detection system is solely a camera-based system. As analyzed in this project, the camera-based system has its inevitable limitations, especially when facing atrocious weather conditions. Therefore a more robust system should rely on the sensor fusion technique. Meanwhile, the proposed lane detection algorithm relies on a predetermined region of interest (ROI), which can cause inaccurate detection results when transferring the system to another platform without manual calibration. Therefore the road area detection and automatic ROI selection will be an

Advanced Driver Intention Inference. https://doi.org/10.1016/B978-0-12-819113-2.00009-9

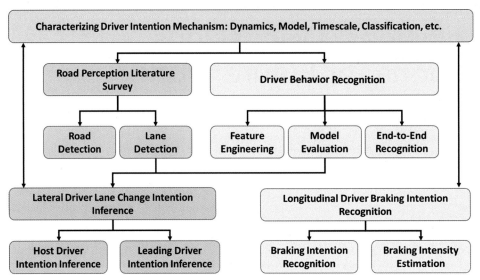

FIGURE 9.1 Main structure summarization of this book.

efficient way to increase the system robustness and transportability. Regarding the sensor-fusion-based road detection system, the cross-fusion network is trained with limited data from KITTI road dataset. No pretrained models are used in this section, which can need more data to train the model and converge the performance.

### Directions for Future Work

The proposed lane detection system is tested with both public datasets and the local Cranfield dataset collected on highways. However, the testing is still not enough to prove the more general performance. The evaluation issue is a challenging task for all the vision-based systems. Therefore to evaluate the system and design a robust driver assistance system, the parallel driving framework is one of the future research directions. The parallel framework will provide more testing dataset to evaluate the performance of the lane detection system. It also enables the online learning and updating of the lane detection system, which makes the system more robust to unseen traffic context. Next, the integration of the lane detection system with other vision-based assistance systems is another research direction. By efficiently integrating lane detection with road detection or vehicle detection, the vision system will be more intelligent to deal with complex traffic context. In terms of the deep road detection network, more datasets are expected in the future to fully exploit the model performance. Another interesting topic is using synthesis data to refine the model and apply domain adaptation methodologies to enhance the model's robustness.

### DRIVING ACTIVITY RECOGNITION AND SECONDARY TASK DETECTION

In this section, normal driving behaviors such as the mirror checking action and multiple secondary tasks are recognized using machine learning methods. The mirror checking behavior is the most important clue for lane change intention inference. For the secondary tasks, it reflects the distraction level of the driver. It is critically important to identify the driver's distraction in normal driving cases. When drivers are being distracted, their behaviors will have a limited contribution to the intention inference. The main contribution of this part includes the following aspects. First, the driving activity and secondary task recognition system is proposed based on the feature evaluation and selection. At this stage, driver head and body joints positions are collected. A few pieces of research in the past consider the impact of driver body joints on distraction detection. The algorithm evaluates the head pose, body pose, driver arm position, and the depth information separately. It is found that the driver body pose features contribute to a more accurate driver status classification. Second, to overcome the complex feature extraction and selection process, an end-to-end deep learning-based classification model is proposed. An unsupervised learning method for driver detection is applied. The deep convolutional neural network model directly takes the color image as the input and outputs the driver status. This method will benefit the design of a low-cost driver behavior detection system. Finally, the driver behavior identification algorithms contribute to a more accurate driver intention inference. It enables the inference system to recognize the driver's intention only when the driver is concentrating on the driving task.

### Algorithm Limitation

Currently, the driver behavior detection dataset contains only five drivers. The five drivers are asked to perform seven tasks, which contains three different secondary tasks. As we all know, the driver can be distracted by different reasons, while the three distraction tasks only contain physical distractions and ignore psychologic distractions. Therefore the algorithm cannot identify the true status of the driver if the driver is psychologically distracted. Moreover, it is not able to collect more naturalistic data in the current stage. Second, deep learning algorithms do not achieve a much better result compared with conventional machine learning methods. The reason for this can be multifold. First, it may be due to the amount of the dataset, which may need to be enriched. Second, the algorithm does not use a precise driver detection algorithm, which makes the segmented image still contain many outliers.

### Directions for Future Work

Therefore future works should concentrate on the collection of more naturalistic data from more drivers. More driver distraction status can also be involved, such as estimation of the cognitive distraction. This requires a comprehensive driver understanding, which needs to take the eye gaze information into consideration. Audio information is another important feature, which is not used in this project. Sometimes audio information carries more distinct features than images, especially when the driver is speaking with other passengers but is looking straight ahead.

Another work is to study the driver behaviors on partially automated vehicles. A driver will have different behaviors when driving automated vehicles; how the driver behavior reasoning system interacts with automated vehicles is a challenging task. Moreover, when using automated vehicles, the driver will have more freedom to perform secondary tasks, which is impossible in conventional vehicles. Therefore the collected dataset would be more valuable for the study of driver behaviors in different situations.

## DRIVER LANE CHANGE INTENTION INFERENCE BASED ON TRAFFIC CONTEXT AND DRIVER BEHAVIOR RECOGNITION

Driver intention is different from the previous driver behavior detection, which should be viewed as a time-series modeling and classification task. Driver intention inference requires to capture the long-term dependency of the temporal driver behavior sequences. In this section, driver lane change intention is inferred based on the studies in the previous sections. As one of the most notable motivations of this book, driver intention is first analyzed according to human intention studies. The generation procedure of lane change intention is analyzed based on the timescale property. The relationship between tactical intention and operational intention is clarified. It is found that traffic context plays the role of intention stimulus, and driver behaviors are the response to the intention. Accordingly, to predict the lane change intention before the maneuver happens, traffic context and driver behaviors are the two most important features. The deep learning-based intention inference algorithm is able to learn the long-term dependence, along with the training data. Based on the statistical analysis of lane change maneuvers from the naturalistic driving dataset, the intention inference algorithm achieved precise prediction results before the maneuver is initiated.

Regarding the surrounding driver intention inference, as it is unable to collect the driver behavior information, a driving trajectory prediction model is constructed instead. The model input takes the easily acceptable vehicle status such as the speed, acceleration, and headways based on vehicle-to-vehicle communication. Then a time-series modeling approach based on the long short-term memory-recurrent neural network (LSTM-RNN) model is used to predict the vehicle trajectory of the leading vehicle.

### Algorithm Limitation

Similar to the existing studies, the dataset in this part is expected to be enriched. The data quantity issue is a common challenge to all the machine learning tasks. To meet the industrial requirement, a mature driver assistance system that can be used in the real world needs hundreds of drivers to be involved. The system still needs a lot of tests before it is open to the public. Moreover, the lane change cases are manually extracted from videos, which is time consuming.

### Directions for Future Work

Based on the abovementioned discussion, future works for lane change inference system design can be summarized as follows. First make the system robust to various drivers by enriching the dataset volume. Next, the lane change cases are expected to be detected automatically based on the lane detection system. If the time stamp of the critical points for each lane change can be detected, the model training and testing can be more efficient. Finally, as discussed in previous sections, the future driver intention inference system should not be working individually. The intention inference system needs to cooperate with other driver status detection systems such as driver workload estimation and driver distraction detection. It is believed that when the driver is overloaded or being distracted, his/her behaviors will be different from those in normal conditions. In the future, the relationship between driver intention and driver workload and the influence on each other will be a challenging research topic.

## DRIVER BRAKING INTENTION RECOGNITION AND BRAKING INTENSITY ESTIMATION BASED ON THE BRAKING STYLE CLASSIFICATION

In addition to driver lateral lane change intention inference, driver longitudinal braking intention is studied in this book. The braking maneuver is analyzed based on a naturalistic driving dataset on a driving simulator. The drivers are required to follow a standard driving cycle to generate the both acceleration and deceleration behaviors. The braking intention is recognized based on two different unsupervised learning approaches using related vehicle status. The purpose of this part is to design a security module for the cyber-physical system of the electronic vehicle. The vehicle can recognize a braking intention in an early stage based on the estimated vehicle dynamics when the braking indicators are under attacked or the signals are lost. Then according to the naturalistic driving data, three different braking levels are clustered using the Gaussian mixture model method. A feedforward neural network (FFNN) is used to predict the braking intensity according to different braking levels. The continuous estimation for braking intensity can reflect the willingness of the braking intention by the driver. Moreover, a real-time braking pressure estimation provides a redundancy-safe module for the hydraulic braking units.

### Algorithm Limitation

Unlike the lane change intention inference system that utilizes multimodal signals to jointly estimate driver intention before the maneuver, the braking intention estimation only uses vehicle dynamic data without any traffic context and driver behaviors such as the foot gesture. Hence, the braking intention can only be recognized after the pedal movement rather than being predicted before the maneuver and recognition delay exists. Moreover, the continuous braking pressure and intensity are estimated based on FFNN, which cannot learn the temporal pattern and dependency with the signal and this can generate an imprecise estimation of future prediction.

### Directions for Future Work

According to the aforementioned discussion, future works can be proposed toward a more accurate and advanced prediction of braking intention. First, along with the vehicle states, driver behavioral signals, such as electroencephalographic signals and foot gesture images, can be involved to provide an early clue about driver behavior. Also, the traffic context can be used to calculate the driving behaviors of the leading vehicle to predict the potential braking maneuver based on the time to collision and headways. As the FFNN model cannot capture the temporal pattern of the braking pressure signal and generate a reasonable prediction according to the past states, time-series models such as hidden Markov model and RNN are expected to be developed for this task. The long-term and short-term dependency between the past sequence and the future sequence can be studied a little bit further. Moreover, the temporal dependency as well as the relationship between other vehicle states and the braking pressure deserve to be further exploited so that a more robust cyber-physical system-based vehicle security system can be designed.

## CONCLUSIONS AND FINAL DISCUSSIONS

During the development of this project, it was found that the intelligent driver assistance system is no longer an isolated function. It requires the fusion of sensors, algorithms, and systems. The studies performed on different aspects should be connected organically. Therefore future works on driver studies should focus on the dynamic integration of different units. Driver intention has been studied in conventional vehicles and it has been proved that the intention can be inferred before the driver takes actions. However, the human intention mechanism inside the human brain is still unclear. Whether a human intention model can be described in a mathematic form is one of the most challenging works for future driver intention studies. If this question can be answered, future intelligent vehicles will be smart enough to prevent accidents from happening.

Despite the great potential of these technologies, more efforts are required to transfer the theoretic research to the practical side. Future transportation systems will have their own challenges and chances. Future transportation systems should adopt the common existence of conventional vehicles and autonomous vehicles, which need a holistic perspective to study the cooperation between different vehicles. How the human driver influences the surrounding autonomous vehicles and how the autonomous vehicles respond to human drivers? As far as we can see, the development of an intelligent transportation system still needs new technology to motivate.

# Index

*Note*: Page numbers followed by "t" indicate tables and "f" indicate figures.